MODERN FORESTS

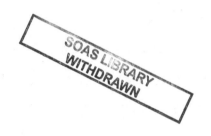

Modern Forests

Statemaking and Environmental Change in Colonial Eastern India

K. Sivaramakrishnan

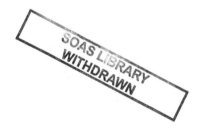
Stanford University Press

Stanford, California

1999

Stanford University Press
Stanford, California

© 1999 Oxford University Press

Originating publisher: Oxford University Press
First published in the U.S.A. by Stanford University Press, 1999

Printed in the United States of America.

ISBN 0-8047-3563-8

LC 99-72814

This book is printed on acid-free paper.

This title will be distributed exclusively by Stanford University Press
in North and Central America, including Canada, U.S. dependencies, and
the Philippines.

For
Bala and our daughters

Contents

Illustrations

Tables

Preface

This book has been prepared from the first volume of a two-volume historical and ethnographic study on statemaking and forest management in Bengal, completed as a PhD dissertation for the Department of Anthropology, Yale University. It draws on extensive historical and archival research carried out in Bengal, Delhi, England, and the United States during the period June 1992 to December 1994, to document and analyse the emergence of rational forest management, related administrative structures, and development regimes in the woodlands of colonial Bengal. The book covers a period of about 180 years that begins around 1767, soon after the first East India Company officials made tentative forays into the jungles to the west of Midnapore town in southwest Bengal, and ends around 1947 when India became independent.

The eastern Indian region has been traversed in diverse ways, and with varying degrees of comprehensiveness, to empirically sustain the main arguments offered by this book. First, to show the influence of regional political and social history on environmental change and forest management I have delved deep into the local particulars of southwest Bengal—an area that comprised adjoining parts of contemporary southern Bihar and western West Bengal. I have discussed this forest-savannah-farmland complex as a political zone of anomaly and an environmental transition zone. Bio-physical processes of environmental change imposed their own logic on resource management strategies and contingent processes of statemaking. My object is not to claim an autonomous sphere of natural agency in the physical world. I suggest, instead, a dialectical relationship between patterns of human engagement with non-human environments and patterns of change in those environments as they influence social-political outcomes of human endeavour, even those of powerful agents like governments, scientists, and elites. This second objective of the book is achieved mainly through a detailed discussion of north Bengal forest management.

A third purpose of this book is to examine the distinctive ways in which a contested history of *sal* (*shorea robusta*) forest management emerged in the colonial period. There was an important conjuncture between the growth of modern forest science in Europe, the exigencies of establishing British empire in India, and field foresters of the Bengal Presidency struggling with the ecological consequences of changing the disturbance regimes under which sal forests originated, were established, and came to dominate particular landscapes. This part of the story takes the book most widely across eastern India and into parts of the United Provinces (present-day Uttar Pradesh). From the eastern end of Assam to the western-most foothills of the Himalaya; from the lower reaches of Nepal and Sikkim montane regions to the upper elevations of the Gangetic valley; and all over the undulating landscape of the Chotanagpur plateau, extending into the central Indian plateau region, sal trees of differing provenance came to dominate the canopy of the forests where social-ecological and edaphic-climate conditions were favourable.

In following the vicissitudes of land management across the eastern Indian region, where a similar forest type was the ground on which diverse statemaking regimes were figured, I have remained within the confines of the Bengal Presidency. Areas that British administrators, explorers, scientists, and foresters came upon and chose to designate as the natural forest estate of empire, with assistance from some residents of those estates and in the face of sustained opposition from others, were always managed landscapes. Nothing brings this home more vividly than the frequent observations of British travellers, surveyors, and explorers about abandoned settlements and young sal regeneration in close proximity to each other. These accounts would note the presence of desired woody vegetation and the absence of undesired forest farmers in the same passage without pausing to reflect on the possible relationship between new tree growth and newly vacated villages. Forests are formed by disturbance and most disturbance, even in the most remote areas, may be traced to human engagement with the bio-physical world. This basic insight was reinforced in my archival forays across Bengal Presidency and so I have chosen to speak of the areas described and discussed in this book as woodland Bengal.

I have incurred numerous debts in the process of carrying out the research. The time has come to place them on record. After nine hectic, highly enjoyable, and immensely educative years in the Indian Administrative Service (IAS), I returned to academic pursuits with encouragement from Anil Agarwal, Kamla Chowdhry, V.B. Eswaran, Lina Fruzzetti, Ramachandra Guha, the late Lovraj Kumar, Naresh Saxena, and Ramaswamy Sudarshan. My interest in the role of forest management in national rural development endeavours blossomed during the late 1980s when working for the National Wastelands Development Board in New Delhi. But it was Ramachandra Guha who had the clarity of vision to suggest that my research could uniquely combine training in history, public administration, forestry, and anthropology by investigating forest management in Bengal.

In Delhi, Ramachandra Guha patiently introduced me to the National Archives, while in London, Richard Bingle was always willing to listen to my endless simple-minded queries and point me in the right direction. Lionel Carter at Cambridge made my brief sojourn at the South Asia collections very productive. Various officials of the West Bengal State Archives were immensely helpful in my search for rather scattered materials on forestry in the nineteenth- and twentieth-century records. The District Magistrate, Midnapore and the Sub-Divisional Officer, Jhargram extended full cooperation to my forays into the dark and dank confines of the record rooms at Midnapore and Jhargram. The Divisional Forest Officer, West Midnapore, and various people in his office, the Bandharbola Range and the Forestry Training Institute in Jhargram were similarly courteous and helpful.

In addition I am grateful to the staff at the Bodleian Library, Oxford; School of Oriental and African Studies Library, London; British Museum, London; Oriental and India Office Collections, London; Central Secretariat Library, New Delhi; Nehru Memorial Museum and Library, New Delhi; National Library, Calcutta; Anthropological Survey of India Library, Calcutta; Centre for Studies in Social Sciences Library, Calcutta; Centre for Science and Environment Library, New Delhi for all their help in finding various materials used in this dissertation. On the subject of libraries, I leave for the last, and a special thank you, the Yale University Library system which, to my delight, kept coming up with material I never imagined would be there. Joseph Miller at the Forestry Library was an encyclopaedic source of information and an astute adviser. The friendly, ever cheerful, people in the Seeley Mudd Library tirelessly brought to me in New Haven books, serials, reports, compendia, colonial documents, and unpublished dissertations from all over the USA and even from Europe.

Arun Ghosh provided invaluable and wide-ranging research assistance in Calcutta. Sukhomoy Sengupta and his family provided me a home in Jhargram when I needed it. They also tirelessly introduced me to the people I had to meet, the Raja of Jhargram, local intellectuals and politicians, the mercurial resident poet, Bhabhatosh Satpathy and many others who, in myriad ways, moved my research speedily forward. B.K. Bardhan Roy, who was then Principal Chief Conservator of Forests, West Bengal, liberally shared unpublished data and many government documents. S.B. Roy, and others at the Institute for Bio-Social Research and Development gave me crucial clues on how to proceed. K.C. Malhotra of the Indian Statistical Institute was full of good advice and information based on his own work in Midnapore as were Ranabir Samaddar and Pashupati Mahato. Gautam Bhadra was a font of wisdom on archival sources in Calcutta, helping me to locate obscure references with his characteristic enthusiasm.

The Centre for Studies in Social Sciences gave me institutional affiliation in Calcutta. Partha Chatterjee was my local mentor there, a patient listener and insightful guide. Anjusri Chakrabarti at the Centre prepared maps of the field sites from topographic base maps and sketches I supplied to her. Kali Babu in the library and Simanti Banerjee as a patient copyist of statistics were invaluable in helping me find what I needed and getting it copied.

In Delhi the Centre for Study of Developing Societies was a generous host. I am grateful to Ashis Nandy for taking interest in my project, and Shiv Viswanathan for his energetic guidance. During my stints in Delhi, Ramachandra Guha took good care of my intellectual and social life. Neeladri Bhattacharya, Prabhu Mohapatra, Suranjan Sinha, Ravi Vasudevan, and Mahesh Ranga-rajan discussed my project at length and offered valuable guidance. Syed Rizvi, then at the Ministry of Environment and Forests,

Arvind Khare, then at the Society for the Promotion of Wastelands Development, and Sunita Narain, still at the Centre for Science and Environment, freely gave of materials and information in their possession. Kanak Mittal, then with the Indira Gandhi National Centre for the Arts, was most generous with advice, and readily parted with copies of rare publications on east Indian tribes, especially the Santhals of Bengal, on whom she has also done research.

In addition to all the intellectual sustenance it could absorb, this project was liberally blessed with financial support at every stage. Initial research trips to India and England in 1992 were funded by the Mellon Fellowship, Department of Anthropology, and the Program in Agrarian Studies, Yale University. Dissertation research in the USA, India, and England during the period June 1993 to December 1994 was made possible by a grant from the Joint Committee on South Asia of the Social Science Research Council and the American Council of Learned Societies with funds provided by the Andrew W. Mellon Foundation and the Ford Foundation. Financial support for this project was also provided by the Wenner-Gren Foundation for Anthropological Research, New York; the Center for International and Area Studies, Yale University; and several Yale University fellowships from 1991 to 1995.

This book has emerged out of revisions and refinements to the first part of a long PhD thesis that I started writing in spring 1995. During the following year, 1995-6, I was privileged to hold the Robert Leylan Prize Fellowship while I completed the dissertation. The book manuscript was prepared while I was a Fellow, Program in Agrarian Studies, Yale University, from September 1996 to August 1997. It was finally revised in 1997-8 while I held a faculty position at the School of Forestry and Environmental Studies, Yale University. A wonderful group of students took my classes there, specially the graduate seminar in Political Ecology, and stimulated amendments that surely benefited the book. I thank my students, the Program in Agrarian Studies, and the Forestry School for giving me time and inspiration to finish this work.

This manuscript has profited greatly from the careful reading and detailed comments that my draft chapters received from a wide group of supervisors, friends, and colleagues. Throughout the time that I have been writing Bill Kelly and Jim Scott have been twin rocks of support and inexhaustible sources of encouragement. Their gentle direction and perceptive comments have incalculably enriched my work. I am grateful to Sugata Bose, Vinay Gidwani, Ramachandra Guha, Sumit Guha, Ivan Katz, David Ludden, Nancy Peluso, Ajay Skaria, and anonymous readers for Oxford University Press and Stanford University Press, for carefully reading and providing detailed comments on the manuscript. Pamela McElwee provided a close editorial reading and gave invaluable assistance as I completed manuscript revisions in the first part of 1998.

Others who have helped me along by reading chapters, or freely discussing ideas with me, include Thomas Abercrombie, Arun Agrawal, David Arnold, Mark Ashton, Mark Baker, Christopher Bayly, Indrani Chatterjee, William Christian Jr, Michael Dove, Margaret Everett, James Fairhead, Richard Grove, Ruhi Grover, Akhil Gupta, Robert Harms, Angelique Haugerud, Karl Jacoby, Dharma Kumar, Melissa Leach, Larry Lohmann, Prabhu Mohapatra, Donald Moore, Witoon Permpongsacharoen, Gyan Prakash, Ravi Rajan, John Richards, Vasant Saberwal, Sanjay Sharma, Helen Siu, Thomas Summerhill, Peter Vandergeest, and John Wargo.

Different parts of the manuscript benefited from being presented at the Cultural Studies Seminar, Centre for Studies in Social Sciences, Calcutta, May 1994; the Association for Asian Studies meetings, April 1995; the Social Science Research Council Conference on Environmental Discourses and Human Welfare in South and Southeast Asia, December 1995; and the Association for Asian Studies meetings, April 1996. Sections of Chapter 2 have appeared in an earlier form as part of my article in *Indian Economic and Social History Review* 33(3): 243–82, 1996. An earlier version of sections from Chapter 5 were included in my article in *Journal of Asian Studies* 56(1): 75–112, 1997. The parts in Chapter 7, on fire and forest regeneration, are abbreviated and revised versions from my article in *Environment and History* 2(2): 145–94, 1996. A slightly different version of Chapter 8 will be published in the January 2000 special issue of *Development and Change* on forests. All these materials are reused, in their amended form, with the permission of those who first published them and respective copyright holders. The actual production of this book was supervised with great care and competence by my editors Bela

Malik at the Oxford University Press, New Delhi, and Muriel Bell at the Stanford University Press.

My research could not have been completed without another, more crucial, kind of support, namely, the love, care, shelter, and nourishment that family and friends lavished upon me over the last few years. Shauri and Ranu Banerji welcomed me into their home in Calcutta. In Delhi my parents took good care of me. Life and work in the US would not have been possible without the wonderful family support of Leela Rangaswamy, Martin Hart, Kamala and Ben Prasad. But above all I shall remain, forever, ineffably grateful to my wife, Bala, for simply being there. She managed alone for the entire duration of my overseas research, and drew upon her great reserves of strength and energy to keep our spirits up when the going looked difficult. Over the years, my daughters have been very indulgent of my preoccupation with the thesis and, then, the book. Their happy voices, music, and occasional complaints combined to constitute a largely congenial polyphony of sounds ringing through the house as I wrote.

This book would not have been possible without the inspiration provided by my grandfathers, K. Anantharaman and S. Jayarama Ayyar, who introduced me to the joys and responsibilities of intellectual pursuits.

Glossary

adalat	court
adivasi	autochthone, original resident, used as non-pejorative term for tribes
amin	land survey assistant of revenue officials in the district administration
arkathi	labour contractor
bankar	forest tolls
bathan	cattle enclosure
bazé	waste
bazé zameen	wasteland
begari	forced labour
bewar	swidden agriculture practised in central India by Baigas
bhuinhari	a form of land tenure granted to those recognized as first clearers of lands in southwest Bengal, usually among Mundas and Oraons
bigha	measure of land area roughly equal to one-third of an acre
bonga	spirit
bund	small earthen dam
burkandazé	armed constable, guard or escort
chakeran	a form of service tenure-lands held rent-free for rendering service to the landlord
chauki	police station or outpost
chaukidar	watchman
chuar	variously describing a bandit, the Bhumij people of the jungle mahals, levies of paik sardars, the lowest rung in zamindari militia

daffadar	labour contractor or rural police
daftar	office
dal	lentil
daroga	head of a police station
dhuna/dhammar	an oleoresin collected by tapping the mature sal tree; used for incense, medicinal and other purposes; a valued trade good in the tribal economy
digwar	landlord militia-man; used for policing duties; sometimes paid through a special service tenure of land in the landlord's estate
diku	outsider or foreigner—usually refers to a non-tribal person
diwan	chief revenue officer
fakeer	ritual specialist
faujdar	head of police court
ghatwal	guardian of passes, warden of marches; also a form of headmanship
goala	milch cattle-keeper
hat	weekly market
hool/hul	rebellion
jamabandi	determination and record of land revenue demand
jhum	shifting cultivation, swidden fields
jote	cultivated field
khandait	swordsman
khas jungle	estate forest
khas khamar	personal demesne or threshing ground
khas mahal	government estate
kheddah	elephant chasing, trapping and capture
khudkasht	occupancy tenant
khuntkatti	tenure of members of the lineage who first reclaimed the land for cultivation
kodo	millet
kumri	a form of shifting cultivation practised in south India
kurao	a form of swidden agriculture practised by the Paharias of the Rajmahal area in Bihar

mahajan	moneylender; usurer; urban shopkeeper
mahal	administrative unit; roughly equivalent to county or subdistrict
mahato	headman (among Oraon)
majhas	land grants or lands held on privileged tenure by a majhi
majhi	headman (among Santhals)
mal	land held by headmen, and other special tenure holders, on which revenue was required to be paid
malguzary	rent roll
mandal	village head in Mahato village in Midnapore; later became a form of village tenure
mofussil	countryside or hinterland
mouza/moujah	village
mouzawar	village-level assessment of land revenues
muhrir/muharrir	writer or clerk
naik/naek	chief
nizamat adalat	chief criminal court
pahan	village priest in Munda and Santhal village
paik	armed attendant or village watchman; holder of special service tenure requiring military and police service to zamindars in the jungle mahals of Midnapore
paikasht	non-occupancy tenant
panchayat	elected self-government body at village cluster, development block, and district levels
pargana	unit of revenue administration equivalent to subdistrict or county
parganait	Santhal headman of a cluster of villages included in a pargana
patni	intermediate rent-collecting tenure in Permanent Settlement areas of Bengal
podu	a form of swidden agriculture practised in Andhra Pradesh
pottah/patta	certificate of landholding
pradhan	village headman in Bhumij village of the jungle mahals
raiyat	peasant

rajbari	zamindar's palace
rakha	protected or kept for a specific purpose
rakhabundy	measurement
saheb	officer
sardar	headman
shikari	hunter
tahsildar	revenue official
tanti	weaver
taungya	system of forest plantation in which tree seedlings are planted along with annual crops like cotton and coarse grains; annual crops are removed from these tree plantation areas after three-four years once the tree crops are established
terai	foothill areas in the eastern Himalayan tract of north Bengal which were characterized by fine-grained, poorly drained, almost swampy soils
thana	police station
thikadar	revenue farmer or rent contractor
zamindar	landlord; person responsible for paying revenue to government under the Permanent Settlement of 1793

Abbreviations

Agri	Agriculture
Asst	Assistant
B&O	Bihar and Orissa
BAAS	British Association for the Advancement of Science
BBORC	Bengal Board of Revenue Consultations
BC	Board's Collections
BCJC	Bengal Criminal Judicial Consultations
BD	Bengal Despatches
BDO	Block Development Officer
BDRM	Bengal District Records, Midnapore
BFAR	Bengal Forest Administration Report
BJCC	Bengal Judicial Civil Consultations
BJP	Bengal Judicial Proceedings
BNR	Bengal Nagpur Railway
BORP	Bihar and Orissa Revenue Proceedings
BOR	Board of Revenue
BPC	Bengal Public Consultations
BRC	Bengal Revenue Consultations
BRP	Bengal Revenue Proceedings
BWAER	Bengal Ward's and Attached Estates Reports
Cal	Calcutta
CF	Conservator of Forests
COD	Court of Directors
Coll	Collection(s)
Collr	Collector
Comml	Commercial
Commr	Commissioner
Corr	Correspondence
CS	Chief Secretary
CSACL	Cambridge South Asia Centre Library
DC	Deputy Commissioner
DCF	Deputy Conservator of Forests

DG	District Gazetteer
Dept	Department
Dy	Deputy
EIC	East India Company
For	Forests
FSO	Forest Settlement Officer
Genl	General
GG-in-C	Governor-General in Council
GOB	Government of Bengal
GOBO	Government of Bihar and Orissa
GOI	Government of India
Govt	Government
GSI	Geological Survey of India
IGF	Inspector-General of Forests
Jt	Joint
Jud	Judicial
l/d	letter dated
LP	Lower Provinces
MFP	Minor Forest Products
Misc	Miscellaneous
MSS	Manuscripts
NAI	National Archives of India, New Delhi
NMML	Nehru Memorial Museum and Library, New Delhi
Offg	Officiating
OIOC	Oriental and India Office Collections, London
Progs	Proceedings
Res	Resolution
Rev	Revenue
SDO	Sub-Divisional Officer
Secy	Secretary
SR	Settlement Report
SWFA	South West Frontier Agency
US	Under Secretary
Vol	Volume
VP-in-C	Vice-President in Council
WBSA	West Bengal State Archives, Calcutta
WBSFP	West Bengal Social Forestry Project
WBSRR	West Bengal Secretariat Record Room, Calcutta
WP	Working Plan

MODERN FORESTS

1

Statemaking and Environmental Change

Considerable areas of these terai forests are of recent alluvial origin, and while the soil is then often light and suitable to the growth of sal, the ground is occupied by very inferior species only . . . we should here step in and assist nature, and by these means add greatly to the value of the estate.
—E.P. Dansey, Conservator of Forests, Bengal, December 1890.[1]

The Importance of Diversity

When the words in the epigraph were penned formal forest conservation in Bengal was twenty-five years old. Yet these were decidedly ambitious words written in momentous times. Over the next two decades the rate of 'managed change' in the forested landscapes of Bengal accelerated at a rapid pace. From the Darjeeling hills to the mangrove swamps of the Sundarbans, from the eastern edge of the central Indian plateau to the tenacious chiefdoms of the Chittagong Hill Tracts, a network of working plans for forest management conjured up regimes of tree and soil manipulation that would add value to the six thousand-odd square miles of Bengal's official forest estate. Taken together, these plans and the attempts to implement them gave expression to the enormous and swift professionalization of modern forestry in colonial Bengal. Foresters in eastern India, working through these plans, participated in the emerging worldwide ideas of rational forest management for the maximization of resources in which dominant social groups had immediate interests.[2] Arguably, the spate of forest working plans that were quickly formulated after the initial Darjeeling plan of 1892 were drawing upon, and disseminating, a sense of technocratic confidence that paid little heed to the ecological complexities of the landscapes they were representing, even as they proposed sweeping interventions in them.[3]

At one level these plans participated in articulating the incipient coherence of a body of technoscientific knowledge that has widely been discussed and named as scientific forestry in the literature on India and southeast Asia.[4] At another level, as the geographic diversity of Bengal intuitively suggests, forest management could not possibly conform to standard continental or colonial models. Landscapes were neither uniformly transformed by, nor similarly obdurate in defying the 'assistance to nature' that became so abundantly and persistently available by the twentieth century, from foresters and their cooperators in Bengal. There could be several measures of these impacts of places (understood as social-ecological geographic entities) on scientific forestry, and how their variations in time and space limited the homogenization of state resource control. One was the frequent changing of forest working plans that became the bane of Bengal forestry. Large areas also remained under management without plans.

By examining these variations closely, this book not only highlights the ecological and social peculiarities of regions,[5] but, more importantly, shows how the manner in which culture, nature, and power are spatially constituted and expressed, influences processes of statemaking. Moreover, the Bengal experience instructs us that environmental change, understood as the interlinked transformation of social and bio-physical landscapes, is better described as 'discordant harmonies' than the customary depiction of an alternation between system-maintaining tendencies and their occasional disruption by cataclysmic events.[6] Such homoeostatic thinking has been powerful in American environmental writing, with the growing influence of conservation movements, that has promoted notions of paradise lost and regained.[7] In contrast the newer focus on complexity tells us that order is going to be more difficult to locate and describe, while any such order will also have unruly elements of indeterminacy. We now have also to recognize that science has as many schisms, conflicts, dissensions, and personality contrasts as any other human activity.[8]

Finally, the Bengal case emphasizes that changes indicated by terms like deforestation, territorialized resource control, and agricultural intensification were not linear or predictably in conformity with a single dominant pattern through most of the colonial period.[9] Some central Indian evidence has been assembled in support of this point.[10] But in India, as in many other parts of the tropical or subtropical world, the dominant account of environmental change is that rapid deforestation in the last one hundred years has been caused principally by population growth and destructive forest clearing by environmentally uncaring farmers. The work of Melissa Leach and James Fairhead is part of a small corpus of work on Africa that has offered careful challenges to these ideas, showing how changes in forest use, and thus the place of forests in agrarian landscapes, vary across space, time, and social groups.[11] Stated more abstractly, these studies support the argument that 'change in nature, or in society, is never unidimensional: it always comes about from interplays—additive or multiplicative, synergistic or antagonistic'.[12]

Grounding these perspectives in woodland Bengal I have examined colonial forestry in at least three guises—first, as a set of material technologies imposed on trees, grasses, and wild animals; second, as a legal regime aimed at appropriation and monopoly in the extraction of natural rents; and third, as a system of rational knowledge that, ironically, became the site of a struggle among technocrats who vied for professional recognition at the upper levels of bureaucracy. I demonstrate that forestry as technology never consolidated its hold on the most valued tree species nor was it able to codify the requirements for their rapid regeneration; forestry as legal regime never achieved consistency in application—sanctions and regulations were often relaxed, and exclusive claims were forsaken when they could either not be enforced or when bureaucratic agencies contested them; and finally, forestry as a system of professional knowledge failed to garner respect or authority because many other colonial elites remained unconvinced of its scientific pedigree. In part this was because woodland Bengal—its people, flora, and fauna—clearly emerged as an agent able to confound foresters and resist their ambitious schemes.[13]

Statemaking and Political Society

This book offers a comprehensive history of the changing culture of forestry and politics in Bengal. It does so by dealing with agrarian and environmental issues as interrelated and mutually constitutive. A fairly rich historiography of rural Bengal has almost without exception neglected these connections. As a result we are placed in a situation where academic curiosity has not extended into areas beyond river valleys, the Gangetic delta, and the plantation economy of the northern parts of the region. Vast areas of Bengal, that geographically are best described as a forest-savannah transition zone, are thus scantily covered by existing scholarship.

Transition zones remind us of the distinct domains that they often separate. In this book I have also examined the making of distinctions that have placed environment in the public domain and made agriculture a private sector activity. I document the emergence of these distinctions as a historical process inflected by localized struggles in arenas where wider separations like that between public and private were created and contested. The partitioning of landscapes and social spheres came to characterize large aspects of modernist state formation.[14] These processes were central to the stabilization of colonial rule, even as they came to be fundamentally altered by issues of race, other forms of difference, and direct subjugation by foreign powers, that were unique to the colonial experience. My study thus advances an important dimension of this complicated picture into the spotlight by examining the creation of distinctions not only within society, and the society-nature combine, but the contested delineation of states, their relations to entities called societies, and the role of such powerful imaginings and institutional arrangements in fashioning environmental change.

One of the most important distinctions that becomes salient here is the one between state and society which is at the heart of modern state formation. By demonstrating the constitution of state and society as complex colonial entities I offer a theory for their mediation by a historically transformed political society. This book shows that the scientifically desirable was always entangled in an argument about the locus of effective governance in colonial woodland Bengal. It contends that through conflict and cooperation between a differentiated society and a heterogeneous colonial state in the making, rural social relations and colonial power were mutually transformed in Bengal. To facilitate the discussion I have introduced the concept of statemaking that is now briefly described.

Statemaking is fundamentally about defining the forms and legitimations of government and governmentality.[15] Statemaking is simultaneously about the making of civil society. Indeed the institution and management of that distinction between state and civil society is one principal object of my inquiry. As Tim Mitchell has recently pointed out, 'the state should not be taken as a free standing entity, whether an agent, instrument, organization or structure, located apart from and opposed to another entity called society.'[16] The production of the state–society distinction, which I discuss as statemaking, must itself be studied.

But there is even more to statemaking. Although the state-society distinction may remain contested, at specific points in history, in particular places, a discernible entity called the state emerges with several internal hierarchies. Statemaking in that context also refers to the ideological and organizational power of the central government to penetrate society, exact compliance, and invoke commitment. This power rests on 'a delicate balance between autonomy and control in the relationship between state and society.'[17] This tension between central direction and local autonomy is at the heart of my argument, which then traces the processes through which a dialectical relationship is established between society and state in the distribution of power.

My project endorses Anna Lowenhaupt-Tsing's argument that 'village politics contribute to making the state; the categories of state rule are actualized in local politics',[18] but it goes further in two respects. First, I analyse statemaking to discover the pathways and modes through which locality, region, nation, and transnational bodies become spatial and cultural entities. Then I employ the concepts of public spheres, counter-publics and multiple spheres to identify the media of political linkages between these entities.

If the constitution of political society describes the institutional domain of statemaking, its cultural domain is understood in my project through the various categories of knowledge. Colonial knowledge, local knowledge, and other forms of hybrid knowledge become crucial because they allow intellectuals, leaders, policymakers, district officers, and field foresters to straddle communities of expertise. The contours of power—or the limits of statemaking—are shaped by structures and processes of knowledge acquisition and dissemination. Forest management was not only predicated on requisite scientific knowledge but on techniques of validating or valorizing certain knowledge while discounting others. Thus was expertise constituted. The struggle over what knowledge was designated as expertise, who generated it, how it was certified, where it was located, and by whom it was practised also became integral to statemaking.

Some recent scholarship on the state has begun to offer useful ways of approaching this shifting boundary between state and society. We are learning to think about state and society not as naturally formed discrete units but as complex entities that emerge out of the boundary drawing that goes on between those who wield legitimated power and those who are subject to such exercise of authority. A discussion centred on statemaking is about the very practices of such boundary work. When we focus on modern statemaking we also find that the boundary work takes on a distinct character influenced by cameral sciences, taxonomy, and rationaliz-ation in the sense of Max Weber's use of the term. Many able scholars have commented on these points. I am suggesting that we have to integrate the description of regimes of cultural represent-ation and political organization with an analysis of how laws, rules, practices, and everyday forms of power are constantly reformulated in the light of experience.

Before proceeding to discuss what newer theories offer as illumin-ation, in our quest for a theory of statemaking, older formulations need to be summarized. Political theory remained occupied with pluralism and double-market-place analysis for a long time, suggesting a stable competition between interest groups rather than structural domination of any particular one. In response, the new state theory argued that pluralism could not explain structural imbalance or violent conflict. This body of work treated stable equilibrium as an empirical impossibility, and a facade for suppressed conflict. It focused on the organizational characteristics of the state and how it might be manipulated. The new state theory saw groups in an unequal struggle to dominate a powerful state apparatus, rather than a game of consensual competition for easily divisible stakes. Lastly, the new state theory saw the state as a relatively autonomous actor.[19]

Curiously, the systems theory of society (*pace* Parsons) and most Marxist theories of state, which have been the dominant paradigms for the study of politics and state formation, have both denied any autonomous role or sphere to the state.[20] If political systems analysts found the state contained in the social system (notwithstanding Poulantzas or his caveats about the effects of state power and its institutional structure), Marxists and neo-Ricardians were ultimately interested in the manipulation of the state by capital.[21] Alternatives were offered which reaffirmed state autonomy on grounds of distinct geopolitical interests and unique military power and bureaucratic organization[22] or argued that power needed to be separated from the state and considered as an autonomous productive force in society.[23]

What all these approaches tended to sacrifice was the historical perspective on statemaking and the constitution of power through the play of structures and strategies. So they failed to offer an explanation for how micro-social diversity leads to macro-social order. They promoted an undifferentiated concept of the state and society that in turn made it difficult to comprehend how different elements in each exert contradictory pressures and form strategic alliances to generate unanticipated patterns of domination and their transformation.[24] While recognizing the prevalence of global strategies of state formation we should see them for what they become as they work to reduce the complexity of social relations and fix them in temporary, unstable, and provisional ways.

Thus a surplus of meanings and practices is always available for articulation into new strategies and power relations which can also exploit the polyvalence of dominant patterns. Such an approach to the study of statemaking provides a way of thinking about the relations between structures and strategy. Its basic elements may then be laid out as follows:

- the state is a specific institutional ensemble with multiple boundaries, no institutional fixity and no pre-given formal or substantive unity;
- formal unity can be established around time, money, space, and law, but this requires articulation with substantive norms

and can happen only in the context of specific state projects and through struggle;

- as a strategic terrain the state is located within a complex dialectic of structures and strategies, each shaping the other;
- class forces and interests do not exhaust the range of forces and interests involved in the state. In addition to other forces and interests constituted outside the state system, the state system itself engenders political interests;
- the power of the state is the power of forces acting in and through the state, including state managers, class forces, gender groups, regional interests and so on. State power also depends on the forms and nature of resistance to state intervention.[25]

Unified views of the state stress its rule-enforcing bureaucratic character, when in fact we could perhaps more profitably study the formulation and transformation of goals in their implementation.[26] Therefore, following the new scholarship and its critique of both Marxist and non-Marxist ideas about statemaking, we may argue that the distinction between state and society is under constant production and this very process should engage our study.[27] Leading us in a similar direction, Crawford Young says, 'state and civil society are indissolubly bound in an unfolding relationship of conflict and cooperation.'[28] Drawing these insights together, it is possible to conclude that statemaking is the process of drawing lines between state and society, something that becomes possible only through significant levels of cooperation at the local level between state agents and political communities, especially their leaders.[29]

But statemaking is not only about defining the state in relation to society. It also refers to the creation and maintenance of complex hierarchies and divisions within the state as a structure, and therefore should be addressed as an effect of detailed processes of spatial organization, temporal arrangement, functional specification, supervision, and surveillance. The conflict between different parts of the state is important for the elaboration of this structure as 'such conflict is an important indication of the permeability of state boundaries . . . it enables one to trace how wider social differences reproduce themselves within the processes of the state.'[30]

Statemaking appears then to be a matter of organizing political subjection within a defined territory (the spatial form of power) and imbuing this subjection with legitimacy. Since the degree of legitimacy influences the costs of maintaining any particular arrangement of subjection, a war of words, meanings, and symbolic exchanges occurs in this terrain of statemaking.[31] Two aspects are important here. First, this process of statemaking is about centralized knowledge, its means of collection, and role in shaping the social identities of citizens.[32] Second, the symbolic struggle that defines the social and cultural space of statemaking takes place on a common ground shaped by bureaucratic practice and popular attitudes, an articulation of local-level interests within the bounds of the acceptable.[33]

In other words, if bureaucratic governance is about the reification and management of taxonomies, we must attend to the liability of these classifications and their recasting in the familiar terms of local experience through which people use and reshape them.[34] Thus, for instance, by reinterpreting categories of rule, rural elites and other groups in woodland Bengal were 'both subverting and reconstructing state power.'[35] They, as well as innovative bureaucrats, invoked their tactical power within the determinate settings offered by structural power.[36] This brings me to the question of the political space in which statemaking occurs.

We have defined the state as an ensemble of institutional orders and heterogeneous social relations whose precise contours, relative unity and distinctive dynamic are never pre-given but are always constituted in and through contingent social practices. We have also argued that statemaking happens through institutional transformation and symbolic exchanges in the public domain. What is this domain? It must be derived from theories of civil society by defining political society (separate from state and political economy) and the public sphere.

In current discussions civil society has come to mean the domain of economic and cultural production—an aggregate of institutions—which exerts pressure and control on state institutions.[37] This definition has been forged in critiques of earlier writing on the subject in various European traditions. For Marxists, civil society was nothing more than bourgeoisie domination under capitalist modes of production shored up by a state that was itself epiphenomenal to the relations of production. This highly reductionist reasoning does violence to rich traditions of political discourse where civil society has meant a search for ways to control and render legitimate state power.

In contrast, Hobbes argued that civil society was a creation of the strong state, while Locke insisted that the constitutional state made civil society fully formed from existing potential. Tom Paine and other libertarians held that civil society existed as a natural condition of freedom and cooperation that was constantly under threat of vitiation by the state. Hegel theorized civil society as a historically produced sphere of ethical life positioned between patriarchal household and the state. It comprised groups and institutions governed by civil law, not directly dependent on the political state.[38]

Hegel is the most important theorist of civil society because he sees the link between civil society and state as one of mediation and interpenetration. He provides us with the fundamental categories for analysis, such as legality, privacy, plurality, association, publicity, and mediation.[39] But the German tradition from Hegel to Marx bifurcated everything sharply into civil society and state. It did not have room for an independent domain between them with distinct institutions and dynamics. The French tradition derived from Tocqueville established political society alongside civil society and state. The Italian tradition going back to Gramsci identified political society with the state.[40]

Foucault presents a far more relentless critique of civil society than his predecessors. Like Marx, Foucault finds that civil society does not circumscribe domination, but supports it. He builds on a core Marxian insight that modernity involves the emergence of new and pervasive forms of domination and stratification.[41] But he goes further than Marx in understanding the social bases of power by delinking power from economy and identifying the subtle pervasive workings of power behind and beyond juridical frameworks. Foucault argues that modern disciplinary society is the source of the power he is describing, not just the juridical liberal-democratic state with its documenting projects. Thus schools, hospitals, and convents are as much responsible for producing power and forms of social control as the state bureaucracies and censuses and prisons. However, he fails to acknowledge how new forms of publicity, association and rights emerging in civil society are key weapons in the hands of collective actors seeking to limit the reach of state and other societal forms of disciplinary power.

Therefore, the place of the state and its interests is somewhat more ambiguous in Foucault's analysis. While he insists that the state is not the sole origin of disciplinary power, he admits that the state is a key actor in generalizing disciplinary power. So for Foucault (unlike Marx or state autonomy theorists) it is not the logic of the economy or the state that penetrates or colonizes civil society. The institutions and practices of civil society generate the technologies of power that are then taken up and globalized by the state and the bourgeoisie. He is ultimately arguing that the state is one locus among many for disciplinary power, suggesting thereby a homology between society and state as distinct realms that are organized by the same disciplinary technologies of power and strategic power relations.

The limitations of this approach, especially its attempt to provide a pervasive account of power and its sources, become clear when we examine the patterns of ramification, persistence, and accommodation between institutional nodes or loci, that are integral to the manifestation and experience of power. To illustrate, let us briefly consider the role of special land tenures in modern Bengal. In Chapter 2 I have introduced a discussion of a series of special tenurial arrangements through which pre-colonial rulers established relations with forested regions in southwest Bengal. Any consideration of these tenures from a Foucaultian perspective would describe their continuation into the colonial period as somehow constituting the expanded universe of indirect rule. But it is my case that the adaptation of these tenures to local government in southwest Bengal was a measure not only of the ramification of disciplinary power but simultaneously indicative of a different kind of power that worked in certain situations as a limitation of modern colonial state power.

Clearly this would be a more complete understanding, for instance, of *ghatwali* and other such tenures, for it would remind us that power is never more than the ways in which it is experienced and transmitted. We can then avoid providing a partial account of things like special tenures by looking closely at the history of their institutional formation and interrelated functioning amongst other state and social institutions, and their location in political society. As several different critics have argued, Foucault's theory of power in its all-consuming aspect fails to disaggregate the spatial, temporal, institutional, and structural variations of governmentality.[42] Rather than treat civil society as a product of power, it is more usefully understood as a sphere of social interaction between economy and state, composed above all of the intimate sphere (especially the family), the sphere of associations, social movements and forms of public communication. Modern civil society is created through forms of self-constitution and self-mobilization. *Political society* arises from civil society and includes parties, political organizations and political publics, its actors being directly involved with state power. Political society mediates between state and civil society.[43]

My use of political society hence refers both to the local structures of government and to those aspects of civil society that have extra-local political reach. In colonial Bengal this would include headmanship, police and judicial roles of landlords, and labour contractors. Such a formulation of political society becomes possible through Habermasian theories of civil society and their dualistic character. As Nancy Fraser points out, this dualism is manifest in a fourfold treatment of family, state, economy, and public spheres.[44] Additionally, 'the idea of the double character of the institutional make-up of civil society is a real gain because it goes beyond a one-sided stress on alienation or domination (Marx, Foucault) and an equally one-sided stress on integration (Durkheim, Parsons)'.[45]

This double character may also be understood as the combination of political institutions and social movements. Public spheres as concrete expressions of political society are thus historically produced in statemaking processes.[46] This conclusion emerges from the explication of public spheres. 'We speak of the political public sphere . . . when public discussion deals with the objects connected with the activities of the state.'[47] Habermas traced the emergence of what we now discuss as associational politics, the formation of social institutions structurally invited by the expanding world of capitalism and the territorial division of nations. He thus re-examined human agency in the formation of political institutions in European bourgeois society, but always remained aware of the role of capitalism in forming the field of agency, and thus provided a dialectical theory of the public sphere where statemaking takes place.

What were the public spheres in colonial woodland Bengal? The police and other judicial powers of landlords, headmanship, institutionalization of forestry as technical expertise, zamindari networks, the district bar associations, tea planter clubs, the select cadre of civil servants endowed with special powers of administration in the jungle mahals and other exceptionally governed districts, these were some of the aspects of governmentality that created elite publics. Ordinary people entered these arenas at times. They also created their own publics through kinship and regional culture and through the travails of collectively subsisting in a harsh and dangerous landscape. By scrutinizing the interaction and mutual jostling between these publics, and by examining the distinct cultures of governance they expressed and shaped, I bring politics into the study of environmental change.

Critical Environmental Histories and Substantial Political Ecologies

At the broader, comparative level this book engages themes pertaining to environmental history and political ecology. Of these themes, the question of doing environmental history, particularly south Asian environmental history, has led to some introspection among its leading practitioners.[48] Reviews written over the last few years reflect the fast burgeoning nature of one of the most exciting and important subfields of south Asian history. But environmental history also remains caught up in two constraining legacies. At the risk of simplification, these may be stated as the enduring influence of nature–culture dualism in conceptualizing environmental history, and the preoccupation with colonialism as a watershed in south Asian environmental history. Both these features, ultimately limiting the creative scope of environmental historians of south Asia, are well illustrated in the work of Ramachandra Guha. His pioneering study of the western Himalaya, for instance, is based on the premise that fundamental environmental changes in the region began only after the 1878 Indian Forest Act came into force.[49]

More recently, this author has opined that writing environmental history requires taking the 'ecological infrastructure of human society' into account. He says, 'while the ecological infrastructure powerfully conditions the evolution and direction of social life, human intervention itself tries to reshape the natural environment in its own image.'[50] In making this simple separation between nature as infrastructure and society as superstructure, Guha follows a dominant strand of US environmental history which analyses nature and culture in ways that rarely examine the cultural politics of constituting the two categories.[51] Much US environmental history also provides a single framework for the analysis of European expansion. This framework—whether it be tilted toward Alfred Crosby's biological imperialism, Eric Wolf or Carolyn Merchant's versions of an emergent world economic system, or William Cronon's analysis in terms of the spread of utilitarian ideas—suggests a focus on fully formed European structures and ideas that violently disrupted colonized natural and social ecologies.[52] This two-cultures, two-systems approach was very much in keeping with reigning ideas about ecology, mainly studied in Clementsian terms as a closed, steady state-seeking model.[53]

While it would be ill-advised to dismiss the claims of these influential scholars, it deserves to be said that their assertions—which locate Europe as the metropolis and grant it an astonishing capacity to disrupt and colonize peripheral societies and ecologies—appear to me, on the basis of the Bengal study, to be overstated. An amended approach, of the kind I offer in this book, not only problematizes 'nature' and 'culture' as social constructs; it additionally relinquishes the large-scale systems approach to human ecology and history in favour of a mosaic approach that stresses the stochastic effects of local ecological factors, social conflicts, events, and culture. The history of colonialism rapidly dissolves into histories of colonial statemaking that, in terms of control over resources and people, fluctuated between notable success and abysmal failure.

We have already noted how in the last two decades, ecological theory has become more attentive to disorganization and complexity in natural processes.[54] Anthropologists and historians of the environment are displaying the same attentiveness to social complexity and processes through which these complexities are reorganized. This has meant a revision of our understanding of older civilizations and their environmental impacts. Jack Goody, for instance, has challenged the idea that pre-modern societies were more conservation minded, stating that 'all societies are engaged in a struggle with nature. The destruction of wildlife, the denuding of forests, the problem of erosion, these are to be found in simpler as well as complex societies'.[55] Revision of received wisdom has also meant a closer look at pervasive social inequalities, skewed power relations, and the uneven patterns of economic change that have always mediated and influenced environmental outcomes. In the past, even a sophisticated historian such as William Cronon was susceptible to a 'holism that washes out the human diversity of experience and identity.'[56] Exciting new scholarship has begun to unpack the opposed social categories of native and colonialist, or traditional modes and modernization, analysing the intermediate and hybrid categories that came into existence as different people contended and cooperated in their struggles with nature and each other.[57]

So, rather than ponder the question of a distinctive south Asian environmental history, I would argue that the Bengal case illustrates the need for less determinist and linear environmental histories of the sort that are now being written in the USA. These would be histories more attentive to micro-processes, more influenced by the regional particular. Region, in my usage, signifies both distinct geographies and social organization, varied ways of producing a spatial and political identity that marks out a place.[58] I prefer the term region because it allows me to range across different spatial and social scales of analysis. In that sense it encompasses the reference to distinctive social formations suggested by terms like locality and the delimitation of self-contained ecological relationships implied by the terms ecosystem or watershed. But my discussion of regionality tries not to forget that unity within a certain arbitrarily nominated scale of analysis should not be presumed. Where we find unity, or tightly interwoven networks, the processes that produced such an effect require examination. Finding the impress of non-unitary, yet particular places on more general things is the challenge.

We have to attend more carefully to categories like colonizer and colonized, conservation and degradation, because their shape and occupants are not readily apparent. Thus, in southwest Bengal, landlords are at times conservationist, and at other times exploit forests to their permanent detriment. The same can be said of peasantries, bureaucrats, and professional foresters. In very different contexts, therefore, I shall take apart congealed categories, a task in part of cultural analysis, to analyse both social forces and their representation.[59] I propose, then, not a third world environmental history that would be juxtaposed against contrary first world ones, but a careful examination of the way comparable and unique elements are always intermixed in the environmental history of anywhere. How these elements come together, what the blend looks like, is the story to be told.

The gains from such analyses are now obvious. Histories and politics, especially those of colonialism, imperialism, and uneven regional development have been inserted into accounts of various peoples' struggles with nature. Anthropologists and geographers have reasserted the importance of dynamic and contested accounts of cultures and places, and thus the need to examine the cultural politics of landscapes, and their representation, use, and management.[60] They have combined to fashion the field of political ecology which brings our attention to processes of global capitalism that shape heterogeneous local histories, cultures, and societies while reminding us of the micro-politics of peasant struggles over resources and the symbolic contestation that constitutes these struggles.[61]

We now have a growing body of work on the forest histories of India, especially in the colonial period. But with a few exceptions,[62] such work has concentrated on the Himalaya and other hill and mountain environments, to the exclusion of the vast central Indian forest region. As a result a complex history, and rather different trajectories of experience in the colonial period have been subsumed largely into national-scale narratives of colonial forestry based mostly on the montane forest zones. While recent research on central and western India has begun to correct this tendency, eastern India has remained neglected.

Not only does my study correct such neglect, but by focusing on both Himalayan Bengal and southwest Bengal regions adjoining the Chotanagpur plateau, it also compares and contrasts two distinct bio-regions. This book explores the consequences for forest policy, state formation, and social change when a standard management regime is introduced in very different ecological zones. Introducing ecological processes more directly into the historical narratives, the book will offer a valuable comparison of dry and moist forests, and the consequent tropical and temperate forest management that emerged under a common government. To reiterate, this book

offers theoretical innovations in the study of colonial state formation in India. By doing so through the case of forest management—
a subject peculiarly suited to the study of the colonial state—it also
enlarges and extends the historiography of rural Bengal as well as
charting new directions for the environmental history of India.

Further, as a study of the past in the present, my project is also
an historical anthropology. This intellectual tradition, gaining
prominence from the 1960s, participated in and seriously revised
earlier approaches most commonly captured under the rubric of
ethnohistory. A review of historical anthropology as it developed
through the 1980s, through confluences in cultural history and
anthropology, will not be undertaken here as many such exercises
exist.[63] But what does need noting is that by the end of the 1980s
historical anthropology had developed certain predilections and
assumptions. First was the assumption that anthropologically
informed history must focus on Otherness, the irreconcilable
differences between cultures. This led to a second methodological
inclination to privilege knowing the Other through certain sources
considered more authentically representative of a 'native point of
view' or the 'subaltern voice'. Principally oral traditions, folklore,
mythology, and religion served as these privileged sources. In short,
historical anthropology and anthropological history alike developed
an idealist bias, often coupled with structuralist conceptions of
how culture influenced human agency.[64]

Even where historical anthropology was practice and political
economy-oriented, very often the anthropological proficiency of
the research was demonstrated through an analysis of textualized
indigenous knowledge recovered from realms of oral praxis,
maintaining a problematic distinction between culture and practical
reason.[65] In all this work, there appeared to be two discrete worlds,
one inhabited by local culture and resistance, the other by universal
reason and state power. Up to a point, especially in the study of
resistance, protest and defiance, this dichotomization has worked
well. But what if we wish to study cultures of power, the everyday
forms of statemaking and the shifting, conflicted practical modes
of government across space and through time? I would argue that
such a project must bring the past and present together rather
differently.[66] Neither 'invention of tradition' nor 'structure of
conjuncture' provides a satisfactory framework.[67] An alternative

approach that I found myself taking may be summed up, following J.D.Y. Peel's recent formulation, as a historical anthropology that deals with a triangle of relations: first, those between ethnohistory (or history as representation) and the past it represents; second, those between this past and the social forms of the present that are its outcome; and third, those between the present social form and the representations of the past for whose production the present provides a context.[68]

Narrative as the most spontaneous form of historical representation has a central role in this complex of relations. But narratives also reveal the constrained choices through which they are constructed. By the study of narrativization, we may begin to understand the interplay of constraint and choice across asymmetrical power relations in space and time. Narratives are a means to appreciate all social formations and identities, including the processes of statemaking. A number of recent studies have amply demonstrated the immense possibilities of such an approach.[69] But a further problem that cropped up for me had to do with two elements in my project. First, the combination of fieldwork with archival work to study a two-hundred-year period threatened to generate an incommensurable body of narrative evidence, including varieties of colonial and national official narratives, the historical memories of peasantry and other rural people, and the textualized orality of myths, traditions and cosmology referred to earlier. Second, a focus on everyday aspects of statemaking ensured that for the nineteenth century and before, there was a severe imbalance in the availability of state-mediated sources versus sources not so filtered.

Ajay Skaria has grappled with this problem in his work on western India, and resolved it by according ultimate primacy to textualized orality, thereby recovering what he considers the authentic—if discrepant (from the point of view of a colonial historiography)—history of Dangi subalternity. He presumes, as an enabling condition of his conclusion, that there was a tremendous unity of purpose and clarity of vision in 'the colonial project'. The Comaroffs, in contrast, have carried out a splendid ethnography in the missionary archives. They bring out the differences and disagreements through which colonialism emerged, but have chosen to recreate the African side of the story through non-narrative sources and symbolic analysis, tending thus to expel time and agency from that part of their account.[70] Either way the problem that creeps back into historical anthropology is the separation of a category which may be called 'indigenous narratives', usually of resistance, from the narratives of power. This problem may be avoided by stressing the constant interaction between different narratives, their mutual construction and modification through this process.[71]

In the context of studying governmentality, which is what I started with, such a historical anthropology must then historicize state formation, colonialism and domination.[72] As Terence Ranger puts it, we need 'a fully historical treatment of colonial hegemony.'[73] In African studies, Bruce Berman provides an example by discussing the ambivalence of the colonial state, the possibility for the colonial state to be autonomous of the metropolis or capital, and the necessity for it to achieve a measure of legitimacy with the colonized, given its limited resources in terms of coercive power. Berman develops an argument about some field officers infused with a profoundly patriarchal mission, who claimed special local knowledge but worked through shifting and superficial alliances with certain African interests. He calls this paternalistic authoritarianism.[74] I have found uncanny resemblances to Berman's Kenya in the Indian case I have documented by examining the everyday forms of statemaking for forest management.

This kind of historical anthropology requires us to understand the very making of the categories through which we apprehend our objects of study and their constitution in the several conflicting narratives to which archives bear testimony. It also proceeds from the recognition that archives hold not merely the narratives of dominant groups, but the discordances within them. As John and Jean Comaroff emphasize, the archives always contain sets of arguments, partaking of diverse genres of historical heteroglossia that voice complex patterns of social stratification.[75] Through these internal differences of colonialism, we can know the processes through which hegemony was secured, and power was expressed, limited, and distorted, and thus begin to comprehend how statemaking occurred.

Narrative Sequences

With these ideas in mind, the book is organized in two parts. Each part deals with specific themes that illuminate the shaping of governmentalities in woodland Bengal. Overall I am most interested in the development of forms of managerial consciousness and practices in their political, cultural and material settings—taking politics to include the systems of colonial power, culture to include science, and materiality to include both economy and forests.

The introductory chapter provides a theory of statemaking. Building this theory required specification of analytic concepts that would allow delineation of the spheres in which contention and negotiation take place as regimes of forest management stabilize and then transform in a dialectical fashion. This chapter identifies political society as the arena for mediation and interpenetration of society and state, themselves understood as unstable constellations of solidarities, interests, and structural formations.

The rest of the book, alternating between genealogy and chronology, historicizes the production and transformation of political society in southwest Bengal. The chapters in Part One document the ways in which, through conquest and experiments in government, the East India Company raj initiated modes of knowing and disciplining political society. Dealing largely with the first ninety years of colonial rule—the period of Company raj—these chapters describe the schemes of governance devised in woodland Bengal, even as it became a site of statemaking. Here I trace British penetration into the forested areas of Bengal, the stabilization of colonial government in these regions and the development of mechanisms of frontier administration. Various modalities were at work here, notably travel, observation, survey, and historiographic. The latter has been well described by Bernard Cohn, who writes:

starting in the 1770s, in Bengal, the British began to investigate, through what they called 'enquiries', a list of specific questions to which they sought answers about how revenue was assessed and collected. Out of this grew the most extensive and continuous administrative activity of the British, which they termed the land settlement process.[76]

The early administration of woodland Bengal emerged as a series of exceptions and anomalies within the overarching standardizations undertaken in the land settlement process. Well into the nineteenth century, forests were unruly, uncivil wastes in most of Bengal, awaiting the civilizing touch of the plough wielded by a settled cultivator. Several cultures of governance were elaborated, with some borrowing on pre-colonial forms, frequently mixing with Eurasian innovations, and occasionally thrusting a chiefly British preoccupation into the managerial landscape. What unified them was the urge to render the landscape of woodland Bengal productive and secure for the formation of stable village communities.

A major concern here was hunting and game management. We need to distinguish types of hunters and phases. There were differences between pre-colonial and colonial era indigenous hunting; hunting for subsistence and trade; hunting as an adjunct of agriculture or a subsidy from other forms of frontier enterprise; elite hunting for the thrill of the chase, and so on. Few historians have appreciated how wildlife resources could determine the pace and character of the frontier process. Wild meat subsidized a variety of intrusions, from missionary to military and whaling expeditions, to mining and railroad construction. Game reduced reliance on long supply lines.[77] Livestock predators like jackals, hyenas, wild dogs, large cats, wolves, leopards, tigers, and bears were classified as vermin, in part becoming so because their natural prey had become scant due to hunting by British of various descriptions. During this period, barring the few isolated teak plantations in southwest Bengal, the urge to redesign the landscape, make it legible and secure, rarely manifested in the orderly regeneration of timber trees. Environmental change revealed the impact of these cultures of governance.

Part Two spans the period 1860–1947. In the latter part of the nineteenth century, changes were made with the introduction of formal forest management, the elaboration of a system of scientific forestry and finally the assertion of a strongly centralizing state forestry regime that encompassed both production of large volumes of timber, and other forest revenues, and the conservation of the environment. Chapter 5 discusses the regional variations that emerged in regimes of restriction and exclusion, constructed through policies of forest reservation and protection, as they encountered rural society and polity in different parts of Bengal. Not only do we get competing political publics as a result but spatially and temporally variegated patterns emerge in the production of political society. Chapter 6 considers examples of forest management that illuminate its continuing entanglement in agrarian economies.

Chapters 7 and 8 take up the classic themes of fire, forest working plans, and silviculture, to identify the procedures and disciplines through which scientific forestry was constructed as a regime of governance. Through the rhetoric and imperatives of conservation ideas, scientific forestry also emerged as a development regime. While nation-building and conservation came to dominate the rhetoric of governance, the idea of 'development' being the legitimate mission of the national state was continually reinvented.[78] The several transformations of this idea, and the force it exercised on forest management, are the storyline that holds these chapters together. In my approach 'development' has a long history precisely because it is not a received doctrine. I illustrate its continuous production and negotiation through processes of statemaking in the forests of Bengal. By arguing for a deeply historicized account of development, in my case the discussion of scientific forestry as a development regime, I am concurring with scholars who have shown that post-colonial Indian development discourse was shaped by the colonial governmental forms that emerged in the later part of the nineteenth century.[79]

My study emphasizes the constructed quality of development discourse and reveals its production in particular historically defined locations. To that end my analysis seeks to critique and move beyond recent post-structuralist and 'systems of knowledge' approaches to the anthropology of the development concept.[80] This task has already been taken up by some historically minded scholarship of development that has emerged in the aftermath of the post-structuralist wave of work. For instance, Fred Cooper does venture into the recent history of development discourse in west Africa. Partha Chatterjee similarly discusses the early-twentieth-century history of the rise of development planning ideas and their subsequent adoption in the first Indian government under Nehru.[81] But this limited historicization still excludes the scientific, conservationist, and social reformist aspects of development discourse which have a deeper and more contested history. Arguably, concentrating on these neglected facets of development discourse will reveal more the impress of struggles at various levels on the processes of statemaking.

This part of my book, therefore, offers two important critiques of debates surrounding forest management and colonialism in Asia. First, historians of modern forestry have tended so far to underestimate the persistence of regional variations in forest management. This has in turn led to generalized conclusions about deforestation and hence blanket prescriptions for conservation and restoration of forest health.[82] Second, scientific forestry has been treated as received doctrine, either emerging from the intellectual history of western science, or from the imperial, exploitative project of colonialism. Such consolidation of the identity of scientific forestry over time and space obscures the relationship between governance and knowledge production as it unfolds in specific historical locations.[83] By examining and showing the conflicted constitution of scientific forestry through divergent experiences in north and southwest Bengal, the second part of the book lends agency to a reconstructed nature, as management was thwarted, often not only by the resistance of disgruntled inhabitants of woodland Bengal, but by recalcitrant and poorly understood ecological processes as well.

Thus the chapters in Part Two examine the elaboration of formal structures of forest management under regimes of control distinguished by their creation of reserves and protected areas. Within each regime I study its character, its technologies of power, and the social ecology of the Bengal regions in which they came to be maintained. We can then see how patterns of statemaking are influenced by social formations and regional geographies. I argue, additionally, that these social and environmental 'forces' are not simply external elements acting upon statemaking, but are constructed through varied cultures of governance. Regional diversity in statemaking is a product of natural, social, and political processes that are both interrelated and imagined. Understanding these diversities and the manner of their production enables our scholarship to move beyond impoverished dichotomies like peasant versus state, indigenous versus western, national versus imperial, agriculture versus forest. Even more importantly, we can begin to discern and appreciate the making of these antinomies in particular historical settings. It is to this making, starting with the first tentative campaigns of British sepoys in the jungles beyond Midnapore town, that I turn in the next chapter.

Notes

1. OIOC P/3871 BRP (For), May–Jul 1891, A progs 98–121, Jul 1891, head 4, coll 2, no. 266DF-G, Darjeeling, 30 Dec 1890, E.P. Dansey, CF Bengal, to Secy GOB Rev, p. 13.

2. Similar trends marked the emergence of the Forest Service in the United States of America, under the leadership of Gifford Pinchot and Bernhard Fernow. See Worster 1994: 269; as also Pinchot 1947; and Hays 1959. For comparable African cases, see P. Richards 1988; and various essays in Anderson and Grove 1989.

3. We are now more acutely aware of the inextricable links between representation and intervention, and should be wary of easily dichotomizing the two processes of cognition and action. A fine discussion of these issues as they pertain to analyses of colonial landscape management may be found in Moore and Vaughan 1994: xxi–xxii. I have, therefore, tried to 'read' representations through interventions—the local politics of forest management—as much as I have inferred particular interventions from systematized and power-imbued representations exemplified by plans and their transformation.

4. Ramachandra Guha 1989; and Peluso 1992.

5. I recognize, with Rangarajan (1996a: 6), that such refinement of existing environmental historiography of India is necessary.

6. I am drawing here upon new writing in ecological sciences that is now more attuned to processes of micro-variation, cumulative impacts, and contingent change in the world it studies. This has a parallel in similar shifts towards open-ended analyses in social sciences, as all realms of human inquiry become increasingly sceptical of systems analysis. For relevant discussions of ecological thought see Pickett and White 1985; Botkin 1990; and Hagen 1992.

7. Worster 1993: 9. We have to consider the vicissitudes of US traditions of environmental historiography carefully because they have indelibly marked the field worldwide. The emergence of more regionally distinctive traditions of environmental history has required debating with and differentiating from the hegemonic paradigms of the New World. Interesting considerations of these issues may be found in Beinart and Coates 1996; and Arnold and Guha 1995.

8. Worster 1994: 411–20. Suggesting the need to revise unified theories of ecological imperialism, Grove (1995a: 7) has similarly argued that early colonial contexts were characterized by heterogeneity and ambivalence. I shall show that colonialism remained plural and contentious even when it matured in Bengal.

9. The main effort to depict forest loss in colonial India as a unidirectional trend may be found in the cluster of writing emerging from the quantitative analysis of land use changes attempted by the Duke University Project. See for instance, Richards et al. 1985; Richards and Hagen 1987; Richards and McAlpin 1983; Tucker 1989; and Richards and Flint 1990. An important west Indian revision has already been offered by Sumit Guha (forthcoming).

10. Sumit Guha (forthcoming); Rangarajan 1996a.

11. Leach 1994: 99; and more generally, Fairhead and Leach 1996.

12. Smil 1993: 24.

13. I am indebted to one of the anonymous readers of my manuscript, for Stanford University Press, for suggesting this bold formulation of the main arguments of the book.

14. A comprehensive theoretical and comparative analysis of these processes may be found in Scott 1998.

15. I use these terms as defined and discussed by Foucault 1991; and Colin Gordon 1991.

16. Mitchell 1991: 95. Also see Migdal 1988; and Crawford Young 1994.

17. Siu 1989: 8. For another east Asian case of nineteenth-century statemaking that exemplifies a similar argument, see Kelly 1985.

18. Lowenhaupt-Tsing 1993: 26.

19. Cerny 1990: 12-15. See also, Martinussen 1994: 230-1; Bright and Harding 1984: 2.

20. See Mitchell (1991) on systems theory. Marxist writing on the state went through many refinements, notably in the contributions of Poulantzas and Laclau. But as Bob Jessop has recently pointed out, for Marx, the state was parasite, epiphenomenon, instrument of class rule, or a set of institutions that emerge with the possibility of public power and government. Jessop (1990: 27-9), concludes, 'nowhere in the Marxist classics do we find a well-formulated, coherent and sustained theoretical analysis of the state.' Still, Marxists, especially when they discuss the state as a system of political domination with specific effects, do point the way in directions that we need to explore further. See in this respect Poulantzas 1978; Laclau 1977; and Miliband 1983.

21. Jessop 1990: 31.

22. The best presentation of this view is Skocpol 1985: 3-37; and Mann 1986: 109-36. See also Cerny 1990. Such scholarship followed Weberian theory calling for the analytic separation of society and state. See, for the classic argument, Bendix 1977.

23. Foucault 1990: 94-101.

24. Migdal 1994b: 7-36.

25. This formulation is adapted from Jessop 1990: 269-70; and Migdal 1994a: 3-4.

26. Such formulations follow one stream of Weberian thought that stresses a state monopoly of authoritative, binding rule making in a territorially demarcated area. See Weber 1968: 64.

27. Mitchell 1991a.

28. Crawford Young 1994: 24.

29. Joel Migdal (1988) makes this case comparatively, while Nugent (1994) provides a case study in confirmation. See also Sayer and Corrigan (1985: 16) who assert, 'the central state capacity in England was from the start . . . based on a high degree of involvement of local elites in the exercise of governance.'

30. Mitchell 1991a: 88, 91-5.

31. These ideas combine a reading of Philip Abrams and Max Weber. See Abrams 1988; and Parsons 1964; Bendix 1977: 286-98. My discussion of territoriality follows Sack 1986: 26.

32. Sayer and Corrigan 1985: 124. Recent scholarship on African state formation has reiterated this point, and as a counterpart to south Asia in terms of colonial statemaking, offers interesting comparisons. For the African case see Crawford Young 1994; Peel 1983; and most importantly, Bayart 1993: 179–80.

33. Herzfeld 1992: 10–15. These popular attitudes are not autonomous, they are formed in relation to the power of the state. Nugent (1993) is an elegant little study exemplifying this argument. At the same time they bring to statemaking the sharp edges of a marginality born of discontent in the context of regional underdevelopment. The concept of marginality utilized here is from Lowenhaupt-Tsing 1994.

34. Herzfeld 1992: 68; Douglas 1986; Handelman (1990) alert us to this point in different ways.

35. The phrase is taken from Verdery 1995: 239.

36. This important distinction was suggested by Wolf 1990; and is elaborated by Ghani 1995: 31–48.

37. Keane 1988: 14; see also Keane 1984.

38. Keane 1988: 32–59.

39. Pelczynski 1984; Avineri 1972.

40. Cohen and Arato 1992: 38–78; see also Perry Anderson 1976–7.

41. Cohen and Arato 1992: 255–95; see also Foucault 1979, 1991.

42. Cohen and Arato 1992: 277–86; Sangren 1995: 17–26; Habermas 1981; Fraser 1989: 32–7.

43. My discussion here is influenced by Cohen and Arato 1992; and Bayart 1993: 163–77. Unlike Bayart, however, I use the term to suggest a diversity of possibilities rather than the consolidation of elite control.

44. Fraser 1989: 123.

45. Cohen and Arato 1992: 525.

46. Tilly 1984: 310–15.

47. Habermas 1989, 1974: 49; cited by Eley 1994: 297–335.

48. As attested to by Ramachandra Guha 1993; Arnold and Guha 1995; Rangarajan 1996b; and Grove 1997.

49. Ramachandra Guha 1989.

50. Ramachandra Guha 1993: 125.

51. The topic is, of course, debated in US historiography. See Demeritt 1994; and Cronon 1995.

52. Crosby 1986; Merchant 1989; Wolf 1982; Cronon 1983.

53. Worster 1996; Hagen 1992.

54. See for instance, Botkin 1990; and Maser 1988. See also Hagen 1992; and Worster 1996.

55. Goody 1996: 262. See also Atran 1990.

56. Alan Taylor 1996: 7. He has recognized this and admitted the deficiencies that crept into environmental history as a result of such writing. See Cronon 1990.

57. Important examples are Faragher 1986; Boag 1992; Alan Taylor 1995; Bunting 1996, and Hurley 1995.

58. Some essays in Godlewska and Smith (1994) demonstrate the value of being attuned to geographic variability. Regional geography has long insisted on this point. See for instance, Pred 1984; and Urry 1987.

59. Arnold (1996) has pointed the way towards such analysis, but all too briefly.

60. Some important contributions have been Wilson 1992; Croll and Parkin 1992; Milton 1993; Massey 1994; Hirsch and O'Hanlon 1995, and Robertson 1996.

61. A comprehensive review of these ideas and their illustration through a southern African case may be found in Donald Moore (1996); and Donald Moore (forthcoming). He is refining arguments that, in African studies, have been suggested by Mandala 1990; Fiona Mackenzie 1991; Berry 1993; Leach 1994; Moore and Vaughan 1994; Fairhead and Leach 1996; and Rocheleau et al. 1996. Peet and Watts (1996) is the best introduction to recent writing in the political ecology genre across the world.

62. For instance Skaria 1999; and Rangarajan 1996a.

63. See Sivaramakrishnan 1995b; Peel 1995; Skaria 1992: ch. 2; Abbott 1991; Silverman and Gulliver 1992; Kellogg 1991; Hunt 1989; Dirks 1987; Cohn 1987a; Carmack 1972.

64. Notable figures here are Sahlins 1981, Parmentier 1987, Behar 1986, Kaplan 1990, Lederman 1986. Many contributions to Ranajit Guha (1982–9) reveal the same tendencies. I have discussed this in detail in Sivaramakrishnan 1995b.

65. For south Asia, this would hold for all the six recent and excellent works of anthropological history that come readily to mind. See Dirks 1987; Hardiman 1987; Ortner 1989; Prakash 1990a; Sundar 1995; and Skaria 1999.

66. Examples and formulations that I draw upon heavily in making this assertion include those offered by Rosaldo 1980; Kelly 1985; Siu 1989; Feierman 1990; Roseberry 1989; Donham 1990; and Haugerud 1995.

67. While Terence Ranger has handsomely acknowledged this point recently, Marshall Sahlins has been more reluctant to concede it. See respectively, Ranger 1994; and Sahlins 1993.

68. Peel 1995: 583.

69. Stone 1979; Hayden 1980; White 1980, 1984; Baker 1985; Carr 1986; Trouillot 1989; Bhabha 1990; Norman 1991; Comaroff and Comaroff 1991; Stoler 1992; Dipesh Chakravarty 1992; Duara 1995; Boyarin 1995; Amin 1995; Peel 1995.

70. See Skaria 1999; and Comaroff and Comaroff 1991. Peel (1995) makes the same point.

71. Peel 1995: 592–606. Also see Feierman (1990), for another example of the interpenetration of narratives, and the role of Gramscian organic intellectuals in mediating the dialectical relationship between narratives.

72. I am particularly influenced here by Dirks and Cohn 1988; and Sayer and Corrigan 1985.

73. Ranger 1994: 24.

74. Berman 1990.

75. Comaroff and Comaroff 1992: 34.

76. Cohn 1996: 5.

77. Beinart and Coates (1996: 17–25) make this point very well for both Africa and the USA in different periods.

78. Theoretical support for this point may be found in Cowen and Shenton 1995.

79. The best exposition of this approach is Ludden 1992a.

80. Notably the work of Appfel-Marglin and Marglin 1990; Banuri and Appfel-Marglin 1993; Ferguson 1994; Escobar 1995.

81. See Cooper 1997; and Partha Chatterjee 1993: 200–19.

82. In India, though based on regional studies, colonial forestry starting from the late nineteenth century has been depicted as reproducing a national and successional scheme. See Ramachandra Guha 1989; Gadgil and Guha 1992; Rangarajan 1996a; Prasad 1994. In other parts of Asia, national histories are perforce undifferentiating. See Peluso 1992; Totman 1989; Menzies 1994.

83. The origins and consequences of scientific forestry in India have recently inspired some debate, within the limited and unproblematized terrain I have indicated. See Grove 1993a, 1995a; Ramachandra Guha 1983, 1989; Skaria 1998; Prasad 1994.

Part One

Intimations of
a Governmental Rationality
at the Margins of Empire

The next three chapters deal chiefly with the first ninety years of colonial rule in Bengal—the period often designated as that of Company raj. They trace the social processes, environmental conditions, and the tentative construction of colonial representations of society and nature that marked this period. Political conquest on the forested frontiers of an emerging empire was lengthy and uncertain. In pursuit of a stable agrarian order, Company raj in southwest Bengal and other central Indian forest regions worked to dismantle forest polities and thus rid itself of the endemic problems that characterized frontiers.[1]

Chapter 2 explores the making of this political frontier during times when the forests were largely seen by their new British rulers as a refuge for politically recalcitrant and stubbornly backward people.[2] During the period 1770 to 1820, the East India Company struggled to stabilize its rule in the jungle areas.[3] The struggle was marked by protracted campaigns against jungle landlords; the construction of these people, their followers, and the local peasantry as primitive peoples; attempts to consolidate and enhance land revenue; the *chuar* (rural militia) disturbances; and the establishment of a police and judicial administration acceptable to British notions of rule of law prevalent at the time. In Chapter 2 these aspects of statemaking are examined to draw out the ways in which the idea of a political frontier persisted. The consequences for forest management are examined in the following chapters.

This chapter uses two organizing concepts of 'discourse of frontiers' and 'zones of anomaly' to understand the conflicted emergence and maintenance of political frontiers under changing logics and institutions of government. The notion of a 'discourse of frontiers' speaks to the formation of colonial ideas about the intractability of certain areas and their populace in forest Bengal under emerging systems of governance. The concept of 'zones of anomaly' is used to elucidate the social-ecological factors in the jungle mahals that shaped these ideas of intransigence.[4]

In late-eighteenth-century Bengal, as Company raj expanded and intensified, a specific model of rural governance was conceived and executed in the form of the Permanent Settlement of 1793. The Permanent Settlement proposed a separation of economic production from political administration, defining spheres of social autonomy and state control. This proposal provided the framework for much of the nineteenth-century statemaking in rural Bengal.

Zones of anomaly are geographic spaces in the terrain targeted by the Permanent Settlement where its application was thwarted. Woodland Bengal is an appropriate location for the study of such zones because there the Permanent Settlement model was swiftly repudiated. The resultant modification of administrative arrangements in southwest Bengal, particularly the jungle mahals, had long-lasting effects. A word of caution is needed here. It is not my intent to argue that forests were somehow inherently a site of primordial resistance to colonial government and therefore a natural zone of anomaly in colonial schemes. In Bengal forests alone, such a claim would be easily disproved. Rather, certain forest areas, at different stages in colonial statemaking, in diverse ways, and on changing terms, became zones of anomaly.

Through a rapidly growing body of scholarship on south and southeast Asia, we know how classificatory and spatial forms of knowledge, in the shape of maps and censuses, became technologies of modern state formation. The jungle mahals exemplify well these processes in the case of British expansion into southwest Bengal, through successively more sophisticated survey and documentation of the region and its peoples. In fact the jungle mahals were brought effectively under a standard British rule of law only in the early years of the twentieth century when a cadastral survey and plot-by-plot enumeration of agricultural holdings was completed there in 1915, along with a corresponding demarcation of forests into reserved, protected and private wooded estates. Clearly, surveying as a powerful means of knowing was integral to establishing British rule in India.[5]

At the same time, perceptions of the surveyability, or survey-worthiness, of any particular region—a function both of technology and estimates of profitability—were formed by administrators on colonial frontiers. Thus James Rennell, surveying Bengal in the 1780s, simply excluded the jungle mahals, for they lay neither conveniently along navigable rivers nor bore repute of great resources.[6] One reason for the jungle mahals becoming a zone of anomaly in an expanding British imperium was that the omissions of Rennell were not rectified for a hundred years. Stable represent-ations of landscapes could not be established in such areas. Disputes about the shape of legitimate and modern governance in woodland Bengal then turned on what might become the dominant mode of landscape designation.

Chapters 3 and 4 take up the story of emerging governmentality in colonial forest Bengal through themes of information and empire.[7] A vanguard of military surveyors, botanists, commercial travellers, and emissaries to Indian regional states fanned out across the subcontinent. They encountered and reported on forests as a wild place. Some among them, notably amateur naturalists in the army corps of surgeons and frontier administrators who were fond of hunting, also described the ravaged forest, forecasting the destruction of potentially valuable resources. They initiated some of the earliest debates about environmental hazards caused by forest loss.[8]

Travelling, surveying, and fighting wild animals, officials of the Company raj worked to identify, control, and conserve landscapes of production in the forests of Bengal. They initiated or accentuated processes of topographic demarcation and cultural differentiation among places and tribal peoples in Bengal that culminated in long-lasting and deeply transforming outcomes like the sedentarization of shifting cultivators, vermin eradication, and the creation of tribal homelands. Not only did the experience of forests as refuge and wild place lay the ground for an incipient colonial ethnology that

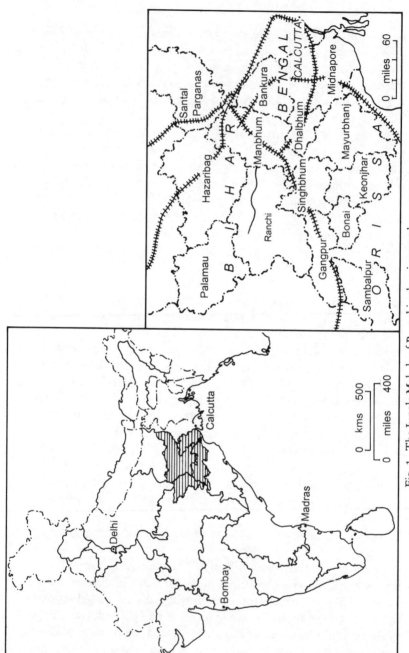

Fig. 1. The Jungle Mahals of Bengal in the nineteenth century

would introduce irreducible differences between caste and tribal society in India, it subsequently produced a hierarchized sociology of tribal primitivity that shaped the fortunes of forest-dwelling populations. Thus Chapters 3 and 4 document the contradictory nature of colonial discourse in forest Bengal, and show how the politics of establishing empire in these regions generated diverse descriptions of landscape and varied strategies for their control.

Notes

1. See for instance Guha 1996; Skaria 1999.

2. Even in pre-colonial times tribes were on the frontier of regional states, though this neglected history of their relations with Mughal and successor states is only recently being written. See Singh 1988; Guha 1995.

3. Das 1973: 2. This volume along with his other book, Das (1984) are the only detailed accounts of British entry into the jungle mahals. Others writings on the topic include Jha 1960, 1961, 1957, 1966: ch. 1; Basu 1956a, 1956b; Chatterjee 1954; Das 1971; Price 1876.

4. Unlike the case of most premodern states, such zones of anomaly were more troublesome to the rationalizing, centralizing and individuating tendencies of modern statemaking. In earlier times, as Perry Anderson says, 'different juridical instances were geographically interwoven and stratified, and plural allegiances, asymmetrical suzerainties and anomalous enclaves abounded.' See Anderson 1974: 37–8.

5. Edney 1990.

6. A useful review of early surveys is provided by Markham (1878).

7. In some ways all the chapters in the first part of this book are powerfully influenced by two ground-breaking works that were completed recently. The first is Bayly 1996. The second is Scott 1998. I would like to believe, however, that my work advances the project initiated by both these inspirational books in new directions by putting them into creative conversation.

8. The authoritative work on this subject is Grove 1995.

2

British Entry into the Jungle Mahals

Introduction

Early in the twentieth century, one of the longest serving district officers from Midnapore and Santhal Parganas described Midnapore thus: 'while the west and the north were hundred years ago given up to savagery, the east and the south were civilized.'[1] He went on to add that policing in the 'wild west' was done by *paik*s (rural militia) who held land on special tenures on condition that they would turn out to fight for their chief when required. Their existence signalled to him that 'this region had formerly been a frontier.'[2] What distinguished Midnapore, and adjoining Chotanagpur, into two zones was a combination of landscape and people. The populations of these frontier tracts, mainly of the Kolarian linguistic group, were described by their chief colonial ethnologist as 'the most unimprovable ... who finding the old country is becoming too civilized for them, fly from the clearances they have made, hide themselves in the hill forests, and relapse into their condition as savages.'[3]

The country too was unremarkable for it yielded little in terms of images of production, prosperity, or landscapes well endowed with resources. As Colebrooke and Lambert had written more than a hundred years before Carstairs, the district officer quoted above, 'the elevated tract occupying the southwest of Bengal ... is in the views of commerce and finance, of little note for any production it now affords.'[4] The very appellation 'jungle mahals' that got attached to these areas was, like 'bush' in Africa and

Australia, a place sharply demarcated as being beyond settlement and not sharing the destiny of cleared lands.[5]

Imagined in these ways, the jungle mahals (see Figure 1) became zones of anomaly in the emerging governmentality of colonial rule.[6] These areas revealed a terrain where fields and farmers were not readily visible. They marked out places where commerce was most vulnerable to predation. Revenue was collected in these zones through tribute rather than taxation. They were also places where the turbulence of zamindars did not yield swiftly to standard pacification strategies.[7] Such was the condition of the jungle zamindaris of Midnapore.

Enduring tropes of wildness that we have briefly sampled in the colonial official writing about forested zones of anomaly in southwest Bengal had ramified by the end of the nineteenth century into what we shall discuss in this chapter as a *discourse of frontiers.*[8] Zones of anomaly were a physical counterpart to the ideas and practices embedded in the discourse of frontiers. These concepts remind us that statemaking occurred within places even as it produced them. For statemaking established a palpable nexus of institutional structures and practices centred on government, while always being a historically constructed and contested exercise in legitimation.[9] Everyday routines of government defined and created certain kind of identities while denying others and these regimes of regulation were the loci of power.[10] Regional variations in these routines spatialized the operation of power. Zones of anomaly and the discourse of frontiers were manifestations of such spatialization.

This politics of space and placemaking refers to the ways 'in which spatially extensive fields of economic and political power are mediated through historically defined conjunctures of social interaction specific to localities.'[11] A concept of place helps us understand the spatiality of social life and power, by exploring the relations between time, space and regionalization, and between temporal and spatial matrices. We realize through it that economistic notions of location, geographical notions of locale and anthropological understandings of community are complementary.[12] Given our perspective on statemaking via issues of community construction and control, being aware of the role of place can remove a lot of the problems that have arisen from the confusion surrounding the notion of community. The term community has been ill-served by its ambiguous legacy. Most of the time people use a notion of community in ways that rather than distinguishing its two connotations—a morally valued way of life and the constitution of social relations in a discrete geographical setting—have more often conflated them.[13]

In this chapter therefore, the politics of establishing Company raj in the jungle mahals of Bengal will help us understand the importance of place and spatialization in analyses of statemaking. The main argument offered in the chapter is that during the first five fractious decades of British rule in the forested regions of southwest Bengal, the entire tract was largely defined as a zone of anomaly. Such a zone had a cultural geography different from river valleys and plains. There were differences in landscape, people, and polity. Company officials initially encountering the region most often called it 'impenetrable hilly jungle'. The people of these jungles were characterized in early administrative accounts as 'extraordinarily primitive'.

In Bengal the combination of 'primitive' people and illegible places offered little evidence of stable polity, and thus no convenient locus of governance to the Company officials chosen to administer them. The differences in human geography that they perceived were generated in contrast to the large agrarian villages of the adjoining plains which presented a legible landscape of rice fields, orchards, ponds, streams, and people comprehensible as Hindu or Muslim. Initially, the troublesome forest regions were a frontier zone, but over the years they became not the periphery of settlement but islands of disturbance that Ajay Skaria in the context of western India has discussed as the formation of the inner frontier.[14]

Skaria has described both the creation and dismantling of the inner frontier in western India. Increase in colonial information about forest areas, demilitarization of forest chiefs, sedentarization of forest tribes, demarcation of reserved forests, are forms of statemaking that certainly had the effect of restructuring frontiers. But the impact of these processes was uneven and modulated by regional variations. Variations are important because they were reinforced through the formation of anomalous zones of governance and the discourse of frontiers. This chapter will focus on their production in initial colonial encounters with woodland Bengal. Different manifestations of the zones of anomaly will be taken up in subsequent chapters.

What was the discourse of frontiers? How were zones of anomaly described? What were the patterns of incorporation by exclusion that defined them?[15] Characteristic features of representations and governance should become evident as the discussion proceeds. But I shall anticipate in the next few paragraphs the main tracks the chapter will take through the forest of evidence about the conquest of the jungle mahals, the emerging perceptions of the place, the politics of chuar unrest, the introduction of standard procedures of land administration and policing, and finally the creation of exceptionally administered pockets in the Company raj.

We shall see the creation of *thana*s (police stations), the *daroga* (police) system and other police arrangements; their failure, partially as a consequence of terrain; and the shifting alliances between paiks, chuars and dispossessed zamindars that emerged in the jungle mahals in the early years of the Company raj. These processes illuminate patterns of statemaking that undermine any mode of analysis which assumes the unitary, strong, centralizing colonial state as a *fait accompli*; and which then treats zamindars as either a stooge of the central state, or describes them as being in an unholy alliance with moneylenders and merchants. Such homogenized depictions of the colonial state facilitate a complementary account that gives peasant insurgency an essential territoriality.[16]

In contrast I would argue that territorialization and de-territorialization are processes encompassed in statemaking. They took particular forms in nineteenth century Bengal as they engaged a politics of placemaking that provides a crucial prehistory to spectacular outbursts of strongly place-centred rural revolt like the Santhal rebellion, which did take on, during its brief life, the character of class war. Earlier cases like the Kol and Bhumij revolts in woodland Bengal, and the protracted chuar disturbances, also do not lend themselves to analysis that does not study them in contexts of placemaking.

Spatiality becomes particularly important when we note that the Company raj was most tentative in its early years. As Sanjay Sharma has pointed out, 'till the 1830s the colonial state had not triumphed as the supreme political institution, given several political and fiscal constraints. Its ideological ambitions were not sufficiently

backed by a developed infrastructure and administrative structures.'[17] This was particularly true of woodland Bengal, and zones
of anomaly more generally, for they emerged as blank spots in the
cultivated vistas of British sovereignty. Objectification of the colony
in the light of certain kinds of knowledge was crucial to establishing
colonial power. Zones of anomaly were areas of darkness.

By not considering these variations and limitations of colonial
state power and its tenuous relationship with regional elites or
landlords, the negotiated quality of statemaking would be lost to
our analysis. We are then likely to describe the politics of regional
autonomy in terms of primordial ethnic territoriality without
recognizing how these forms of insurgent discourse are historically
produced through interaction with discourses of rule.[18] Preserving
a revenue regime of quit rents and police powers for zamindars
was a way of asserting regional autonomy, and the cross-class
alliances of raiyats, paiks and zamindars in the chuar, naik and
Bhumij revolts nurtured the discourse of frontiers. This is the
precise significance of noting that the chuar attacks were carefully
planned on outsider rent collectors.[19] More important to my
analysis is the way these attacks were not on all outsiders but only
those likely to establish and fortify a state system that disempowered
the insiders. Through a discussion of the chuar disturbances we
shall identify how colonial interventions in the structures of local
power and authority in the jungle mahals were attempted and
modified.

With the creation of the South West Frontier Agency (SWFA)
in 1833, the jungle mahals were placed in a special administrative
category of regions left largely under the charge of individuals
working paternalistically and directly with a local people, 'wild,
imperfectly civilized and occasionally disturbed', in hilly and jungle
areas. Here the forest was integral to the production of a colonial
stereotype, namely, the simple, innocent, tribal people who needed
protection from venal petty bureaucrats, non-tribal merchants and
moneylenders.[20] Through the interaction between colonial legal-
administrative procedures and socio-cultural stereotypes on the
one hand, and local political-ecological history on the other, the
zone of anomaly continued to persist as a combination of particular
people and their place. As an idea it ramified from the formation
of the SWFA, to inform forest management in Bengal in different

ways at different times and places. Exploring those processes in the second quarter of the nineteenth century in forest Bengal, this chapter will conclude by sketching the relevance of zones of anomaly to the rest of this book.

The First Campaigns and Descriptions of Terrain

In January 1767 the emerging Company raj took its first steps to settle the jungle zamindaris.[21] After receiving charge of revenue collection in Midnapore, the East India Company had been preoccupied with the eastern part of the district, where settled agriculture, an accessible countryside, and its apparently well established zamindaris were quickly controlled.[22] But Midnapore district straddled a geographical transition zone, where the 'semi-aquatic rice plain' of the Ganges delta rose through rolling upland country into the rocky escarpments and gneissic tablelands of the Chotanagpur plateau.[23] The British soon discovered the large western jungle tracts which were in the hands of zamindars who had paid little revenue since the Maratha incursions of 1740–50. In 1767 the Resident at Midnapore was instructed 'to reduce zamindars to the westward' and then demolish their mud forts, other than those required for the protection of the country.[24]

Caught up in what was to become an incompatible vision of rule through zamindars presiding over stable peasant communities, the political conquest of woodland Bengal continued over the next few years.[25] During his travels through the jungles a few years later, George Vansittart, the second Resident at Midnapore, noted the extent of mulberry trees and the potential for silk trade. He made unsuccessful attempts to persuade *tantis* (weavers) to settle there in great numbers.[26] Rather than enter into settlement, not sharing this optimistic view of future relations with British power, most jungle zamindars in the first instance just withdrew into the remote and dense jungles of their *pargana*s.[27]

In the early years British forays into the jungle parganas were limited. Spurred by the quest for valuable commodities like precious stones, European traders were travelling through the interior of central India soon after British accession to the Dewani of Bengal.[28] These travellers tended to avoid forests for fear of tigers. By the

1790s, some journals did occasionally mention encountering teak, sal and *sissoo* (*dalbergia latifolia*) forests, all Indian timbers by then widely recognized as commercially valuable.[29]

One such memoir was penned by William Hodges, who travelled in the Rajmahal Hills with Cleveland in 1782. He wrote, 'the appearance of this part of the country is very singular, having immense masses of stone piled on one another; from the interstices of which very large timber trees grow out—in many of these rocks I found the teek (*sic*), a timber remarkable for its hardness and size.'[30] Hodges was also an accomplished artist and his landscapes are a valuable visual record of the country from the late eighteenth century (Figures 2–5). Many paintings depict the areas around Rajmahal as a patchwork of forests, grasslands and groves of fruit trees.[31] The 'impenetrable jungle' of military campaigns and revenue administration was, in fact, made up of managed forests and fields. A more legible and benign landscape became visible on closer examination.

Often these travellers recount narrow escapes from the numerous tribes of the central Indian forest region. For example, one traveller passing through Gond territories noted that the Banjaras, an itinerant trading community that had established links with certain forest tribes leading to a barter trade in lac, iron ore and other forest products, were also often plundered by their tribal partners in trade.[32] The same traveller notes, describing the region adjoining

Table 2.1. Revenue Demand, in rupees, of Select Jungle Estates, 1767–1817

Zamindari	Number of villages in 1817	Previous demand	Ferguson's settlement	Prevalent in 1817
Ramgarh	25	126	616	672
Lalgarh*	-	-	879	-
Jamboni	72	84	516	670
Jatibhunni*	145	-	703	4000
Jhargram	33	238	400	236

* included earlier in Ramgarh and Jamboni, respectively.
Source: The table is adapted from figures provide by Das, *Civil Rebellion*, p. 37; and OIOC P/55/13 BRC, 1 July 1808, progs 71, l/d 2 Jan 1808, Collr Midnapore to BOR.

Fig. 2. A view of a hill village in the district of Baugelepoor

Fig. 3. A view of the pass of Sicri Gully

Fig. 4. A view in the Jungle Terry

Fig. 5. Mountains of Rajemahal where they descend to the Ganges

Chotanagpur, that the forest was a mixture of thick woods and grasslands exhibiting no sign of cultivation and contained many kinds of trees dominated by sal and Boswellia. The forests did display signs of annual firing for pasture.[33] The extensive practice of collecting sal resin (*dhuna/dhammar*) by girdling meant that the trees were mostly young. Myrobalans (*emblica phyllanthus*) and *chironji* nuts, along with resins and starch prepared from wild tubers dug up from the forest, were part of a vigorous trade from Chotanagpur to Banaras where these products were exchanged for salt and tobacco.[34] Jungle zamindars, lords of the transition zone between forest tribes and regional states, participated in this complex trade and raid economy, with its complicated delineation of jurisdictions.

During this period, the East India Company pursued expansion of the revenue base through subjugation of jungle zamindars as a way to compensate for losses in the plains caused by remissions during floods and droughts.[35] But even that limited purpose seems not to have been served. In the decade 1760–70, the revenue from Bengal went up from 116,925 rupees annually to 902,149 rupees, though the western jungles accounted only for 2.5 per cent of this amount. The revenue demand was nevertheless sharply increased in the jungle zamindaris, as Table 2.1 shows using figures for the first five estates settled in the Midnapore jungles.[36] Having made the settlements, officials remained sceptical about the ability of the zamindars to pay when the area remained turbulent and sparsely tilled. In 1769, George Vansittart, the first Resident at Midnapore to travel through the jungle areas extensively, wrote, 'the whole western part of this district is overrun with jungles in which there are scattered some trifling villages interspersed with few cultivated fields.'[37]

In 1773, Edward Baber, the Collector of Midnapore, provided the first detailed description of the Jungle Mahals:

the western jungle is an extent of country of about 80 miles in length and 60 miles in breadth. On the east it is bounded by Midnapore, on the west by Singhbhoom, on the north by Pachete, and the south by Mayurbhanj. There is very little land cultivated in this whole extent and very disproportionate part of it capable of cultivation. The soil is very rocky, the country is mountainous and overspread with thick woods which render it in many places impassable. It has always been annexed to the province of Midnapore but from its situation was never

greatly regarded by the Nabob's government. The zamindars sometimes paid their rent or rather tribute and sometimes not . . . these
zamindars are mere freebooters who plunder their neighbours and one
another and their tenants are banditti whom they chiefly employ in
these outrages. These depredations keep the zamindars and their tenants
continually in arms for after the harvest is gathered in there is scarcely
one of them who does not call his ryots to his standard either to
defend his own property or attack his neighbours—the effects of this
feudal anarchy are that the revenue is very precarious—the zamindars
are refractory and the inhabitants rude and ungovernable.[38]

The Collector went on to propose a geographical and ecological
argument for the intransigence of the region, stating that these
people not being 'under better subjection after having been reduced
so long' was principally due to the nature of the country which
from its woods and mountains was rendered almost inaccessible.
Baber's commentary found a wider resonance. In central India,
regional magnates like the Raja of Berar had long settled for a
regime of 'annual tribute and imperfect obedience' from the forest
tribes in their territories, faced with a strategy of flight into
inaccessible forests when any more systematic and regular control
was attempted.[39]

After reaping the annual harvest, the jungle chiefs would retire
to their forest-enclosed forts with the grain and their cattle to
elude sepoys unfamiliar with the terrain. The jungle zamindars did
not of course see themselves as fugitives from a fair revenue assessment of their lands by the Company raj. Some of these zamindars
described their country as jungle and their rents as a kind of quit
rent collected from paiks and chuars who were service tenure-
holding militias.[40] In making this claim, they were suggesting both
that the country was incapable of yielding more and that the
organization of tax collections was not rigorous. To some extent the
problems of jungle zamindars were part of the wider oppressiveness
of British rule that George Vansittart had acknowledged as early
as 1769. He had written:

since the government has fallen into the hands of the English every
part of the country is visibly in decline. Trade, manufacture, agriculture
are considerably diminished, many of the inhabitants have been driven
in oppression from their homes and the collection of revenue becomes
every year more difficult.[41]

The Company, however, continued to seek a more stable regime of control which would place these territories firmly under their power. The object of political control was pacification and revenue from land; the colonial authorities would ensure law and order to enable subjects to prosper and generate revenue. Ratnalekha Ray has described this process as reducing zamindaris from a polity to a unit of economic production.[42] That was certainly the intent. But in the jungle mahals this reduction remained more an aspiration, and less a reality, till well into the nineteenth century.

Surveying the outcome of these aspirations forty years after they began to materialize, the landmark Fifth Report on the affairs of the East India Company observed in 1812, 'it was essentially as revenue collectors that the English entered into the actual occupation of the country . . . the exigencies of revenue service . . . compelled them to elaborate a system of government.'[43] So curbing the freedom of these jungle chiefs and restraining their authority over the landscape necessitated creating more discernible areas of cultivation, wastes and forests. Therefore, jungle chiefs and their militia were divested of police powers while ecological warfare was waged on the landscape.[44] Felling trees, and thus obstructing the march of sepoys, was a common enough strategy for the paik sardars and their men, but the large-scale clearing of forests during military campaigns that the eighteenth-century states and the British embarked upon, was a political strategy that sought to 'break up the unstable concentrations of power on the fringes of the arable.'[45] Goals of political control and economic profits were pursued in conjunction with each other and forest clearing appeared to serve both purposes well.

The clearing of wastelands for cultivation was given high priority with the creation of a separate *bazé zameen daftar* (wastelands office). From the beginning the large extent of bazé zameen was considered the main problem.[46] Though, ironically enough, the leading agrarian historian of Bengal has found that the Bazé Zameen Regulation of 1788 actually retarded agricultural recovery after the famine of 1769–70 since this restricted the zamindars in making rent-free land grants.[47] Zamindari power and authority rested in a large measure on the ability to grant revenue-free holdings to service tenure holders like militia, police, palanquin bearers, ritual specialists, and *chaukidars* (watchmen), among others.

The attack on bazé zameen constituted an attack on this privilege. In many districts the resumption or assessment of bazé zameen included service tenure holdings like paikan and *chakeran* (personal servants) lands.[48] Zamindars who acquiesced more readily in new arrangements often provided paiks and chuars in the campaign against more recalcitrant ones, like the Raja of Jamboni did against Jagannath Dhal of Ghatsila.[49] In these instances the forces of new collaborators were turned against obdurate foes in the spirit of wartime realignments that might mark any process of conquest. But the particular configuration of alignments, the sequence in which estates were acquired, was also influenced by the feuds and historical rivalries between various jungle zamindars. The ultimate consequence was that the addition of local forces often provided vital advantage to the Company battalions finding their way through the jungles. But the formal conquest of the western jungles did not yield much dividend as manpower was not available to clear the wastelands and extend cultivation. People of the adjacent lowlands did not rise to inducements to settle in the jungle mahals.[50]

The intractability of the problem is brought out by Price, the colonial historian of Midnapore who wrote:

even in 1800 nearly two-thirds of the district consisted of jungle the greater part of which was uninhabited and inaccessible. Where it was otherwise, and the soil fit for culture the want of water, the want of bunds, the extreme disinclination of the lower orders of the natives to settle in the jungles and of the higher orders to engage in any undertaking attended with expense and risk for remote advantage seemed insuperable obstacles to any great progress in clearing the jungle.[51]

By the turn of the century it was becoming evident that profit must be sought in the raw materials either actually in the country or capable of being immediately produced by the cultivation of the soil. One official perceived the opportunities in the following terms:

the first of these is wood, the cutting down and disposal of which is an object of importance as it brings them ready money, tends to clear the country, makes it more healthy and its proceeds will gradually induce the people to settle themselves in fixed habitations and turn them more and more to cultivation. Charcoal might be made and firewood procured in any quantities, very good boats have been and any number might be built. I am told that farther in the jungles and on the banks of the Subarnarekha larger timbers might be procured . . . iron ore abounds in many parts of the country in question, and formerly many ironworks were established . . . this is an article doubly valuable as it would give employment to a number of people and the wood charcoal required in preparation would greatly tend to clear the country of jungle . . . stick lac, dammar, beeswax and honey are articles at present produced for sale . . . but all this is clearly a poor second choice to extending cultivation . . . .[52]

The report of the Assistant Magistrate was reiterating a fond hope expressed by the higher echelons of the Company raj, for the wide propagation of indigo. The Board of Trade was already aware that Midnapore was particularly inhospitable to cotton because of its poor laterite soils in the jungle tracts.[53]

As the political frontier expanded through expeditions into other remote parts of the empire, officials surveying the areas dominated by forests and tribes consistently generated a similar picture of unavailed potential in nature and the desirability of tapping into this potential. Settled cultivation remained, in the colonial imagination, the best way to attain these ends. For instance, after reporting in 1848 that half of Porahat parganas was jungle, Captain Haughten proposed a threefold classification of the land based on whether it was capable of retaining water for the cultivation of rice.[54] Another report surveying the tracts of Darjeeling recently ceded by Nepal, described the Morung areas inhabited by the Cooch tribes:

at present there would seem to be some defect in the system of cultivation, the land being left fallow for some years after yielding a cotton crop. In fact the cotton is grown on fresh lands just recovered from the jungle, by burning the trees and underwood. On this a good crop is obtained, but rice is sown the next year, and afterwards the land is left till it is again covered with jungle. Why such a course should be necessary here when it is not found so in similar tracts of zillah Beerbhoom, or the skirts of hills, which are gradually being brought under cultivation through the labour of Santhals, it is not easy to understand. If the agency of Dhangurs, who come from the Beerbhoom jungles were brought into play in these hills, probably the system of cultivation might be improved and rendered more permanent and effective. The only reason for resorting to the fresh soil of the forest is that it is a less troublesome method than the improvement of land already reclaimed.[55]

Zones of anomaly, marked by these ecological and social distinctions, were constituted in woodland Bengal, by 1850, at the boundaries of settled agriculture. But soon thereafter they were opened to reconfiguration by the spread of forest reservation. While that story is taken up later, in the next section we shall return to Midnapore and the people encountered in the jungly terrain, to learn more about the social formations that fleshed out the frontier and came to be recognized through a discourse of marginality and difference.

People of the Jungle Mahals

The Bengal countryside the British began to traverse after 1765, was devastated by acute drought in 1768 and 1769, culminating in the famine of 1770.[56] The western districts suffered the worst and even ten years later many formerly inhabited parts of the districts remained impassable jungle. The highway of imperial armies had become overgrown and wild to the extent that a small band of sepoys could not force their way through it in 1780, and the Company was forced to offer handsome rewards for killing tigers.[57] The area was thinly populated compared to the alluvial plains of eastern Midnapore, and the inhabitants, to Vansittarts's eye, were mostly paiks who were negligent of cultivation.[58] These paiks had a hierarchy of *sardar*s (headmen), *khandait*s (swordsmen) and chuars. The chuars, who were mostly Bhumij people, lived mainly in the hills between Ghatsila and Barabhum and were probably the original inhabitants of the jungle mahals.[59] These chuars were the source of zamindari militia and held lands in service tenures that we have already noticed as paikan, ghatwali and chakeran tenures.[60]

Underlining their importance and standing as a rural elite, Henry Strachey, the Midnapore magistrate around 1800, pointed out that the sardar paiks were considered talukadars (revenue officers) who had acknowledged the zamindar as their chief. Their ancestors had possessed the lands currently occupied by them.[61] But these paiks obviously did not preside over a stable peasant community in the jungle parganas. In the last quarter of the eighteenth century, forest rapidly encroached on abandoned fields and the jungle zamindars were hard pressed to organize cultivation through migrant labour from the east. For such outside assistance they paid a premium. In western Burdwan, the *khudkasht* (those cultivating their own lands) raiyats, reckoned as aborigines, paid a higher rent for their lands than the *paikasht* (those brought in seasonally to cultivate) raiyats who were considered of superior rank. Among the paikasht, the *nij ganji* paikasht who cultivated lands in their own villages paid more than the *bazé ganji* paikasht, who cultivated land at a distance.[62]

When environmental changes sharpened hill–plain hostilities, they had the geographical consequence of widening the forest-savannah transition zone that buffered settled agriculture from their forest-based raiders. The Santhals, who came to occupy these fringe areas, were recent settlers early in the nineteenth century.[63] Forest clearing by Santhals for agriculture was part of a migratory process that gathered momentum through the first quarter of the nineteenth century in the jungle mahals, bringing together forest peoples like Paharias and Kharias, Santhals and the Bhumij peasantry from which the chuars, paiks, ghatwals and jungle zamindar were drawn.[64]

The increased surveying of the early nineteenth century added many hilly and forested areas to the growing list of frontier landscapes. Concurrently the chuars as a people became part of the wider category of wild tribes like the Kols, characterized by their quickness to a violent display of anger, who would bring out the arrow and the axe at the slightest provocation. They did not allow any outsiders other than weavers and *goala*s (cowherds) to settle in their midst, preferring to keep large areas of village land as 'waste'.[65] As peasants these people exhibited, to the British administrators keen on observing a healthy attachment to land, a singular lack of tenacity in settling on identified lands and converting them into productive farms. As one account written in the middle of the nineteenth century has it:

the cultivators in these jungles formerly held their fields in some part without leases . . . they brought their whole produce to the zamindar, who gave them means of support during the year . . . if they were oppressed . . . a whole village would literally in one night up-stick (their huts were made of sticks and leaves) and off to another zamindar whose general character promised better treatment.[66]

Such acute scarcity of labour meant that neighbouring zamindars

were often in competition for the same labour force, the mobilization of which was left to the paik sardars and chuars, who sought
to command both loyalty and terror. They remained influential
longer in the jungle zamindaris, which were the only ones by 1802
that still had armed retainers.[67] The relationship between zamindars,
paiks, and chuars was close. Most jungle zamindar families were
related by marriage.[68] The paiks and chuars were mostly of the
same ethnicity as the zamindars, generally Bhumij, and known to
intermarry.[69] The principal leaders of the chuars were always
connected with each other by ties of blood and affinity. For example
Haroo Digwar, the main agent of disturbances in 1780–5, was
related to Chattar Singh, zamindar of Baugree; father of Achul
Naik, and father-in-law of Govardan Digpaty, the prime movers
of unrest in the 1798–1815 period.[70] In a rare study of caste
composition and mobility of agriculturists in southwestern Bengal,
Hitesranjan Sanyal argues that the Sadgops, pastoralists from
Burdwan, became agriculturists by moving into the jungly lateritic
tracts of western Midnapore.[71] They imposed themselves over the
existing forest tribes and may have coined the contemptuous term
chuar for the Bhumij.

On the contrary, Ratnalekha Ray's analysis of landholders and
peasantry in Midnapore at the commencement of British rule reveals
that in the jungle parganas the zamindari of Midnapore alone passed
out of the hands of a tribal raja into Sadgop hands, before the
Permanent Settlement of 1793; and after 1793, only three other
parganas were bought by Bengali speculators, and retained, out of
at least seventeen distinct estates.[72] The patriarch of the Sadgop
family that came to own the Midnapore zamindari had worked as
the diwan of the Khaira (*sic*) raja, whom Ratna Ray calls a tribal
chief.[73] He was an exception. Most jungle zamindaris remained
with the families that owned them and fought to regain them
when temporarily dispossessed, like Durjan Singh of Raipur and
the Raja of Pachet.

Fiercely enforced ties of kinship and ethnicity allowed jungle
zamindars to successfully resist the transfer of their estates to
outsiders. While these ties facilitated a strong loyalty, they did not
interfere with a loosely structured administration. Even the Midnapore zamindari under Sadgop ownership, the only one with a
strong and extensive control over substantial areas in both the

jungle and riverine areas, followed the practice of leaving jungle parganas on small quit rents with paik sardars. When these paiks were displaced they often paralysed the revenue collection.

The paiks were retained through rent-free grants of land for which they performed a variety of police and revenue-collecting functions. Not surprisingly, the periodic disturbances that occurred from 1767 through 1817 were led by disgruntled paiks as the British rule sought to dismantle their military fiscalism. The Company government began to realize this by 1800 after the chuar disturbances had gone through their most violent and widespread phase in the last years of the eighteenth century. As one senior official concluded:

the disturbances . . . have been principally excited by . . . paiks who were formerly employed by zamindars in services connected with the police of the country and more particularly for its defence against the incursion of chuars, a predatory race of people inhabiting the hills and woods on the frontier of the district. In consideration of these services the paiks were permitted to enjoy at a small quit rent extensive tracts of land in Midnapore . . . paiks were resumed and the jumma of their lands incorporated in that of different zamindaris in which they were situated . . . the paiks not reconciled to this change . . . abandoned their lands and entered into league with chuars.[74]

If the paiks and chuars were characterized derogatorily as savage and prone to violence, the jungle landlords fared no better. They were described as rude and illiterate, thereby incapable of comprehending anything but the simplest and easiest regulations, while possessed of a character that made them 'ignorant, savage, vicious, selfish and corrupt.'[75] The jungle zamindars were also deeply averse to attending court proceedings as it involved travelling larger distances and they hesitated to move far from their jungle strongholds. To one commentator, this was at least partly due to every zamindar having been at some time a criminal in the eye of the law.[76]

Henry Strachey, the Magistrate of Midnapore, who was instrumental in the return of police powers to zamindars in Midnapore and the restoration of paikan tenures, wrote in 1802 of the decay of the typical stone-and-mud forts of the jungle rajas. This to him was proof of their decline and the passing of their estates into the hands of urban adventurers.[77] Ratna Ray argues against this picture by pointing out that the transfers that occurred were more to zamindari servants, employees of the collectorate and local business-men. The point remains that they were Bengali and not related to the original jungle zamindars.[78] The chuar disturbances that are discussed below targeted all such transferees as outsiders, attacking thereby the transformations of the local state system by zamindari settlement and daroga policing.

The transformation of ownership in landlord rights was partly thwarted in the jungle parganas. But the gradual extension of cultivation by forest clearing after 1800 introduced new groups into the peasantry and spread a system of *mouzawar* assessment (village tenure) that strengthened village heads. They were called *mandals*, *pradhans* and *majhis* depending on whether these were Mahato, Bhumij or Santhal villages. They often organized raiyats effectively against *jamabandi* (revenue assessment).[79]

The eighteenth century witnessed a rise of these mandals, a process accelerated by the famine of 1770 which destabilized the rural economy by increasing wastelands, creating labour scarcity and shortages in capital. Their rise was facilitated in the fifty years following the famine, as the paik sardars and other landed elites of the jungle areas came to depend on paikasht raiyats who through ownership of plough animals and other capital required for agricultural extension established a migratory system of cultivation where they retained larger shares of land rent than the less enterprising and poorly endowed khudkasht raiyats.[80] Zamindar–mandal tensions can be noticed by late in the eighteenth century, and the first decades of the nineteenth century. Since labour remained the limiting factor of production in the jungle parganas even after 1850, mandals, with their control over labour and large uncultivated lands, remained eminent in the rural social structure through most of the nineteenth century.

Over the first fifty years of the Company raj a mixture of old and new ethnic groups rolled back the frontier in the jungle mahals in a gradual process that by and large worked through zamindari encouragement. New people were constantly becoming native to these places in the forest-savannah transition zones of southwest Bengal. These close to the ground transformations of agrarian society, however, exacerbated elite conflict, and played into a series of confrontations between regional chiefs and Company power. These frequently violent struggles articulated a politics of place in the manner already suggested in the introduction. The next section will examine in detail the logic of chuar unrest and evaluate its impact on colonial statemaking in the jungle mahals till the formation of SWFA in 1833.

The Chuar Disturbances: A Politics of Place

About two decades ago, one historian of Bengal suggested that the rural disorder of the late eighteenth century had not been studied in detail yet.[81] This omission in the historiography of Bengal remains. The subsequent work of John McLane on banditry also dwells on the early nineteenth century and the Bengal lowlands.[82] Though the period from 1770 to 1800 is particularly marked by unrest and statemaking in the highland frontier of western Bengal, McLane, like Chatterjee, commences his discussion with the Permanent Settlement. His account explains the emergence of banditry in terms of the impact of this settlement on rural social relations, and the centralizing tendencies of colonial statemaking that he traces from this 1793 watershed. In his scheme, chuar unrest became the last stand of independent chiefs against British conquest and early-nineteenth-century banditry the outcome of such conquest which led to the alteration of tenant-landlord relations and the reorganization of police and judicial administration.[83]

Binod Das and Jagdish Jha conform to this line of reasoning. They argue that the chuar disturbances were a direct response to the sale of jungle zamindaris in arrears of revenue and the abolition of paikan tenures.[84] Thus they too give great importance to 1793 and the impact of Cornwallis on the forested landscape. We have noted earlier that there was a wider process of consolidation of village oligarchies against the backdrop of crumbling superior administrative structures, and especially shrinking zamindari power. This was partly hastened by natural calamities like famine and British policies of demilitarization of zamindars. Ratna Ray seeks to explain chuar disturbances as an anarchic consequence of this breakdown.[85]

But chuar disturbances always worked through an alliance between jungle zamindars and their faithful militia. It is difficult to read them as emergent class conflict between rising rich peasants and fading magnates. In that sense there is still a missing history. It is not my purpose to provide this missing history of late-eighteenth-century rural unrest in the Bengal uplands. But by examining certain aspects of it I wish to propose that the politics of the jungle mahals illuminate a special terrain of statemaking in colonial history which we have already identified as the zones of anomaly. The politics of chuar unrest and its aftermath set in motion a logic that became a recurrent feature of colonial rule in the jungle mahals. The persistence of the discourse of frontiers in statemaking defies a historiography that suggests the inexorable subsumption of frontiers in the expansion of British rule and its stabilization.

As a solitary illustration of the way these aspects of state reorganization of administrative and legal arrangements in these newly defined zones of anomaly in turn reproduced and perpetuated the discourse of frontiers, consider the following opening description given by a Bengal civilian writing around 1900 about the Santhal area:

Here Bengal, Behar and Chotanagpur meet. Thrust in between them like a wedge is this debatable ground, a tangled mass of hill and jungle peopled by uncouth aboriginal races, standing like the furthest outpost of barbarism to face the highest civilization that Hindu and Musulman successively planted at its gates. Its primitiveness untouched by the stir of passing events, it has looked down with unresponsive gaze upon all that was best and highest and furthest advanced in the India of the day.[86]

Chuar unrest participated in the production of these primitive places through their influence on colonial imagination, ethnology and administrative policies. To understand that genealogy let us return to the immediate issues of chuar unrest in the jungle mahals in the early years of British political expansion over the region.

In their discussion of the spread of banditry in lower Bengal, McLane and Chatterjee point to the leadership of *chaukidar*s and other displaced service tenure holders who lost rent-free lands to the combined onslaught of resumption by the state (as it took over police functions) and landlords (seeking enhanced rents to meet exorbitant revenue demands).[87] The other point is that these dacoits were complicit with their patron zamindars.[88] But so too were the chuars from the very beginning. So rather than treat chuar unrest as an upsurge of traditional banditry in a period of eighteenth-century anarchy, or as the assertion of essential tribal solidarity and independence, we may more usefully note the constant negotiation of structures of local authority and governance that becomes particularly visible during these moments.

By the early 1800s, the jungle mahals witnessed a return to zamindari control of local policing.[89] This marked out the region as an administrative frontier where the consolidation of daroga and magistracy control over police and judicial administration was retarded.[90] The tenuous hold of Mughal authority on the jungle mahals has already been noted. In most of Bengal the local administration had emerged as a complementary arrangement between mughal *faujdars* (police officers) and zamindars.[91] But the jungle provinces were the areas where this complementarity was most attenuated with jungle zamindars and their militia (paiks and chuars) most dominant. So any discussion in terms of the collapse of Mughal police arrangements and the consequent efforts of British rule to introduce the daroga system are of little relevance to these areas.

That is why it becomes important to examine the instances of chuar unrest so carefully reported by Price in his narrative of the troubles in the jungle provinces.[92] By looking closely at the types of violence perpetrated, and the careful selection of targets in chuar attacks, we may identify the production of a politics of place and a concerted attempt to preserve local structures of authority that conflated economic and judicial functions. While the increase in revenue demands certainly provides an immediate edge to conflicts, the struggle in the jungle provinces seems to have been more over who collected revenue and how the demand would be ascertained.

Deposing before the faujdari court of Midnapore, one Anantram Nandi of Anandpore stated in March 1799, 'I and Cochil Dey my brother-in-law went to Dhobasul to bring sixteen arrahs of paddy . . . hundred robbers came and plundered four bullocks loaded with grain . . . they said . . . by doing so . . . the resumed paikan lands would be restored.'[93] By disrupting the grain trade, attacking outsider rent collectors and drawing a line in the forest that was hazardous for anybody to cross, these local elites were using terrain and place to define the limits of administrative innovation.[94] Price also reports the refusal of *pottah*s (certificates of landholding) by raiyats in the jungle areas, clearly an attempt to prevent the neat categorization of wastelands as cultivable and uncultivable.

By the beginning of 1796, the intensified depredations of chuars in the hilly tracts of Midnapore compelled local officers to seek military assistance from Calcutta.[95] The chuars traversed the district in bodies of five to eight hundred, driving cattle and anything else that moves before them. As one military report from Pachet observed, 'vast herds of cattle have been driven off by the malcontents and have been traced to Manbhoom.'[96] Among the items reported stolen were clothes and some utensils. A message, originating as much from jungle landlords as from the chuars, was thus sent to land purchasers and their adherents that they and their commerce were unwelcome.

The chuars concentrated their attacks on officials appointed for revenue collection, causing them to abandon their posts and raiyats to flee to more peaceful estates like Baminbhoom and Burdwan.[97] In early 1799 the Collector of Midnapore reported that the chuars had come to a village six miles from Midnapore and left a notice for Raghunath Pal, the *tahsildar* appointed for resumed paikan lands, asking for the supply of rice, *dal* (lentils) and other supplies failing which these demands for food taxes would be enforced.[98]

Numerous accounts are given, in the cases filed by thanadars and *burkandazés* (constables) in the *nizamat* and *faujdari adalat*s (criminal and police courts), of chuars attacking whole villages and burning them. It is important to note that the villages were generally those where revenue collection by outsiders had been possible.[99] In Salbani the village accounts were burnt after killing Bhaktaram, the petty revenue official, leading the District Collector (the top East India Company official) to despair about the *mofussil* (country) settlements of the jungle parganas. Price notes that the 'chuars were never to be completely suppressed as long as they could take refuge in the wilds and fastnesses to the west of the district.'[100]

Several decades later, in Singhbhum, seeing a proposed road being marked off, the Coles (Kols) deserted their homes and took to the jungles, which is not only a token of fear but is generally amongst such wild tribes, token of proposed resistance also, and they only returned when the marks had been removed.[101]

The jungle as an uncivil refuge for the tribal people is a colonial construct that emerged from this period of strife and lodged in the discourse of frontiers.[102] The chuars had by this time entered Company discourse as vicious, desperate mobile banditti who lived by rapine, pillage, extortion, and murder, using the forest as their sanctuary and base of operations. In a perceptive report of 1804, Ernst, the former Collector of Midnapore and Judge of Circuit Court, noted that the chuars were dismissed paiks who clearly used their banditry and ability to strike fear in the raiyats to renegotiate their terms of employment with zamindars.[103] Their participation in more localized forms of statemaking was clearly disrupted by Company intervention. While in some situations, the zamindars conveniently blamed poor revenue collection and hence the need for a lower revenue demand on the depredations of chuars, on other occasions they made common cause with them. Yet the restoration of police powers to zamindars and the policy of using paiks to catch chuars was evidently predicated on the belief that zamindars, though rude and ignorant, commanded the loyalty of their raiyats, which qualified them to make police administration effective.[104]

The zamindars and paiks, having secured the withdrawal of darogas from the jungle mahals, did not prosecute their task of apprehending and liquidating chuars with any vigour. In this way they also frustrated Bengal government efforts to minimize reliance on the army in maintaining order in the jungle parganas. Since the tracking of people, as of game, was possible only after leaf fall and burning of the forest floor in March–April (*Chaitra–Baisakh*), military contingents were detained on several occasions for this purpose, upsetting the schedule of troop movements to accommodate zamindari policing arrangements.[105]

Henry Strachey recommended the return of zamindari policing in the jungle mahals in the hope that by degrees the tranquillity of the jungle mahals would come to depend more on laws and less on the character of those entrusted with police powers. Another practical cause for an exceptional local delegation of policing was the unhealthy conditions of forests, especially in the rainy season and high losses of sepoys and non-local burkandazés to forest fevers and other ailments.[106] But the transition to rule of law proved elusive. Magistrates from Strachey to Hodgson, over the period 1800–15, reported the reluctance of victims and witnesses of chuar atrocities to giving evidence.

During the period 1770–87 the chuars were uniformly styled rebels and insurgents. From 1795 to 1815, the chuar disturbances were more particularly attributed to various factors: resumption of service tenures that deprived ghatwals and paiks of the means of subsistence; hereditary feuds among jungle landlords; the essential lawless habits of chuars; and the new proprietors created by auction sales of estates failing to cooperate with the magistracy and daroga establishment in securing law and order.[107] Disagreeing with descriptions of the chuars as criminals, bandits, and rebels, the Court of Directors held that the chuars were contesting the locus of governance by challenging definitions of what was just. 'The strength of the country, the nature of the contest, and the peculiarity of their cause' led to a dreadful war that ultimately was about the enforcement and rejection of a system of justice. In those turbulent times the administration of justice was often the vehicle of government as it searched for the order and tranquillity that would become the foundation for prosperity and enhanced yields of revenue from land.[108] The struggle, and the repudiation of magistrate-daroga raj, was particularly protracted in some places but in the rest of the jungle mahals police regulations were introduced and withdrawn in the space of seven years in the last decade of the eighteenth century. Turning to that wider story at this point will facilitate grasping the larger implications of the discourse of frontiers.

Jungle Landlords and the Police

Through the first two decades of their control over the forested estates, the British were exasperated with the protracted turbulence and disorderly conduct on the part of zamindars in the jungle mahals. Regulation XXII of 1793 established a police administration by the government of the East India Company. In the process it took policing work from zamindars in Bengal and gave it to thanadars (police station heads) to be appointed by the government. These thanas had jurisdiction over a twenty-mile radius and all manner of village watchmen were placed under the jurisdiction of their darogas.[109] These darogas were placed under the direct supervision of the District Magistrate, effecting the separation of revenue and judicial functions between Company bureaucracy and landed elite that has been referred to earlier.[110]

That the arrangements largely failed in the jungle provinces is clear from the continuous disturbances that occurred in the next seven years, reaching a peak in 1799. March 1799 marked the murder of Rasiklal Ghose, daroga of Thana Janpur, one of two thanas set up under the new police regulations in Midnapore. The Collector of Midnapore, J. Imhoff was quick to report the ineffectiveness of the daroga system of police for a populace largely made up of savages, and requested special arrangements. He also highlighted the frequent failure of auction purchasers when trying to take possession of jungle estates, as in the case of Durjan Singh of Raipur. In the Collector's words, restoration of paikan lands alone would prevent the country 'becoming a harbour of robbers and wild beasts'.[111]

Imhoff further recommended that a corps of rangers be created following the example of Bhagalpur Archers formed by Augustus Cleveland from amongst the Paharias in the Rajmahal Hills. Jungle zamindars and their paiks could restrict chuars driving stolen cattle through their estates, while the government would not have to rely on sepoys unfamiliar with the terrain and susceptible to jungle fevers.[112] The Board of Revenue endorsed this report, noting that resumed paikan lands in Midnapore had been assessed at twice their earlier quit rent. These demands had not been collected after the first year. What seemed to clinch the issue was the perception, epitomizing the discourse of frontiers, that 'the people are extremely wild and ignorant and to the last degree pertinacious of their customs. Sound policy would seem to suggest indulgence in those customs is best calculated to preserve the peace of the country.'[113]

Briefly skipping ahead of the story, it is useful to note that this view contained the germ of the idea that restored the zamindari and ghatwali police, leading to the ultimate deregulation of the jungle mahal area in 1833. The Board of Control, taking serious note of the violence and its apparent origin in the resumption of paikan lands, ordered that the fixed settlement be entered into with paiks and a better description of the jungle mahals be provided in the process.[114] A police committee, in its report to the Governor-General in Council, recommended a district system of police with joint responsibility to restored paiks and darogas.[115] This did not appeal to the Collector of Midnapore and others who held the chuars were bandits, and felt, therefore, that settling with them was 'making treaty with the open enemy'. Such a policy was bound to encourage other banditti to seek advantageous settlements through rebellion.[116] But in their deliberations on the system of police, the Company authorities were more concerned with securing the conciliation and cooperation of jungle tribes through their zamindars. So the government approved the restoration of zamindari control over policing, through paiks and ghatwals, giving them joint responsibility with darogas.[117]

The restoration of police powers to zamindars occasioned an extensive official intercourse with zamindars and from this Strachey, the architect of the policy of zamindari conciliation and involvement, argued

the weakness and inefficiency of the former system of police in the jungles arose . . . from the zamindars not possessing power in proportion to the interest they have in the peace of the country, that is to say, generally speaking in proportion to their property.[118]

He went on to recommend this system to the rest of jungle zamindaris. In giving the jungle zamindars the responsibility of maintaining law and order in their estates, terrain was clearly an issue, since from the vantage of their jungle fastnesses these landlords always harassed troops with impunity. Another matter of concern was the creation of a buffer zone between British Bengal and the Mahratta raiders who came through these jungle territories from the southwest (Orissa and Central Provinces). Recognizing the impracticability of deploying armed forces in a scattered fashion, the Governor-General had ordered military commanders to work with district magistrates, who in turn were instructed to secure the effective cooperation of hill and jungle zamindars 'in such measures that may eventually be pursued to repel any hostile invasions.'[119]

But the question of local knowledge and information was equally important and in many ways linked to the mysteries of the wild lands and difficulties of a somewhat incomprehensible geography. Local knowledge, invaluable to maintaining law and order was available to the superseded village watchmen who had caste and kin ties to the chuars. This was soon recognized by the Company administration as its still uncertain and scant judicial machinery began to deal with the flood of reports of violent rural crime that flowed in with the spread of chuar disturbances. One judge concluded, 'the prevention of gang robbery must arise entirely from the vigilance of village watchmen . . . assisted by villagers . . .

who know the robbers.'[120] He thus articulated the paradox that the frontiers of both society and state were the heartland of vital information transactions. Typically the excluded tribal people were the ones who knew the forest, its people, paths and products; they purveyed this information to get leverage in the state system. As Christopher Bayly says, centralization of rule by the British, or their decision to rely on superficial structures, was tempered in specific locations by their ability to collect such information.[121] The pertinent question, from the standpoint of information and its accessibility, in the jungle mahals of the early nineteenth century was, where else should the special dispensations made in Midnapore be extended?

With the formation of the Jungle Mahals district in 1805, the zamindari police system was extended there.[122] When the Permanent Settlement was extended to Dhalbhum in 1800, it became another province of Midnapore, though not in the Jungle Mahals district, to be part of the zamindari police arrangements. But the Company government was anxious to confine these special provisions to a small area:

We believe it to be true that the local power and influence of the zamindar would give them peculiar advantage in suppressing disorder and depredations. We nevertheless think that there would be serious grounds of objection to the general adoption of that principle. The perpetual property conferred on the zamindar in the lands must gradually strengthen that power which they possessed before from office and prescription. The progressive improvement of the country which may be expected under the present system of land tenancy will increase the wealth and independence of the owners of the land. We cannot but doubt if it would be prudent to invest the same persons with the additional authority which the administration of police would bestow and we have understood that zamindars themselves have connived at if not conducted depredations against neighbouring districts and even their own ryots. We think it possible too that they have been indisposed to the success of any system of police which did not vest the power in their own hands.[123]

With these words the Board of Control seemed to limit the special nature of arrangements to regions characterized by peculiar problems. The separation of powers, especially economic functions from those of governance, was thus reversed in the zones of anomaly.[124]

Withdrawal of daroga policing was followed by the withdrawal of regulations for sale of estates.[125] The Board of Revenue also suggested that regulations for the partition of estates not be applied to jungle zamindaris, the reason being 'the nature of the country . . . and disposition of the people, attached to their chiefs and in the last degree tenacious of their local customs.' Such allegiance made difficult the interference of government in the internal arrangements of managing land.[126] Even earlier, the issue of pattas (certificates of landholding) and the necessity of measuring lands under the Permanent Settlement had produced a volley of protest. The Raja of Pachet said:

my country abounds in hills and woods, and the villages were never measured, if now I make the measurements the raiyats will run away and the villages will be depopulated. A mahato takes a pottah for a village and then divides the land into 16 parts to raiyats who cultivate and pay revenue . . . there has never been a rakhabundy (measurement).[127]

Nearly twenty years later, the Collector of Midnapore opposed the imposition of Regulation XIII of 1817, in the jungle estates, asserting the establishment of *kanungo*s (revenue officials) in those areas was likely to fail on account of *the natural disposition of the people and obstacles offered by the face of the country*. His words offer one of the best examples of the way a zone of anomaly was defined:

the seclusion in which the jungle populations live tends to separate them both in fact and idea from the inhabitants of the open country and from their attachment to long prescribed customs and local usages, innovation or anything bearing that appearance is particularly hateful to them—the ferocious nature they display when under the influence of passion dictates the expediency of avoiding any reason to excite to irritation a class of men easily thrown into outrage and disorder and very difficult when once roused by wrongs either real or imaginary to be quieted. The nature of the country . . . where in large tracts . . . the true distinction and definition of boundaries . . . would require perhaps greater skill and science than could be met with . . .[128]

Thus the relaxation of direct control over policing was accompanied by a parallel hesitation to closely supervise revenue collection, allowing in effect a degree of local autonomy both in land and judicial administration that was unique to the jungle estates.

At the same time the restoration of police functions to zamindars did not restore status quo ante. Each of the district chiefs framed elaborate rules, inspired by the work of Henry Strachey in Midnapore, and placed the appointment of paiks under the supervision of the magistracy.[129] The police had, by the middle of the nineteenth century, settled into three systems: (a) the usual thana system for regulated provinces; (b) the unchecked and unassisted zamindar as in jungle mahals and Manbhum; and (c) the zamindar assisted by a clerk as in Jhalda and Patcoom.[130]

Zones of Anomaly in Woodland Bengal

In 1833 the Governor-General had agreed to exclude Chotanagpur and the jungly parts adjacent to it in the districts of jungle mahals and Midnapore from the operation of general regulations and they were formed into a separate jurisdiction under the superintendence of the Political Agent for the South West Frontier.[131] This had followed the famous report of joint commissioners W. Dent and T. Wilkinson, appointed to inquire into the affairs of Chotanagpur following the Kol uprising, also known as Gunga Narain's Hungama.[132] In Barabhoom and Dhalbhum, the prime movers of the unrest had been sardar ghatwals, who held on nominal quit rents lands surrounding the *ghats* (hill passes) they guarded. They had also served as zamindari police. These ghatwals were mainly drawn from the Bhumij. The region also had scattered Santhal, Kurmi (Mahato), Kharia (Lodha) and Paharia populations.[133]

Dent's report on the causes of the unrest in Chotanagpur during 1831–3, rehearses all the constituent elements of what we have already discussed as the discourse of frontiers. Describing the dense jungle, meagre cultivation, forest trade and nominal rents paid by paiks, ghatwals and other zamindari militia, Dent concluded that Gunga Narain Singh killed his uncle Madhub Singh because he committed excesses in league with outsiders. That unholy alliance had benefited from the presence of thanas and darogas. By sacking police stations Gunga Narain and his followers revived the logic of chuari.[134] In response Dent recommended direct rule that would safeguard the role and property of jungle zamindars and their privileged tenants.[135]

The campaign against Gunga Narain occasioned a close

acquaintance with the hills and forests of Chotanagpur, revealing a brisk trade in coal and iron, in a landscape often very pleasing to the eye. After taking one stronghold from the insurgents, Dent wrote of it, 'the place was . . . situated in a beautiful glen . . . the houses were large, well built and had an air of comfort which quite surprised me.' Later, observing the industry evident from crops in the fields he conceded the Bhumij had considerably risen in his estimation.[136] But these jungle fastnesses were also the terrain where sepoys had always struggled against disease and strangeness. The jungle mahals field force lost ten times more soldiers and camp followers to cholera and jungle fever than to matchlock wounds inflicted by Bhumij and Kol fighters.[137] When Gunga Narain was defeated and killed in February 1833, while attacking a thana of Thakur Chetan Singh of Kharsawan in Singhbhum, local assistance in the form of village guides and the support of many ghatwals had played a crucial role in the British success.[138]

Disturbances in Singhbhum followed the same pattern. Dispossessed Mankis and Mundas (traditional headmen among the Kols and Mundas) plundered the Muslim *thikadar*s (revenue farmers) who particularly incensed the Mankis by denying them fruits from trees planted by their ancestors. Dent and Wilkinson explained the violence not only as having this material origin but also in terms of Kol psyche, and its 'semi-barbarous state', that inclined them easily to plunder and fuelled a fierce loyalty to village heads.[139] Analysing Bhil disorderliness in western India in the same fashion, John Malcolm said that since the Bhils were 'constantly exposed to danger from their fellow creatures, and from the ferocity of wild beasts, with whom they shared the forests . . . they have cherished predatory habits as a means of subsistence.'[140]

It was the same argument about what we might call 'deep primitivity' or something ordinarily unredeemable. Yet, this condition was at times diagnosed as holding out the precarious possibility of reclamation through the efforts of extraordinary men like Augustus Cleveland, who worked with the Paharias, or James Outram, who 'tamed' the Bhils. Both these young officers of the Company raj in its forest provinces had raised military corps from their 'brigand' charges. In Singhbhum, Dent and Wilkinson reiterated the need to rely on zamindari policing, and suggested that Mankis and Mundas as tenure holders should be the focus of village administration. If their tenures were confiscated for default in payment, Dent and Wilkinson recommended transferring the management of the tenure to a kinsman or a European agent.[141]

Having constituted the legal-administrative entity called the SWFA, the Government of Bengal directed Dent and Wilkinson to identify the precise tracts that would be included there.[142] This exercise admirably illuminates the ways in which the creation of specially governed territories drew on the notion of zones of anomaly. From the Midnapore Magistracy Dhalbhum was identified for inclusion in the SWFA because it was large, jungly and a difficult place to get information about, in other words, inaccessible or illegible.[143] Other places like Patcoom and Barabhoom were included in SWFA because they were inhabited by the 'notorious Bhumij people'.[144] Ramgarh possessed all the qualifications laid out in regulation XIII of 1833 creating SWFA; a jungly geography, tribal sociology, chronic disturbance—both through administration and commerce—under conditions of outsider contact, and favourable response to a system of zamindari policing.[145] Even into the 1850s, only 25 per cent of the SWFA was under cultivation; trade in raw silk, rosins, and buffaloes provided the main revenue.[146]

The zone of anomaly thus became a special place defined by the unique combination of tribal people and forest landscape. For the purpose of governance it was also a space less legible than others surrounding it. Defined by multiple differences from the other spaces of colonial rule in Bengal, zones of anomaly also showed the limits of colonial knowledge, which was already being assembled through compilation of customs and stereotypes, through readings of classical texts in Persian and Sanskrit, and by way of a geographical sensibility that saw field and forest in neat separation. Even if not recorded in these terms, by constituting SWFA and listing the reasons for doing so in the manner it did, the Company raj acknowledged its constraints in certain tribal places and forest spaces.[147] This created an enduring pattern of exceptionalism in the political administration of certain regions of woodland Bengal.

After the Santhal *hool* (uprising) of 1855 this form of rule grew more sophisticated, and in addition to the suspension of civil and criminal law as normally applied elsewhere, these non-regulation provinces (or scheduled districts as they were called after 1874) also had special tenancy law that sought to prevent land alienation from tribal indigenes to non-tribals.[148] In non-regulation provinces district officers exercised tremendous discretion as local specialists, making 'man dominant over machine' for administering 'localities or races having peculiarities, especially those reputed backward, such as for instance the western frontier districts of Bengal.'[149]

These themes of colonial knowledge, regimes of control and their variations in forest Bengal are taken up in the four chapters in Part Two of the book. In the next two chapters we shall deal with the prior processes of statemaking through which woodland Bengal became more amenable to the transformation from wild to manageable. Chapter 3, and subsequent parts of this book, will also highlight the distinctive impact of zones of anomaly, as regional feature and governmentality, on the emerging discourses, practices, and institutions of forest management. Thus, while identifying the other modes of statemaking that worked through categories of community construction, resource control and expertise designation, we shall continue to examine the redefinition of zones of anomaly and their influence on forest management in Bengal.

Notes

1. Carstairs 1912: 188.
2. Ibid.: 197.
3. Dalton 1865: 4.
4. Henry Thomas Colebrooke and A. Lambert, *Remarks on the Present State of the Husbandry and Commerce of Bengal* (Calcutta, 1795), p. 2.
5. For the Australian bush and colonial descriptions of it in these terms see Paul Carter 1989: 144–9. For sub-Saharan Africa and the extensive debates on bushmen, encompassing issues raised here, see Wilmsen 1989: 24–32.
6. Jungle Mahals here, and hereafter, refers to all of the following. Initially, as O'Malley (1911: 195) says it was 'a term applied in the eighteenth century to the territory lying between Birbhum, Bankura, Midnapore and the hilly country of Chota Nagpur . . . in Midnapore the term applied to thanas Binpur, Garbeta, Gopiballabhpur, Jhargram and Salbani.' After being omitted from some of the earliest surveys like that of Rennell (1976[1793]), the nebulous area signified by jungle mahals constantly changed in the next half century. Suchibrata Sen (1984: 2–3) discusses some of these changes in administrative jurisdictions over the first sixty years of British rule. He also notes the use of the term for a larger region contained in the present-day districts of Birbhum, Bankura, Singhbhum, Purulia and Midnapore, while it was simultaneously applied to the western part of Midnapore district. Also see McAlpin 1981(1909): 3–4. The broadest interpretation of the jungle mahals region has recently been provided in a map included in Samaddar (1998) which shows several Orissa

districts of Mayurbhanj, Keonjhar, Sambalpur, and Bihar districts of Ranchi, Santhal Parganas, Hazaribagh and Palamau.

7. Initial resistance to British control probably marked most zamindari areas, but more commonly such resistance was dissipated within a couple of decades of conquest in the populous, cultivated riverine areas. A good example is the case of Bihar discussed by Yang (1989: 53–67).

8. I use the term discourse here in its Foucaultian sense of a complex network of representations and practices. In my case, the entrenchment of ideas is through routines of governance they create and the power of practices to perpetuate official stereotypes.

9. Abrams 1988: 82; see also Alonso 1994.

10. Roseberry 1994: 357.

11. Agnew 1989: 9.

12. Theories of place owe a lot to Giddens and his critique of positivist and radical social science. For an early and extended account of his theories of spatiality that inform more recent discussions of place, see Giddens 1981. See also Massey 1984; Agnew 1987.

13. Agnew (1989: 15–23) provides a useful summary of the development of the community concept in social scientific analysis. Other thoughtful critiques are provided by Calhoun 1978, 1980.

14. See Skaria 1999.

15. This last question is discussed in the context of European colonization of Australia by Paul Carter (1989: 158–71).

16. Ranajit Guha 1983: 225–6, 278–330.

17. Sharma 1993: 349.

18. There is no simple relationship or opposition between local autonomy and centralization in processes of statemaking. Following recent writing on state building in early modern Japan, I shall argue that local politics need to be visualized as component parts of a broader political structure. See Totman 1993; and Ravina 1995.

19. This point is well made by Ranajit Guha (1983: 22, 281–7).

20. These ideas, images and policies informed colonial rule in a wide tribal belt across peninsular India; see Hardiman 1994: 108; Padel 1987; Skaria 1997.

21. Price 1876: 38.

22. For a recent, brief and well-explained discussion of the emergence of zamindari in Bengal and its nature on the eve of British rule, see McLane 1993: 9–15. 'Zamindari was essentially a function of governance of the country and was an integral part of the political and administrative structure of Bengali society,' says Ray (1979: 24).

23. These topographic descriptions are based on O'Malley (1910a: 5, 73).

24. Walter K. Firminger (ed.), *Bengal District Records: Midnapore, Volume I, 1763–67* (Calcutta: Military Orphan Press, 1926), hereafter *BDRM*, v. 1, letter no. 60, 17 Mar 1766, Harry Verelst to John Graham, p. 48.

25. Even in plains Bengal, any homogeneous view of zamindars was fraught with problems, as the Company raj discovered. See Hunter (1894: 29–44). For a detailed discussion of the scepticism certain officials felt, in the tradition of John Shore's dissent in the Permanent Settlement, for rule by standard

regulations in the forest provinces of Chotanagpur, see Jha 1987: 8–66, and Jha 1964.

26. *BDRM*, v. 1, letter no. 124, 14 Feb 1767, Ferguson to Graham; pp. 95–6; Bodleian Library, Oxford, Manuscript Coll, George Vansittart papers, b67, Midnapore Journal, vol. I, p. 7.

27. *BDRM*, v. 1, letters no. 127–30, 16–22 Feb 1767, exchanged by Ferguson and Graham, about the Ameynagar Raja, pp. 97–102.

28. T. Motte, 'A Narrative of a Journey to the Diamond Mines of at Sumbhalpoor in the Province of Orissa' (extracted from 1799 *Asiatic Annual Register*), in Anon, *Early European Travellers in the Nagpur Territories* (Nagpur: Government Press, 1930), pp. 1–50.

29. Daniel Robinson Leckie, 'Journal of a Route to Nagpur by Way of Cuttac, Burrosumber, Donghur Ghur, and the Southern Banjare Ghat in the months of March, April, May 1790 with an Account of Nagpur and a Journal from that Place to Banaras by the Sohagee Pass' (first published London: John Stockdale, 1800) in Anon, *Early European Travellers*, pp. 51–90, 64.

30. OIOC W/2126, William Hodges, *Travels in India* (London: J. Edwards, 1793), p. 87.

31. Hodges, *Travels in India* contains many reprints. See in particular a picture entitled 'The Pass of Sicri Gully from Bengal entering the Province of Bahar,' p. 23. I am grateful to James Lancaster for discussing the art of William Hodges with me.

32. Capt. J.T. Blunt, 'Narrative of a Route from Chunargarh to Yertna-goodam in Ellore Circar,' in Anon, *Early European Travellers*, pp. 91–174, 153.

33. 'A Narrative of a Journey from Mirzapur to Nagpur,' in Anon, *Early European Travellers*, pp. 175–229, 189, 199.

34. Ibid., pp. 175–229, 191–2.

35. *BDRM*, v. 1, letter no. 269, 29 Nov 1767, G. Vansittart, Collr Midnapore, to Richard Becher, Collr Genl at Calcutta, pp. 195–6.

36. *BDRM*, v. 2, Harry Verelst, 'View of the English government in Bengal,' p. 77; *BDRM*, v. 1, letter no. 285, 28 Dec 1767, Richard Becher to George Vansittart, Collr Midnapore, p. 203.

37. Bodleian Library, Oxford, Manuscript Coll, George Vansittart papers, b67, *Midnapore Journal*, Vol III, p. 10; b100 letters to England, Aug 1766–Sep 1769, l/d 9 Feb 1769 to Henry Vansittart.

38. *BDRM*, v. 4, letter no. 163, 6 Feb 1773, Edward Baber Collr Midnapore, to Warren Hastings, Governor, Fort William, pp. 106–7. The evidence of the independence of jungle chiefs in western Midnapore and surrounding forest areas from the Bengal subah of Mughal rule and the subsequent Nawabi of Bengal is scattered but conclusive. Hunter 1877: 346; Blochmann 1873; Ray 1979, chapter on Midnapore; all these authors cite *Ain-i-Akbari* and other Persian sources for arriving at this conclusion. M.M. Chakravarty (1908: 285–6) reports that Mahavira, the founder of Jainism, passed through the jungle mahals and was attacked by forest peoples.

39. 'A Narrative of a Journey from Mirzapur to Nagpur,' in Anon, *Early European Travellers*, pp. 175–229, 195.

40. Walter Firminger (ed.), *The Fifth Report of the Select Committee to Parliament* (Calcutta: R. Cambray & Co., 1812, reprinted 1917), pp. cxxviii–cxxix.

41. Bodliean Library, Oxford, Manuscript Coll, George Vansittart papers, b100, letters to England, Aug 1766–Sep 1769, l/d 29 Sep 1769 to Henry Vansittart, p. 82.

42. Ray 1979: 79; McNeile 1866: 140; Firminger, *Fifth Report*.

43. Firminger, ibid.: iii.

44. In 1768, the zamindar of Dhalbhum fought Company forces by felling trees and barricading passes to obstruct the advance of sepoys, see Binod Das 1973: 39. For similar accounts of ecological warfare during nineteenth-century political disturbances in forested parts of China, see, Menzies 1994: 22–3.

45. Bayly 1990: 108. See also Grove 1993a; Rangarajan 1994: 149–50; Sumit Guha 1995; and Mann 1995; who provide ample evidence of similar ecological warfare in different parts of India during the late eighteenth and early nineteenth centuries. For a similar discussion of wildlands as disorderly space in Imperial China during the nineteenth century, see Menzies 1992: 720.

46. George Vansittart's journal is full of this issue. He notes that in Mouza Agrachour, 3,141 bighas out of 4,824 was bazé zameen. In parganas Margate, though bazé zameen was 2,243 bighas out of 17,997 bighas of land, only 6,451 bighas were assessed in the jummabundy. Bodleian Library, Oxford, Manuscript Coll, George Vansittart papers, b67, Midnapore Journal, Vol II, pp. 2–25, entries dated 6 May and 11 May 1767.

47. B.B. Chaudhuri 1976: 299; the term bazé zameen was applied to jungles considered culturable wasteland. See J.C. Sengupta, 'Introduction', in Asok Mitra and Ranajit Guha (eds), *Bengal District Records, Midnapore, 1777–1800* (Alipore: Government Press, 1962), p. ccxli (hereafter *BDRM* II).

48. McLane 1984: 36, n. 41. For a detailed discussion of inam lands and other forms of land gifts that were crucial to the power of little kings in south India, see Dirks 1987, esp. chs 3 and 4. There too, in the Permanent Settlement areas of Madras Presidency the colonial state curbed the zamindari control of land allocations and thereby sapped their political authority.

49. *BDRM*, v. 2, letter no. 352, 4 Jun 1768, Lieut George Rooke to George Vansittart, Collr Midnapore, p. 74.

50. *BDRM*, v. 2, letter no. 83, 30 Jan 1771, Collr Midnapore to Claud Russell, Collr Genl, p. 48.

51. Price 1876: 34. In 1815 this was still the picture, with jungles being used for cattle grazing. Hamilton 1815: 119–21; Hamilton 1820: 148.

52. OIOC F/4/97, BC 1986, pp. 78–81.

53. OIOC V6151, *Reports and Documents Connected with the Proceedings of the East India Company in Regard to the Culture and Manufacture of Cotton Wool, Raw Silk, and Indigo in India* (London: J.L. Cox, 1836), pp. 44, 328.

54. OIOC V/23/94, H. Ricketts, *Report on the District of Singhbhum*, Selections from the Records of Bengal Government, no. XVI (Calcutta: F. Carbery, Military Orphan Press, 1854), hereafter SRBG 16, pp. 69, 71.

55. W.B. Jackson, *Report on Darjeeling*, Selections from the Records of Government, Bengal, XVII (Calcutta: Thomas Jones, Calcutta Gazette Office, 1854), hereafter SRBG 17, pp. 4–5.

56. Hunter (1868) gives a graphic account of the famine and its consequences.

57. Hunter 1868: 67–8; B.B. Chaudhuri 1976: 296; Ramsbotham 1930. This book reproduces the *Amini Commission Report*, a survey of Bengal agriculture in 1772–6, which reveals the extent of depopulation caused by the famine and thereby the degree to which forest encroached on former farm lands.

58. *BDRM*, v. 2, 10 Apr 1769, Vansittart to James Alexander, Collr Genl.

59. OIOC F/4/1501, BC 58886, extract BCJC of 31 Mar 1834, progs 51, 1/d 4 Sep 1833, W. Dent, Jt Commr Jungle Mahals, to Secy Jud, Fort William, p. 3. Binod Das (1973: 46) quite confidently makes this assertion. Surajit Sinha argues that the Bhumij and Kharia were the first settlers of the jungle villages in the Barabhum area and around, based on genealogies collected by him in the 1950s. See, S. Sinha 1957: 24; also see, Dikshit Sinha 1984. Oral histories collected during my field work also confirm that in the Jhargram area the Mahato settlements were made around the middle of the nineteenth century, displacing Bhumij peasantries who were most often described as 'fleeing to Mayurbhanj'.

60. A more detailed discussion of these tenures may be found in Sivarama-krishnan 1996a: ch 9. The ghatwals were guardians of hill passes, or wardens of marches.

61. Price 1876: 73; OIOC F/4/97, BC 1986, extract BCJC letter, 13 Apr 1800, p. 331.

62. OIOC H/385, Home Misc, p. 16.

63. W.B. Oldham 1894: 2–5. He goes on to observe the practice of worshipping sal trees and preserving sacred groves both among the Santhals and the Mal paharias, whom he calls the real aborigines of the area. The same Paharias are described in another account as 'wretched unmanageable . . . living the life of wild animals which divided the jungles with them.' See SRBG 17, p. 27. On divergent accounts of Santhal migration see, Waddell 1893; and A. Campbell 1894.

64. OIOC F/4/97, BC 1986, 1/d 14 Nov 1799, P. Touchet, Comml Resident at Radnagore to Board, p. 64.

65. SRBG 16, p. 92.

66. Bayley 1902: 16–19.

67. McLane 1984: 24, following evidence in the *Fifth Report*.

68 . Jha 1966: 42.

69. OIOC F/4/1501, BC 58886, extract BCJC of 31 Mar 1834, progs 51, 1/d 4 Sep 1833, W. Dent, Jt Commr Jungle Mahals, to Secy Jud, Fort William, pp. 2–3; Jha 1966, 1964. According to Risley, the Bhumij were hinduized Mundas, who had under the leadership of sardar ghatwals and paiks organized the first clearing of the cultivated lands in the jungle parganas. See, Risley 1915: 75, 96–7.

70. OIOC F/4/606, BC 15044, Report of H.B. Bayley, dated 13 Sep 1815, pp. 135–9; OIOC MSS EUR F-128/154, Strachey Papers, Corr of Henry Strachey as Magistrate Midnapore, pp. 7–8, gives several examples of relations between various jungle zamindars like Durjan Singh, Raja of Phulkusuma and the Dhal family of Ghatsila.

71. Sanyal 1971: 321–6.

72. Ray 1979: 131–73. She notes that in the jungle mahals, 'the zamindars were aboriginal bhuiya or Bhumij chiefs who had assumed the status of Rajput or khandait' (p. 147).

73. Ray 1979: 137. The reference is probably to the Kharia hill tribes of Manbhum, a branch of the Lodhas of Midnapore, and confirms that the currently impoverished Lodhas and Kharias were the original kings of the forest. For similar accounts from the Maratha territories, see Sumit Guha (forthcoming).

74. OIOC F/4/70, BC 1580, extract Rev letter from Bengal, 31 Oct 1799.

75. OIOC F/4/97, BC 1986, letter from Magistrate Midnapore, to Secy Rev, extract BCJC 1 May 1800, p. 303; H. Ricketts, *Papers Relating to the South West Frontier, Comprising reports on Purulia or Manbhum, Chotanagpur, Hazaribagh, Sambalpore and South West Frontier Agency*, Selections from the Records of Government, Bengal, XX (Calcutta: Thomas Jones, Calcutta Gazette Office, 1855), hereafter SRBG 20, p. 7.

76. OIOC F/4/98, BC 1987, extract BCJC, 31 Jul 1800, D. Campbell, Magistrate Birbhum, to Secy GOB, Jud and Rev, pp. 84–5.

77. OIOC F/4/98, BC 1987, extract BJCC 55 of 8 Jul 1802, report of the Magistrate Midnapore to BOR.

78. Ray 1979: 153.

79. OIOC P/51/32, BRC, 25 Feb 1789, l/d 13 Feb 1789, C. Keating, Collr Birbhum, to BOR, pp. 847–8.

80. The Amini Commission, which did a detailed survey of Bengal agriculture in 1776–7, reports much of the information we have on rural social structure and labour process in Bengal agriculture. Ramsbotham (1930) notes that the Amini Report was the first technical explanation of the land revenue system in Bengal, and reproduces the text of the report. Binod Das 1984; B.B. Chaudhuri 1976; Ray 1979; McLane 1984; all note the rise of paikasht raiyats and the role of mandals in clearing and settling wastelands after the 1740s. The mandal in those early years was a head raiyat, holding office with the consent of peers, settling disputes within the village and mediating between fellow raiyats and subordinate revenue collectors. Gradually the office became hereditary and mandals became tenure holders in the nineteenth century. Ramsbotham 1930: 108; Colebrooke and Lambert, *Remarks on the Present State of the Husbandry and Commerce of Bengal*, p. 59.

81. Basudeb Chatterjee 1980: 20, n. 1.

82. McLane 1985.

83. Ibid.: 27.

84. Binod Das 1973: 94–5, 1984: 131; Jha 1966: ch. 1.

85. Ray 1979: 70.

86. Bradley-Birt 1905: 2.

87. McLane 1985: 32; Basudeb Chatterjee 1980: 23.

88. McLane 1985: 35.

89. Nandalal Chatterjee 1954: 75.

90. Basudeb Chatterjee (1980: 27–32) discusses the process for the rest of Bengal.

91. Basudeb Chatterjee 1980: 21; also see Ray 1979: 133–45.

92. Price 1953 (1874); Binod Das 1984: 230–73.

93. OIOC P/128/42 BCJC, 4 Jul 1799–29 Aug 1799, progs 5 of 11 Jul 1799.

94. Compare the excellent study of patterns of rural crime during famine in early-nineteenth-century UP, Sharma (1993: 362–7), which argues that certain crop failures sharpened a politics of place where traders and moneylenders, rather than landlords, were targets of attacks.

95. Bengal BOR Consultations, 3–31 May 1796, l/d 20 May 1796, GC-in-C to Secy BOR; l/d 29 Apr 1796, Secy BOR to GOB, enclosing report of the Collr of Midnapore.

96. OIOC P/128/40 BCJC, 4 Jan–29 Mar 1799, progs 11 of 15 Feb 1799, l/d 1 Feb 1799, Richard Blunt, Acting Magistrate Burdwan, to G.H. Barlow Secy Jud, Fort William; OIOC P/128/38 BCJC, 7 Sep–31 Dec 1798, progs 14 of 12 Oct 1798, l/d 26 Sep 1798, Major D. Marshall, Ramgarh Battalion, to Adjutant General.

97. OIOC F/4/70, BC 1580, extract BRC of 22 Feb 1799, Midnapore Collr's reports.

98. Price 1953 (1874): 231.

99. See Binod Das 1973; Jha 1966; and detailed reports of instances in OIOC F/4/70, BC 1580.

100. Price 1953 (1874): 234–41.

101. SRBG 16, p. 68. The civilizing influence of road construction, an obvious prerequisite for promoting commerce and trade, is always emphasized in the discourse of frontiers, see OIOC F/4/97, BC 1986, 'Report of Asst Magistrate Midnapore', p. 85. Flight to elude such 'civilizing' influences can certainly be discussed as 'avoidance protest.' See, Adas 1981.

102. More detailed illustration of this point, by discussing one particular jungle zamindari, may be found in Sivaramakrishnan 1996b: 266–74.

103. OIOC F/4/606, BC 15044, p. 181–8. Henry Strachey in his recommendations on restoring zamindari police in Midnapore said the same. OIOC MSS EUR F-128/154, Strachey Papers, Henry Strachey to H.G. Tucker, Secy to Govt, Jud and Rev Dept, l/d 9 April 1800.

104. Strachey Papers, ibid., pp. 2–3.

105. OIOC MSS EUR 128/154 Strachey Papers, pp. 5–6. Also, OIOC F/4/606, BC 15044, extract BCJC, note from H. Hodgson, Magistrate Midnapore, to W.B. Bayley, Secy Jud, GOB, pp. 24–5. Wellesley had initiated as Governor-General a concerted effort to release the military from police work and exhorted the civil administration to make other arrangements to deal with chuars, but consistently the magistracy had resisted. British Museum, Western MSS Coll, Add. 13472, Secret Dispatches from the GC-in-C to various officers regarding maintaining the integrity of frontiers from Midnapore to Mirzapore, l/d 16 Jul 1803, I. Lumsden, Chief Secy, to Governor-General; l/d 4 Aug 1803, Lumsden to Col Fenwick at Midnapore.

106. OIOC MSS EUR F-128/154, Strachey Papers, pp. 14–20.

107. OIOC F/4/606, BC 15044, Report of Secy Jud, H.B. Bayley, 13 Sep 1815, pp. 212–35.

108. OIOC F/4/823, BC 21887, Orders of the Court of Directors, EIC, extract jud letter, 13 Dec 1820, p. 193.

109. OIOC H/692 Home Misc, 'Notes Inspecting the Rules of Police in the Lower Provinces of Bengal and an Account of the State of it in 1808, Mar 1811, pp. 173-5; OIOC F/4/97, BC 1986, p. 7.

110. Thanas were set up with a daroga, *muhrir* (writer), head constable and 10–40 burkandazés. OIOC H/692, Home Misc, p. 176.

111. OIOC P/54/1 BRC, progs 33, 15 Mar 1799, l/d 4 Mar 1799, J. Imhoff to President BOR.

112. Ibid.

113. OIOC P/54/1 BRC, progs 32, 15 Mar 1799, l/d 8 Mar 1799, G. Hatch Member BOR to Lieut-Genl Clarke, Dy Governor, Fort William.

114. OIOC F/4/70, BC 1580, Board of Control to BOR, extract BRC, 15 Mar 1799.

115. OIOC F/4/97, BC 1986, extract BCJC of 14 Mar 1800.

116. OIOC F/4/70, BC 1580, l/d 16 Apr 1799, J. Imhoff, Collr Midnapore, to President BOR.

117. OIOC P/54/10, BRC, progs 13 of 14 Mar 1800, l/d 10 Jan 1800, Board to Govt; OIOC F/4/97, BC 1986, Govt to BOR, l/d 6 Mar 1800, p. 143. The decision promptly received the blessings of the highest authorities. OIOC E/4/650/35 Bengal Despatches, 2 Dec 1800–2 Apr 1801, EIC Genl Corr, pp. 418–19.

118. OIOC F/4/98, BC 1987, l/d 18 Jul 1800, Henry Strachey, Magistrate Midnapore, to Secy GOB Jud and Rev, p. 48; also OIOC MSS EUR F-128/154, Strachey Papers, corr of Henry Strachey, p. 31.

119. British Museum, Western MSS Coll, Add. 13472, Secret Dispatches from GC-in-C to various officers regarding maintaining the integrity of frontiers from Midnapore to Mirzapore, 16 Jul–4 Aug 1803, pp. 3-4.

120. OIOC F/4/98, BC 1987, l/d 15 Oct 1800, Camac, third judge of the Calcutta court of Circuit to the Nizamat Adalat, p. 172.

121. Bayly 1993: 5, 16–17.

122. OIOC P/129/18 BCJC, 5–26 Dec 1805, progs 16 of 13 Dec 1805, Reg XVII of 1805.

123. OIOC E/4/653/38, Bengal Despatches, 30 Jun 1802–Mar 1805, on landholders in the jungle mahals, of Midnapore reinvested with the management of police—proposition to extend this arrangement throughout the provinces, pp. 24–6.

124. The terrains occupied by thugs and criminal tribes were other examples of such social and political frontiers. See Singha (1990) for an excellent discussion of those cases.

125. Price 1953 (1874): 258.

126. OIOC F/4/70, BC 1580, l/d 19 Apr 1799, Thomas Graham Secy BOR, to Lieut-Genl Clarke, Dy Governor Fort William.

127. OIOC P/53/14, BRC 82 of 27 Mar 1794, Ramgarh Collr to Board, l/d 17 Aug 1793, enclosing petition about introduction of Permanent Settlement in Pachet.

128. OIOC P/57/28 BRC, 16 Jan 1818, progs 70, l/d 30 Dec 1817, A. Campbell, Collr Midnapore, to BOR.

129. OIOC F/4/97, BC 1986, extract BCJC, 1 May 1800, pp. 360–91.

130. SRBG 20, pp. 11–18; OIOC P/55/13 BRC, 1 Jul 1808, progs 21, l/d 28 Jun 1808, W. Blunt, Magistrate, Jungle Mahals, to George Dowdeswell, Secy Jud GOB.

131. OIOC F/4/15011, BC 58885, progs 3 of 1834, l/d 31 Mar 1834, C.T. Metcalfe and W. Blunt to Court of Directors; report of Jt Commrs Dent and Wilkinson received in l/d 4 Sep and 14 Nov 1833, p. 15.

132. Jha (1964) is an exhaustive study of the causes, outbreak, progress, and consequences of this unrest.

133. OIOC F/4/1501, BC 58886, extract BCJC, 31 Mar 1834, progs 51, l/d 4 Sep 1833, W. Dent, Jt Commr, Jungle Mahals, to Secy Jud, Fort William, p. 2.

134. Ibid., pp. 4–45.

135. Ibid., p. 47. For full details of the arrangements proposed by Wilkinson in Chotanagpur, see, Jha 1964: 225–58.

136. OIOC F/4/1501, BC 58888, extract BCJC, 4 Dec 1832, progs 30, Dent to Offg Secy, GOB Jud, pp. 96–7.

137. OIOC F/4/1502, BC 58889, extract BCJC, 15 Jan 1833, progs 42, l/d 27 Dec 1832, I.M. Cash, Asst Surgeon, 34th N.I. and Artillery, to Lieut Lyons, Detachment Staff.

138. OIOC F/4/1501, BC 58889, extract BCJC, 1 Mar 1833, progs 43, l/d 10 Feb 1833, Capt. Wilkinson, Jt Commr, Jungle Mahals, to W.H. McNaughten, Secy Jud GOB, Fort William, p. 155.

139. OIOC F/4/1502, BC 58891, extract BCJC, 3 Jun 1833, progs 4, l/d 16 Nov 1832, W. Dent and Capt. Wilkinson, Jt Commrs, to J. Thomason, Dy Secy, GOB, pp. 1–46.

140. John Malcolm, 'Essay on Bhils', *Transactions of the Royal Asiatic Society of Great Britain and Ireland*, 1, 1827, p. 89. Quoted in Stewart Gordon 1994: 158.

141. OIOC F/4/1502, BC 58891, extract BCJC, 3 Jun 1833, progs 4, l/d 16 Nov 1832, W. Dent and Capt. Wilkinson, Jt Commrs, to J. Thomason, Dy Secy, GOB, pp. 47–52; progs 65, note, Hazaribagh, 5 Jan 1833, W. Dent, Jt Commr, pp. 57–74. Dent had some reservations on the role of zamindars, but certainly agreed on making the headmen the pivot of administration.

142. Ibid., progs 69, l/d 3 Jun 1833, Secy Jud GOB to Jt Commrs, p. 143.

143. OIOC F/4/1502, BC 58891, extract BCJC, 2 Dec 1833, progs 55, l/d 24 Jul 1833, H. D'Oyly, Magistrate Midnapore, to W. Dent, Commr, pp. 189–90. Illegibility of landscapes as an impediment to generalized planning or administration is very ably discussed by Holston (1989).

144. Ibid., progs 57, l/d 10 Aug 1833, H.P. Russell, Magistrate Jungle Mahals, to W. Dent, Commr, p. 193.

145. Ibid., progs 61, orders dated 31 Mar 1834, Secy Jud GOB, giving full text of Reg XIII of 1833; OIOC F/4/1502, BC 58894, extract BCJC, 12 Apr 1831, progs 35, l/d 14 Mar 1831, S.J. Cuthbert, Magistrate and Collr, Ramgarh, to W. Lambert, Commr, Patna Division, pp. 7–15.

146. Anon, *Reports on the Political State of the South West Frontier Agency*, Selections from the Records of the Bengal Government, no. XI (Calcutta: F. Carbery, Military Orphan Press, 1853), hereafter SRBG 11, p. 13.

147. OIOC E/4/745, Bengal Despatches, Jud Dept progs 4 of 16 Sep 1835, draft 507/1835, pp. 361–83.

148. Jha 1966: 157–60. The development of ethnoregionalism in Chotanagpur and the adjoining areas that are now part of the Jharkhand movement and its claim to a separate state, in interaction with tribal policy in colonial India, is discussed by Corbridge (1986). He suggests that the emergence, timing, and intensity of ethnic and regional politics in the Jharkhand can be understood in terms of a series of redefinitions of region and tribe.

149. Carstairs 1912: 224–5.

3

Geographies of Empire: The Transition from 'Wild' to Managed Landscapes

All these trees mixed with shrubs loaded with parasites give to the forest the appearance of a conservatory in disorder in which individual plants are indistinguishable, and not the majestic aspect of the fine forests of our countries, which bear more resemblance to galleries of gothic columns. . . .[1]

What is evoked here are the spatial forms and fantasies through which a culture declares its presence. It is spatiality . . . as a form of history. That cultural space has such a history is evident from the historical documents themselves. For the literature of spatial history the letters home, the explorer's journals, the unfinished map—written traces which, but for their spatial occasion, would not have come into being.[2]

Introduction

This chapter seeks to show how forests began to take shape as a more distinct domain of management under the colonial government (becoming more than the fringes of, or impediments to, agriculture). It will also examine the relationships between state-making and knowledge production as will the next chapter. As a 'colonial state' was constructed, so was a constellation of systematically related and empowered representations of the colonized people and landscapes that we may gloss as 'colonial knowledge'.

I shall begin by pointing to the failure of recent writing on colonial discourse to explicate the variegated, conflicted, and shifting modes of official knowledge produced by colonial expansion in India. I shall also suggest that identifying evolutionary trajectories in the development of colonial knowledge is not enough. We need to understand regional variations—the spatiality of colonial knowledge—as well.

I shall situate the production of knowledge, in contrast to prosopographical approaches that are utilized by historians of science,[3] in the very modalities of colonial expansion—surveying, vermin eradication, sedentarization of shifting cultivators, botanical and geological explorations, commercial missions, and war[4]—that required and constrained the production of hybrid knowledge. Prosopography identifies key agents, and is used to good effect by Richard Grove and Ravi Rajan for describing the spread of modern forestry and conservation ideas in British colonies. But it explains neither the transformation of ideas into discourses, nor the heterogeneity of discourse. Further, prosopography does not help explain what causes this diversity and what its consequences might be.

This chapter will, therefore, describe the specific, more systematic ways of conquering forests that facilitated British rule in woodland Bengal. I seek thereby to elucidate the modes of knowledge that emerged in their actual, historically grounded, moments of production. Clearly the development of earth and plant sciences, landscape aesthetics, and hunting attitudes in Europe was salient to the production of colonial knowledge. But it shall remain my argument that the institutionalization, dislocation, transformation, and hybridized production of knowledge in diverse colonial situations more fundamentally prefigured the regimes of forest management that emerged in Bengal by the early twentieth century.[5]

Unpacking Colonial Discourse

As we have already seen, political turbulence, a baffling sociology, the apparent lack of a stable agrarian order, and a landscape of 'impenetrable jungle' had combined to shape the early colonial *imaginaire* as the Company raj came to woodland Bengal. Almost concurrently, the production of knowledge about Bengal society had begun.[6] The relationship of knowledge to statemaking, therefore, merits scrutiny. Scholars from diverse disciplines have begun to question the manner in which 'colonial discourse has too frequently been evoked as a global and transhistorical logic of denigration, that has remained impervious to active marking or reformulation . . . a coherent imposition rather than a practically mediated relation.'[7] On the other hand, careful research on the working of colonialism 'reveals competing agendas for using power, competing strategies for maintaining control and doubts about the legitimacy of the venture.'[8]

Colonial knowledge developed in conditions where even mainly 'western' ideas of the eighteenth and nineteenth centuries are not easily explained in terms of a single scheme. European ideas about nature, for example, were caught up in the rivalry between the Cartesian fascination with classificatory sciences and the Romantic sensibility.'[9] These conflicts were only amplified and multiplied by encounters with nature in colonial situations because 'colonial categories were never instituted without their dislocation and trans-formation.'[10] By the 1820s, the expansion of the Company raj in India had deepened and diversified the colonial experience of governance and the forms of knowledge generated in that process.[11] We should then hesitate to propose any master historical narrative called colonial discourse.

Recent studies of colonialism as a culture of representation, influ-ential and persuasive as they often are, have shown heady disregard for such caution.[12] Said's claim that the texts of colonialism 'create not only knowledge but the very reality they appear to describe', by according primacy to representation over historical experience, has allowed a profound lack of historical specificity to enter colonial discourse studies.[13] Colonialism is then treated by scholars, following Saidian modes of analysis, in ways that fail to recognize its fractured quality or how it was racked by internal debate and modified by resistance.[14] Within the field of literary criticism the textualism and idealism of such an approach, which also accepts too readily imperial pretensions of omniscience, have been subject to a compelling critique that we need not rehearse here.[15]

Both in later *Subaltern Studies* and in synthetic historical work like Inden's *Imagining India*, Foucaultian discussions of power and discourse have combined with the Saidian reading of them to achieve the same disappointing effect for the historiography of colonialism in India.[16] A related criticism of Said is offered by David Ludden who says, 'seeing Orientalism in descriptive, literary terms, he makes provocative associations among texts that constitute Orientalism and the dynamics of European power. But the particulars that connect histories of imperialism and knowledge are missing.'[17] I dwell on these observations about Said because it is important to recognize the transformations in colonial knowledge at different periods in colonial histories if we are to study the effects of these transformative processes on colonial policy in any sphere.

One of the important transformations of colonial knowledge in India was the delinking of detailed empirical data collection from its diverse contexts.[18] Aided by the rise of statistical methods in Europe, colonial India produced by the middle of the nineteenth century enormous information on its subject populace. This information was rapidly standardized through the decennial census into instruments of surveillance and compendia of knowledge. Numerous studies have analysed this process.[19] They describe early surveys and empirical inquiries of the Company raj as the experimental or learning phase. The gains of this period were then consolidated, bureaucratized and rationalized through operations like the census, cadastral survey, and land settlements in the Age of Empire (1858–1947), as objectification procedures were mastered and disseminated. According to Appadurai, the post-1870 colonial state unyoked the ecological and physical landscape from the social relations associated with it. For Dirks the massive weight of the late-nineteenth-century documentation projects erased the 'ambivalences and contests within early colonial historicities', political triumphs made possible the illusions of permanence.[20] Most recently, Skaria, in a study of the Dangs in western India, has described this transition in knowledge and attendant administrative arrangements as one moving from forms of potential control to forms of everyday control.[21]

Space in the Forests

Much of the history of colonial forestry in India fits into this account that describes ever-tightening screws of resource control turned by a specialized state agency over the last hundred years of colonial rule.[22] But we still remain above the level of regional variations. Once we begin to examine those, thus interleaving imperial history with the history of spatiality, the transitions are less evident as neat successions and accumulations of knowledge and political practices.[23] Lefebvre's observation, that the history of space does not always coincide with widely accepted periodizations, is amply corroborated in woodland Bengal when we examine regional variations.[24] For example, the work of surveying was to discover and to order. It was thus at once a spatial practice and a representation of space. In woodland Bengal, the work of surveying was never finished, and more than the fruits of exploration, surveying itself became the decisive discovery, a mode of knowledge that coexisted with and shaped the cataloguing, managerial, developmental modes of knowledge throughout the nineteenth century.[25]

Representations—the confluence of ideology and knowledge— achieve consistency only by intervening in social space and its production.[26] Thus botanical surveys, tree plantation, and timber conservancy were measures that entered the colonial repertoire for knowing and ordering the landscape at the same time, by the beginning of the nineteenth century. Where they occurred, persisted, failed, shrank, and enlarged are all aspects that set colonialism off against its 'historical and geographical particularities'.[27] We have to describe, therefore, the production of landscapes as historical spaces and recognize the concomitant production of societies and cultures that takes place. Tribal people were associated with tribal places, and the production of these places had as much to do with the history of these people as it had to do with the geography of their lands. When landscapes were partitioned through colonial survey and policy, their inhabitants were also compartmentalized.[28]

This chapter demonstrates that spatial history can begin to overcome the problem of diffuseness in colonial discourse theory. Such an approach, informed by a critical historical geography, has the merit of transcending the distinctions between metaphorical and material space, allowing us to see that colonial projects were shaped by the diverse and dependent settings through which they worked.[29] Without adopting all of Lefebvre's teleological formulations about absolute space and its transformation into abstract space by the work of capitalism, we should note the patterns of commodification of space and bureaucratization through space that he points out.[30]

Following Deleuze and Guattari, colonialism in woodland Bengal will also be discussed in terms of de-territorialization and re-territorialization, to enable the simultaneous discussion of its historical material procedures and ideological operation.[31] Territorialization is important for two reasons. First, colonialism was very much about the appropriation of space by the creation of state forms that emphasized territorial forms of control. Second, and more important, by studying the patterns of territorial control we can appreciate the variations in regional processes of statemaking and note their production through the influence of local ecology, politics, and history on what we frequently discuss as trans-regional representations and practices.[32]

This brings me to the making of tribal places in woodland Bengal. Tribal places were made, not only by curbing practices like shifting cultivation but by assigning them a specific terrain which was designated the place of tribes unredeemable from 'backward agriculture'. Key administrators, scientists, and other influential imperial agents joined to imagine a significant congruence between the space of primeval landscapes and their human inhabitants.[33] Their scientific criticism of *kumri* and *podu* (forms of shifting cultivation in south India) aimed at regulating patterns of settlement and production such that British control over forests was facilitated and agendas of pacification by rule of law were accomplished.[34] 'Tribe' and 'caste' ultimately became the main categories of colonial sociology in India. By the end of the nineteenth century they were still being parsed by prominent colonial ethnologists exasperated by the blurring of lines between them.[35] Thus Herbert Risley, the authority on Bengal tribes and castes, wrote in 1901:

all over India at present there is going on a process of the gradual and insensible transformation of tribes into castes . . . the main agency at work is fiction . . . I hope the ethnographic surveys will throw a great deal of light on these singular forms of evolution by which large masses of people surrender a condition of comparative freedom and take in exchange a condition that becomes more burdensome.[36]

We have already remarked on how the documentation project, especially through the censuses, sparked and shaped these processes. But the numerous studies of social categorization inadequately deal with the integral spatial components of such colonial discourse.

Both literary critics and historians have studied the objectifying gaze of colonialism, but confined their analysis mostly to its impact on the body, occasionally to consequences for the appropriation of space, but rarely focused on *the body in space*.[37] Even recent studies of tribal sedentarization in India, both in the pre-colonial and British period, do not note the making of tribal places.[38]

I would, therefore, like to emphasize the confluence of projects of people and landscape classification. This is important because

it is now widely acknowledged that what were until recently regarded as hunting and gathering tribes were in many cases reduced within the last hundred years or so to an economic condition in which they are forced to survive by foraging, begging, thieving and such activities.[39]

Such alteration and degradation of subsistence production for certain tribes was accompanied in the colonial period by their confinement to certain spaces through force or voluntary movement and settlement. When it came to forest management, therefore, people were not so much unyoked from ecology as Arjun Appadurai has it, as they were classified by relation to ecology. We may see this clearly in the following discussion about the creation of the Santhal Parganas district.

Santhals and Paharias certainly came to belong in the tribal world of Bengal designated by colonial ethnologists and administrators. But they were differentiated by their place in a regional agrarian economy. Paharias invited greater official opprobrium and suffered gradual impoverishment as the landscape of the new Santhal Parganas district was partitioned into areas fanned and forested. They were the group on the hilltops most obdurately resistant to sedentary cultivation. Santhals in contrast, emerged as the apparently successful pioneers who first extended and then stabilized the cultivated arable. But this was made possible by creating an enclave—far from complete—for Santhals viewed as tribes sufficiently imbued with a developmental spirit. Other Santhals left for a variety of reasons to work in tea gardens and other plantation economies in north Bengal and Assam. They more often became deracinated labour in far-off places. In those situations it was not their potential to become developed (read incipient caste-like qualities) that came to be valorized but their tribal 'hardiness and simplicity'.

Discerning and Making Tribal Places

The work of Cleveland with the Paharias in woodland Bengal marked the first instance where the need for administrative exceptionalism was recognized.[40] Like the Paharias, their neighbours the Santhals are reported to have lived by hunting and raiding lowland fanners. Grain, salt, tobacco, cattle, and goats were generally taken in these raids.[41] William Hodges describes Paharia raiding at its height. 'Like those of all savages, their incursions were merely predatory . . . entering villages by night, murdering husbandmen, drove off their cattle and then secured themselves in the hills.[42] Curbed by British rule in such practices, they were established in their own areas to till in peace. Soon there emerged a clear distribution of Santhals in valleys and Paharias on barren hills, in what became after 1855 the Santhal Parganas district, an area of 5,500 square miles taken from Bhagalpur and Birbhum.[43]

After Cleveland's pacification of the Paharias in 1782, disputes over grazing, forest products and boundaries of the hill people's terrain continued between Paharias and lowland zamindars, leading to John Ward's demarcation of the Damin-i-Koh in 1832. Rapid Santhal migration and settlement in the skirt of the hills followed, as official encouragement thickened an earlier flow. Between 1838 and 1851, Santhal villages increased from forty to 1,473, their population from 3,000 to 82,795, their revenue payments from 2,000 rupees to 43,918 rupees.[44] The Santhal Parganas, marked by masonry pillars, became a particularly well-delineated case of the exceptionally administered district, where government was to rely on indigenous forms of village organization.[45] These forms, characterized by the hierarchy of village *manjhi, des manjhi* and *parganait* (pargana head), were considered sufficient justification to persevere with the policy of a segregated Santhal province as the question of revenue settlements and letting *dikus* (outsiders) into the provincial land markets repeatedly came up in later decades.[46] As one Santhal officer, an ardent advocate of protecting their place, wrote in 1881, 'to let in the dikus for the purpose of breaking up Santhal tribal unity would be like drugging a spirited horse . . . instead of taming him by good riding.[47]

If the Santhal system re-created the Paharia stipendiary chiefs and self-governing councils in some respects it also divided the integral landscape of Rajmahal into hill and plain. Paharias confined to the hilltops soon became impoverished and in the 1880s, the inquiries of the Dufferin Committee revealed as much. Carstairs, the Deputy Commissioner of Santhal Parganas wrote, 'Mr. Grant, SDO Godda, . . . searched fifty Paharia houses and did not find so much as food for the evening meal of the day. They were waiting for the return of women who had carried firewood for sale to the market and would bring back food.'[48] In contrast the Santhals seemed prosperous with their terraced rice fields, pigs, poultry, goats, cattle, and liberal access to small game and the fruit and seeds of mahua (*bassia latifoia*), sal, and kendu (*diospyros melanoxylon*) from the forests.[49]

The divergent futures of Paharias and Santhals through the nineteenth century were built on distinct patterns of deforestation. In the hills, Paharia territory, shifting cultivation persisted but in the foothills and plains of the Damin, forest clearing established Santhal agriculture. Even by the early decades of the nineteenth century, when Buchanan-Hamilton toured the Bengal districts, the view of the Rajmahal hills was one of mostly stunted trees. It was a landscape created by shifting cultivation, firewood removal, sal dhuna extraction, and annual fires for both pasture regeneration and amelioration of the air.[50] Paharia clearly managed forests for what came to be known in the jargon of forest conservancy as minor forest products. They grew sesamum and *kurthi* (a pulse), in clearings around preserved kendu, palas (*butea frondos*), asan (*terminalia tomentosa*), and mahua trees. The flower of the mahua was a valuable food, while the *asan* was used to breed the *tasar* silkworm. Even though ghatwals and other elites had sought monopolies of tasar silkworm breeding by surrounding the activity with rituals of purity (like vegetarian observances), the poor Paharia was still the source of cocoons that were skilfully gathered while foraging in the forest. In addition to cocoons and charcoal, which they made abundantly, Paharias traded millets, sorghum, and pulses grown in their *jhums* for salt, iron and clothes. The northern Paharias also grew some cotton.[51]

The making of tribal places, in which restricting shifting cultivation was a key strategy, was a pacificatory tactic that compelled close observation of landscape use. Even the critics grudgingly noted that cutting down trees was only one among many ways forests were utilized in these places. An account otherwise largely sceptical of tribal villagers' management of forests, especially shifting cultivation, records, with refreshing candour, that villages tended to be embowered in groves of fruit trees containing mango, tamarind, peepul (*ficus religiosa*), and surrounded by extensive forests, providing, shade, food, small timber, and fodder. To ensure a liberal supply of fodder these forests were interspersed with grasslands.[52] In the long run, pioneering settled agriculture differentiated the relatively 'successful aboriginals' from the unsuccessful ones.

One commentator noted, 'though the Kolarians, particularly the Santhals, cherish religious groves, they seem unable to stand the sight of fine old trees and invariably cut them down.[53] Clearly at this point in the history of land settlement, and its conversion to what were colonially sanctioned 'productive uses', Santhals were a class of aborigines appreciated not for their love of trees but their willingness to cut them down in the service of agriculture. It was this aspect of their reputation that earned them their place in British India, carved out in the skirts of the Rajmahal hills. Sutherland, the Joint Magistrate of Monghyr wrote in 1817, after investigating tenures in the Rajmahal area, that 'very extensive forests ... have been brought under gradual cultivation by industrious Santhals.'[54] But the Santhal migration and clearing of the hills was itself impelled by the spread of settled cultivation. E.G. Man, a contemporary chronicler of this process, portrays the Santhal as a primitive backwoodsman who cleared and prepared the land for the advancing Bengali. A cycle of reclamation and dispossession seemed to be at work. The creation of the Santhal Parganas was an attempt to break the cycle and reward Santhal enterprise.[55]

The making of tribal places was also predicated on distinguishing their inhabitants from Hindus. Manbhum, a district carved out of the former Jungle Mahals district, was largely populated by Bhumij cultivators even in 1865. Dalton had classified them as Kolarian and Munda, after observing their ceremonies, language, and marriage practices.[56] Manbhum was thus part of the larger tract running from the hilly areas west and south of Bengal, Bihar and Benaras to the frontiers of Hyderabad, and many of the aboriginal zamindars of the jungle mahals were described as becoming Hinduized because they aspired to Rajput status.[57]

To the first colonial chroniclers of their fortunes, the rajas, mankis, ghatwals and other privileged status holders in Chotanagpur, and its adjoining areas, had attained an exalted standing and prosperity that had followed their Hinduization. Dalton expresses this viewpoint clearly when he says:

left to themselves the Kols increased and multiplied, and lived a happy Arcadian sort of life under their republican form of government for many centuries; but it is said a wily Brahmin at last obtained a footing among them and an important change in the form of government was the result.[58]

What was this change? The chief alteration of aboriginal society was its division into the Hinduized and the others, with the former coming to dominate the latter through service tenures and office. Further, rural hierarchy was introduced by the rajas who appointed revenue farmers and favoured Brahmins and Rajputs with land grants, the cumulative effect being the erosion of Munda and Manki authority as the collective proprietary land system they presided over was dismantled.[59] A related problem was the migration of Santhals and other tribes out of regions where outsider (non-tribal) *jagirdar*s and *thikadar*s (rent collecting intermediaries) obtained rent contracts over them and raised rents. These trends both ousted Santhal *parganait*s (chiefs of parganas) and increased rental demands.

As one exasperated Santhal officer put it, no sooner is a village rescued from its primitive jungle by the labours of the Santhals and brought into a tolerable state of cultivation than it is given in farm to some jobber or other, and the Santhal sent to the wall . . . .'[60] Through this narrative of aboriginal dispossession the 'tribal' officers of Bengal carved out tribal places as zones of paternalist, local knowledge-based direct administration in a region increasingly subject to bureaucratic and rationalized governance through standard laws, codes and procedures.[61] Disturbed in some respects by the permeability of the boundary between tribe and caste, these administrators sought to create and preserve the defining characteristic of tribes, which may be summed up as 'a matter of remaining outside of state and civilization.'[62]

By the 1870s, the southwestern parts of woodland Bengal—the emerging tribal heartland—were taking shape as a landscape where two important elements of policy converged. The spread of forest conservancy reinforced this convergence. One element worked to hasten the gradual transformation of wildlands and wastelands into an ordered terrain of fields and groves. The patchwork of forests and fields, supporting the varied demands made on them, demonstrated the processes whereby villages were formed in the jungle areas. The chequered landscape of embanked fields, uplands, and jungle indicated a transition zone. Forest dependence for the mostly tribal cultivators of these areas had an ecological basis. They tilled poor soil, which meant a high proportion of largely rainfed agriculture was on uplands that had to be fallowed (for three to five years) after every season of cultivation. These lands could yield only coarse rice or millets.

The tenants of these lands often did *begari* (forced free labour) for zamindars since they paid no rent on occasionally cultivated lands.[63] The aboriginal agricultural classes played a key role in this managed transition, and they were least amenable to standard forest rules. They participated in what Dalton saw as a double civilization process of the land and of themselves. In his words, 'to throw into their country a staff of forest conservancy officers' was guaranteed to arrest this civilizational momentum.[64] The other element was a concerted effort by sections of the Bengal government to prevent tribal people from losing their lands and homes. Dalton and other officers were joined in this campaign by missionaries. They were all votaries of administrative exceptionalism in these tribal places and argued that loss of land and traditional polity were destroying aboriginal society and thus undermining British rule itself.

The combination of these themes focused attention on the locus of governance and its relation to complex patterns of transformation in the landscape. Efforts to preserve 'tribal republics' and the patchwork of forest and field considered their 'natural habitat' came to distinguish the administration of the southwestern region in woodland Bengal. An anonymous tract, dedicated to creating the tribal history consonant with these efforts, thus records that

according to traditions the Kols, in their two tribes of Munda and Oraon, were the first inhabitants of Chotanagpur . . . from ancient times they live in patriarchal style under heads of villages and heads of districts (munda and manki) each ruling over one or more villages.[65]

This document goes on to note the types of tenures—*rajhas* (paying rent to the king), *bhuinhari* (rent-free), *majhas* (land granted to majhi or villager in charge of rent collection).[66] When Chotanagpur came under British rule in 1818, the area was opened to 'Mahomeddans, Hindus and vagabonds', who as thikadars and jagirdars began oppression of the Kols.[67]

In the next fifty years British curiosity expanded into neighbouring tributary states. Kols who had migrated away from oppression into the isolated hills and dense sal forests of Bonai and Gangpur were living by girdling sal for the resin dhuna and raising tasar silkworms on its associate, the asan tree.[68] Such images of tribal idylls in the forests only spurred certain Bengal officers to secure the continuation of traditional forms of local governance. They argued that Mankis and Mundas should be restored to their original authority and possessions, and since this was not possible under existing Bengal regulations, special provisions like the Chotanagpur Tenures Act were introduced. In these measures we may find a modified reiteration of the discourse of frontiers discussed in the last chapter.

Arguably, the principle animating these mid-nineteenth-century discussions of what were coming to be distinguished as tribal places was one of first naming. Naming schemes for the landscape expressed a state of expectation.[69] Paul Carter evocatively portrays the explorers of rural Australia who assiduously located objects of cultural significance: rivers, mountains, meadows, plains of promise. Like them the pioneers of colonial administration in woodland Bengal searched for hints of the habitable, the glimmer of exchange value. Their lexicon for the landscapes they discovered was constructed from agricultural and tenurial usage. An aborigine, in this context, was the first clearer of the land for cultivation. Such a claim was strengthened by further claims, in most cases, that these first clearers were previous occupants of the land who had subsisted on forest products. These original settlers, defined by the transformation of landscape that gave them their name, were also mobile. Their sedentarization was both a loss of territory and a territorialization prescribing how they would cultivate and where.[70]

In a fine-grained study of agrarian relations in Chotanagpur, Prabhu Mohapatra has shown that a widely subscribed official view was consolidated during the cadastral survey and settlement operations of the early twentieth century. According to this position, there were customary and community rights over land in Chotanagpur that preceded property and tenure rights granted under the Permanent Settlement of 1793.[71] Missionary anthropologists like Father Hoffman, the author of *Encyclopaedia Mundarica*, were important to the spread of this idea. The *khuntkatti* (agnatic kin-based village settlement) system epitomized this collective land control by tribes and village communities and its variants were found among the pioneer tribes who had cleared the jungle. In these villages, even wood for domestic purposes from forested wastes adjoining fields could be taken only with the consent of the extended family of first clearers.[72] Sacred groves were created to appease *bongas* (spirits) believed to be disturbed by forest clearing, and the *Pahan* (priest) carried out all ritual and religious functions on behalf of the village. Thus there was a separation between secular and sacred leadership between Munda or Pradhan and the Pahan.[73]

The headman (pradhan/munda/manjhi) was the representative of the corporate village group and through him the villagers expressed themselves in economic matters such as reclaiming waste for cultivation or rights in jungle lands. Landlord villages marked the breakdown of the khuntkatti system, but there too ordinary rent-paying cultivators enjoyed usufructuary rights in jungles and the right to reclaim jungles at privileged rates.[74] As recently as the 1930s, a Bihar officer, who came to be renowned as an ethnographer of Santhals and other tribes of the region, recalled his encounter with such first clearers like the Kharias on route to a bear hunt through thick sal jungle.[75] The identification of land clearance with tribal groups entered the expressive forms of tribal unrest, as exemplified in the Tana Bhagat movement of 1918, with its cry that title deeds were embodied in 'my spade, my axe, and my plough'.[76]

It is worth noting that becoming native to a place was always preceded by a movement into that place, voluntary or forced, that is often described as migration sometimes being characterized the original migration. Both recent migration, as in the Santhal case, and long processes of sedentarization followed by stratification, as in the Munda case, established tribal places in southwest

Bengal.[77] But in all cases tribal relations with outsiders were similarly treated by the British. Colonial rulers sought to mitigate the exploitative influence of non-tribals on tribal society. Concurrently they redefined tribal relations with forests, to reduce the diversity of forest use, and thus establish effective control over forest products and management. Both strategies characterized the British making of tribal places.

Extending the Arable: Securing Landscapes from Vermin

Another facet of the landscape-ordering procedures of colonial rule was the systematic redrawing of wildland boundaries. The extermination of carnivora, some herbivorous large mammals, and poisonous snakes, a process known as vermin eradication, reached its highest intensity only in the last decades of the nineteenth century.[78] The classical mode of vermin eradication was hunting. Throughout western history hunting had been regarded as a sort of character-building sport, marking good hunters as potentially good soldiers, pioneers, explorers, and leaders of empire. As a result class formation could be seen in hunting as it entered formal aspects of elite culture. Hunting quickly became ritualized, with the production of manuals and rules of sportsmanship designed to establish the power and civility of the elite hunters.[79]

As a combination of the pragmatic compulsions of extending empire into forest areas and the symbolic merits of affirming imperial rule everywhere in the colonies, hunting took its place among the various techniques of ordering the jungle landscape. Thus a historian of imperial hunting notes, 'the protective hunt, with its symbolic dependence was to be very important in the imperial setting.[80] Colonial officers found that hunting combined pleasure and rule very well. The mandatory touring and camping of district life presented many opportunities for sport. The trusty *shikari* (native hunter) would be sent ahead to scout out game. When his *saheb* arrived he would have details of tiger, buffalo, deer, pigs, and partridge available in the vicinity. As the subordinates pitched camp, *huzoor* (the district officer) would hunt. Tigers, where found, were the first choice because often 'a villager would come with tears to relate how his best milch cow had just been carried off, and that would decide the matter.'[81]

When cast by such supplication in his most dramatically paternal role, the district officer was keenly aware that swift and successful slaying of the tiger was desirable on many counts. In one stroke, or load of shot, he could underline the power of colonial government to protect its domain, expand the cultivated realm, and secure the terrain. These spatial outcomes of hunting cleared the ground for the more territorial and zoned approaches to forest management that followed. By establishing forests and their adjacent villages as equally protected for dwelling and commerce, colonial hunting in the nineteenth century may be understood as a somewhat 'pre-territorial' but distinct mode of statemaking in woodland Bengal.[82]

Not surprisingly then, early British commentators on the Indian hunting scene describe Indian shikar and staged animal fights by aristocrats without any pejoratives inflecting their narrative.[83] In one of the earliest English accounts of hunting practices and animal life in India, Johnson noted the existence of many professional hunters, who protected villagers, stock and travellers from carnivores, and secured crops from herbivores. As guardians of passes they had service grants of land from landlords. They divided the work of dealing with different vermin among different castes, and used a variety of methods. The *machan* was used for defensive or protective hunting.[84] Bishop Heber's narrative describes many tribes who lived by the chase in the hilly regions between Rajmahal and Burdwan. The Paharias among them had perfected techniques of killing large animals, including tigers, with poisoned arrows. They were reported to obtain their supply of poison from the Garos of Meghalaya.[85] However, much of this came to be regarded over the next hundred years as poaching, mindless destruction, or inhumane treatment of wild animals. Such representations abound principally in the memoirs of eminent Indian foresters.[86]

The spread of British control over general administration in forested areas gradually led to their assumption of the vermin eradication role. So villagers turned to them for protection, particularly when subject to a dire menace like man-eating tigers. Earlier this role had belonged to the professional shikari with traps, poisoned arrows and the occasional muzzle loader.[87] An officer successfully killing such a tiger could instantly win a lot of admiration among his colleagues, while advancing his acceptance among the local populace under his jurisdiction. The lasting perception of the tiger as vermin, because they hindered development work, forest management and the spread of cultivation, probably accelerated its endangerment by the 1930s in India.[88] But it was not tigers alone. Elephants, bears, wild buffaloes, bison, boar, and even deer were vermin for most of the nineteenth century, many of them being driven to extinction in the plains and uplands of lower Bengal by the end of the century (Figures 6–9). At the beginning of the nineteenth century Bengal had been described as the paradise of pigsticking and was known for the menace of bears in cane fields. Rhinoceroses were found in the Bengal plains forests till the early nineteenth century, but by mid-century they were not to be found south of the Ganges, north of Purnea and along the Brahmaputra valley.[89]

Writing about Birbhum in the jungle mahals, Hunter says, 'the ravages of wild elephants were on a larger scale, and their extermination formed one of the most important duties of the Collector for some time after the district passed directly under British rule.'[90] This did not happen quickly. In his survey of the district in 1848–52, Sherwill still described it as more or less covered with jungle, its high peaks of gneiss and narrow valleys between laterite ridges allowing limited cultivation.[91] The forests remained infested with snakes and wild beasts. Tigers, bears, leopards, and elephants had pushed back the line of cultivation in the late eighteenth century, precipitating a slaughter of large mammals and carnivora to protect agriculture. This outburst peaked during the period 1808–16, when an average of 172 animals were killed every month. But the situation did not ease till the mid-nineteenth century.[92] Various types of deer and the *nilgai* (Indian antelope) were to be found in the cultivated plains, along with wild pigs and many types of wildfowl. Wild buffalo, *gaur* (bison), deer and carnivora were found in the forest regions.[93]

Government efforts to secure forest fringe areas from wild animals were sporadic through the first half of the nineteenth century. So upland farmers, largely tribes like Santhals, situated their fields between forest galleries that served to protect them from deer and swine, while grassland intervening between forest and field was regularly burnt to easily detect any carnivora sheltering in the savannah. One narrative of their life and country from the early nineteenth century certifies them as good agriculturists who tended to lose a lot of their crop either to wild animals or to moneylenders.[94] Crops were not the only thing forest villagers lost. As farming settlements constricted the forest-savannah transition zone in places like the Bengal jungle mahals, villagers themselves were sometimes carried away by a tiger or a leopard.[95] Right into the 1850s and later, the Santhals of the Damin were in constant danger of attack from tigers, bears, and leopards, while their crops were threatened by elephants, deer, wild boar and monkeys.[96] In the tributary mahals adjoining Chotanagpur, inhabitants of small hamlets in the wild border country suffered for most of the year from shortage of crops, relying on forest products for sustenance. While this exposed them to wild animals, their scanty crops were often further reduced by wild elephants.[97] Writing of Chang Bhukar, in 1870, Dalton still found the country 'a den of wild beasts . . . a herd of not less than 100 elephants is ravaging it and man-eating tigers abound.'[98]

Some proportion of carnivore were slain as sport and for the security of the work of geological surveyors like Valentine Ball. His reminiscences of work in Bengal are filled with accounts of leopards, bear, and wolves shot, in addition to the usual game like hare, duck, geese, teal, and deer.[99] The Darjeeling topographic survey, conducted some years before Ball wrote his memoirs, was carried out in the midst of elephants, tigers, leopards, panthers, bears, buffalo, and deer of many descriptions. An official of the survey reported shooting a boa constrictor measuring eighteen feet.[100] Such survey work was particularly threatened by tigers, which made travel by foot exceedingly dangerous in places. Reports from the Bengal Surveys contain numerous accounts of tigers so bold that men were lost not only at night but also 'in the midday on the line of march.'[101]

These challenges to the knowledge-garnering tasks of empire compelled various officials to launch a more concerted effort to reduce the menace of wild animals to agriculture and official business in the 1860s. The Bengal government sought to take firm managerial control over non-agricultural lands. It adopted a standard policy granting rewards for slaying specified animals, notably tigers, leopards, bears, hyena, and wolves.[102] Initially there were noticeable reductions in deaths caused by wild animals in some divisions of Bengal. For instance, in the Santhal Parganas, the number of

Fig. 6. Hunters going out in the morning

Fig. 7. Hog-hunters meeting by surprise a tigress and her cubs

Fig. 8. Shooting a leopard

Fig. 9. Driving a bear out of sugar canes

reported deaths went down from 494 in 1858 to 125 in 1863. The figures for wild animals killed in the Santhal Parganas, the most acutely affected district, are given in Table 3.1.

Early success pushed up reward rates. In Chotanagpur, for instance, rates for tigers and wolves had tripled and those for bears and leopards had quadrupled by 1867.[103] However, for a period both the problem and the intensity of efforts to contain it grew. Many of these animals identified as vermin were compelled to concentrate in the forests as their habitats in the plains shrank with the spread of settled cultivation.[104] The forest-savannah transition zones, like parts of the Chotanagpur and the foothills of north Bengal or United Provinces, often became the last refuge of the besieged tigers and leopards. If settled cultivation reduced the beat of wild animals in the plains, plantation crops like tea and coffee took from them the higher reaches of hills particularly in south India. The proximity of domestic cattle, the decline in deer population, and the cover provided in certain areas by fruit trees, agricultural crops, increased encounters between villagers, tigers, and leopards. This provided a favourable environment for carnivores to become man-eaters.[105]

Such ecotonal areas were 'conservatories in disorder' in many respects. While they defied classification as field or forest, they also nurtured bestiality and agents of chaos in the form of wild animals. Similarly, Baiga *bewar* (shifting cultivation) created the forest-savannah zone that extended the habitat of gaur and black-buck, along with that of their predators.[106] Cultivators expected to transform these wild lands into productive fields were thus held at bay, or consumed, leaving the landscape to jungle tribes.[107] This outcome did not imply a tribal invulnerability to animal attack but underlined the necessary coexistence of extensive farming systems and wildlife habitats. Intensive farming required a different segregated spatial arrangement of human and wild animal worlds in the woodlands of Bengal. It is in this context that we can understand the acute exasperation conveyed by the words of one officer who exclaimed, 'it (is) a stain on our administration that, at this date, beasts of prey should still contest the field with us.'[108] Yet the sedentarization policies followed by the government in the aftermath of the Santhal *hool* and the Sepoy Revolt of 1857, may have prolonged this contest, as in certain districts the exodus of keen hunters like the Santhals caused the practice of annual hunts to become irregular, allowing faunal populations to grow.[109] The exclusion, or migration, of certain tribes from particular forests complicated hunting of every sort. Even in the early twentieth century, a keen sportsman forester like James Best acknowledged his debt to the Baigas of the Central Provinces, 'who knew to a yard where a tiger would come out in a beat, and entered into the spirit of tiger hunting with great zest.'[110]

Forest protection, vermin eradication and the creation of a sedentarized peasantry were often contradictory processes. While the first required the exclusion of most tribal people from forests, this led to minimized hunting and an increase in vermin. Similarly sedentarization directed communities adept at a mixture of cultivation, gathering and hunting toward forms of specialization in land use that impaired their hunting. Forest protection interfered with annual fires and other means by which villagers had controlled vermin in the vicinity of settlements. Complex interrelated activities of agriculture, forest management and hunting were thus marked off from each other, serving to increase the burdens of government and at the same time rendering it unable to secure safe pursuit for, or enhanced returns from, any of these activities.

Taken with the demographic movements and recomposition of ethnic groupings in the various provinces of woodland Bengal, the incidence of deaths caused by wild animals serves as a good index of the patterns of forest clearing. Midnapore was the southwest Bengal district where cultivation had made the most inroads into the forest-savannah complex. As Table 3.2 indicates, this district suffered the highest casualties.

The system of rewards for hunting, which had been revived in

Table 3.1. Wild Animals Killed in the Santhal Parganas, 1858–63

Animal	1858	1859	1860	1861	1862	1863
Tiger	13	15	39	15	42	26
Leopard	84	75	122	81	113	142
Bear	2	34	64	44	44	32
Hyena	8	16	18	29	42	43
Wolf	8	4	-	3	19	11

Source: WBSA P/5 8 BJP (Jud), A progs 341–2, 18 Jul 1863, pp. 199–200.

1860 after being temporarily abandoned after the 1857 uprising, and the widespread disarmament of local populace that had followed, was soon perceived to be insufficient. Citing the success in controlling the tiger menace in Bombay, by rearming peasantry, the India Office recommended such selective rearming of the border villages in Bengal as well.[111] In some provinces traditional hunting tribes were recruited to the task. The Pardees of Hyderabad, for instance, otherwise regarded as criminals, vagrants, and gypsies, were used to kill wolves using snares that they made from animal or bird sinew.[112] The search for similar remedies in Bengal was taken up. Declaring that 'rewards cannot and will not reduce the number of wild beast', one Bengal official proposed a force of shikaris armed by the state.[113]

In Patna Division a scheme of this nature was approved. Men of the Baheelia caste, considered fond of shooting, were hired and assigned to different police stations, being supplied there with guns, powder, and shot. They were to be paid five rupees per wolf and one rupee for a whelp. In addition they were free to resort to blind wells and strychnine.[114] The system was pronounced a failure in short order, as the hired shikaris appeared to lack the 'special knowledge and experience required to track out . . . and destroy wolves.'[115] Some of this ineptitude was deliberately feigned. Since spilling the blood of wolves in the vicinity of habitation was considered inauspicious by some forest villagers they often threw shikaris off the trail of wolves.

Many initial policies for vermin eradication were conceived within a British elite cultural framework of sport and gamesmanship that did not permit the use of techniques that smacked of foul play or excessive violence. The reduction of vermin and the regulation of native hunting were seen by many colonial officials as opportunities to fortify the authority of British rule and publicize its firm but paternally kind aspects. But the continuing failure of many of these approaches compelled frontier district administrators to rethink their lofty objectives. Killing by stratagem, in the name of making tribal places safe, was introduced in the face of criticism for its unsportsmanlike use of guns attached to springs. But such techniques endangered human and domestic cattle far too often to prove of lasting value. This meant a reluctant return to the rewards system.[116]

Table 3.2. Wild Animals Destroyed and Persons Killed by them in the Jungle Mahals, 1860–6

District	Persons killed by			Wild animals killed		
	Tigers	Leopards	Bears	Tigers	Leopards	Bears
Manbhum	137	20	17	35	16	68
Singhbhum	8	-	1	89	79	48
Birbhum	18	-	1	2	10	14
Midnapore	227	72	-	16	42	33

Source: WBSA P/140 BJP (Jud), A progs 179–80, 7 Feb 1870, pp. 145–6. This table was prepared by extracting information from a larger table in these proceedings.

The Bengal government also encouraged the district administration to sponsor hunting parties that would revive traditional group hunting. This proved inappropriate against wolves, as they preyed more on villages, usually children and the aged, and were unlikely to cross the path of hunting parties.[117] During the phase of vermin eradication, a range of strategies were launched to intensify and concentrate activities in the hands of the British administrators or their native subordinates. These strategies made the task of culling vermin from the woodlands a direct responsibility of field officers. They could hunt such vermin themselves or find ways to put native hunters to work for the imperial cause. But like the group hunting case briefly mentioned, these officers and their attempts failed for lack of local knowledge. So, despite several limitations— the inaccuracies of deaths reported and attributed to animals, the killing of animals for sport that went unreported, and the failure to estimate crop and cattle destroyed by animals—the rewards systems persevered in Bengal. Over the years it was refined to vary the rewards by species of beast and extent of menace (see Table 3.3). Thus a man-eater or a difficult rogue elephant might merit a special reward.[118] In a move that exacerbated conflict between civil and forest officers, the power to give rewards was delegated to subdivisional officers, bringing destruction of wild animals into clash with fire protection in forests, as the latter gained momentum in the last decade of the nineteenth century.[119]

To sum up, it will be useful to underline some important transitions in statemaking here. Vermin eradication was primarily a programme intended to secure agricultural expansion. In its heyday, roughly 1860 to 1890, relations between Indian and British hunting were transformed in several ways. First, if British and Indian hunting had coexisted before 1860, concerted efforts were made thereafter to render the latter completely a servant of the former, or declare it illegitimate.[120] Hunting, separated out of the tribal economy, became a preserve of governmental action subject to its own rules.

Second, in the project of vermin eradication, British efforts to secure monopolies over legalized violence converged with the creation of tribal places. In securing these places for their aboriginal pioneers the colonial government worked to buffer them from predatory carnivora, plains people, and town dwellers. Some Bengal officials justified deviating from codes of fair play and sportsmanship to accomplish this difficult goal. In those situations the role played by hunting in affirming heroic and just district officers was frequently jeopardized at the political threshold where the spatial segregation of agriculture, forestry, and sport became vulnerable. Though, as already noted, the techniques of killing for which the justification of unfair means was offered, tended to fail at a more mundane level of efficacy.[121]

Third, by the end of the nineteenth century, vermin eradication was not being evaluated for success or failure in terms of its own returns of animals killed or people killed. Forests had become important as a woody resource and were firmly subject to a battery of territorial management strategies. This topic is discussed in detail in the second part of the book. It is important to mention it here because we should not lose sight of the way changes in a single aspect of woodland management—like vermin eradication—were influenced by the dynamic development of vermin eradication policies and their relationship to other related policies like forest management or watershed conservation. The resultant procedures, and their accompanying rhetoric of conservation and prudent resource use, entered the landscape of hunting and transformed it again. A clutch of hunting, shooting and fishing rules in reserved and protected forests brought to the surface simmering conflicts between foresters and district officers. Their upshot is discussed later in this section. Before that vermin eradication has to be rounded off with a discussion of herbivorous and reptilian vermin.

Table 3.3. Scale of Rewards Sanctioned in Bengal, 1870 (in rupees)

Division/District	Tiger	Leopard	Bear	Hyena	Wolf
Midnapore	10.5	2.5	2.5	2	-
Chotanagpur	30	10	10	5	20
Santhal Parganas	10	10	2.5	2.5	5
Patna	5	2.5	2.5	2	20
Bancoorah	10	2.5	2.5	2	-

Source: WBSA P/150 BJP (Jud), A progs 126–9, 17 Dec 1870, no. 5512, Fort William, 17 Dec 1870, A. Mackenzie, Offg Junior Secy GOB, to Secy GOI Home, p. 115. This table is extracted from a larger one in these proceedings.

Elephants, unlike carnivora, were taken off the extermination lists soon after the 1860s. As a first measure the Bengal government was advised to regulate elephant hunting. Remarking with strong disapproval on the indiscriminate hunting that prevailed, one senior army officer and surveyor wrote:

at present elephant catchers from the Jalpaiguri and Cooch Behar Duars are permitted to hunt as they please, these men break up the herds, as they hunt only by running down the animals, separating the calves, which when found alone are quickly killed and eaten by the men of the hill tribes . . . all this should be stopped . . . the right to hunt over determined tracts should be sold to approved parties.[122]

Charges of cruelty and wasteful hunting by Indians were, of course, part of the standard vocabulary of colonial rhetoric, comparable to the critique of shifting cultivation.

The comparison may be extended to the way such critique glossed over or rationalized the incorporation of 'local knowledge' into colonial management strategies. With elephants, the case is particularly clear. A long-standing cross-oceanic trade had engendered elaborate methods of capturing and transporting elephants. During the seventeenth century, Indian rulers of Bengal and south India were major importers of elephants from southeast Asia. But by 1750, the elephant trade was often moving the other way, with diminishing demand in India and exhausted supply in southeast Asia.[123] Control of elephant hunting was a means of benefiting from this trade, though later due to the extensive use of elephants in timbering, elephant management got linked to an increasingly centrally regulated forest management. The question of government monopoly in elephant catching was first raised in 1851. The Legal Remembrancer pointed out that law and custom affirmed government monopolies in Arrakan and Cachar, but such claims were unsupported in Sylhet and Chittagong. Preventing the Jaintia raja and others from hunting in Assam, by 1855 the government moved towards rules declaring elephant catching a state monopoly.[124] The only regulation, however, where by statute the government claimed ownership of elephants, was created in 1873 when the establishment of the inner line system included elephants in the list of items requiring permits for trade across the line.[125] This caused hardships to raiyats whose crops were regularly destroyed by elephants in permanently settled estates, while landlords who claimed a right to hunt elephants and tax others who did so in their forests were peeved at the loss of privilege and revenue. In Assam, the costs of operating the monopoly led to its partial dismantling by 1859.[126]

By the 1860s, the Government of India was contemplating elephant breeding and a detailed account of the practice followed by the Karens, Siamese and other tribes who bred elephants was compiled.[127] Soon, through *kheddah* (elephant catching) operations they were being caught and tamed in large numbers. The kheddah consisted of a large enclosure surrounded by a ditch and a paling of timbers. Three to four thousand people participated in drives lasting several days with trumpets, drums, and fireworks to bring the elephants into the enclosure through a narrow tunnel.[128] In the next decade, kheddah use became less, largely as the emerging forestry establishment became critical of elephant hunting. By 1879, Elephant Preservation Act was enacted in India and killing, authorized by the District Magistrate, was confined to rogue animals.[129] The Act was extended to most parts of Bengal, including Midnapore district by 1894.[130]

As Assam and Bengal had become the principal elephant hunting ground, the Act was not well received there. Sanderson, the leading expert in India on elephants, made a powerful plea for a balanced approach to their preservation and capture. Protective laws, exclusion of hunters from forest reserves and government monopoly had raised the price of a baggage elephant from eight hundred to two thousand rupees within the 1870s. Elephants were regularly imported from Burma, as a result.[131] The case of elephant preservation thus created a conflict between foresters ostensibly worried about their destruction, and commissary officials, planters, and others who used them as labour. Vermin eradication was rapidly substituted by conservation ideas in this case as the institutions of territorial forestry were strengthened. But more than this tension between contradictory government and business interests the damage caused to crops by unrestricted elephant breeding meant that pit-falls, poisoned spears, and other modes of hunting that the British wished to suppress returned, perforce, as farmers were left to defend their fields and grain stores on their own.[132]

The other target of vermin eradication was the prolific poisonous snake population of the jungle mahals on which twelve thousand rupees were spent in 1863 alone. But the administrators of these districts despaired of reducing the number of snakes as the low jungle and dry soil, interspersed with sparsely cultivated fields, provided a favourable breeding ground.[133] The villagers tended to make cobras their chief target for remunerated snake killing. Cobras tended to inhabit fields more than forests and thus their killing, while not ridding forests of venomous snakes, made the mandatory securing of fields a profitable task for farmers.[134] As may be seen from Table 3.4, little progress was made in southwest Bengal under the rewards system towards the ambitious goal of exterminating venomous snakes. But the system continued, with both the rewards for cobras and deaths by snakebite increasing annually through the 1870s.[135]

In Bengal, as in the rest of India, hunting regulation arose through the impetus provided by the Indian Forest Act of 1878. The Act included skins, horns, tusks, and bones in forest products.[136] The Elephant Preservation Act that came into force in 1879 has already been noted. This was followed by the Wild Birds and Game Protection Act in 1887, which actually recognized the diversity of prevailing conditions and gave local governments considerable leeway in framing rules.[137] The Wild Birds and Animal Preservation Act of 1912 consolidated the nineteenth-century laws. A variety of hunting and shooting rules, initiated in the 1880s, were standardized by 1915. These classified as 'A forests' areas which were sanctuaries, and 'B forests' areas where seasonal hunting by permits was allowed.[138] These rules also prohibited certain forms of hunting like use of vehicles, traps, explosive and unmanned guns. The Bengal amendment to the Elephant Preservation Act was passed in 1932, allowing destruction of rogues, and by the end of colonial rule, Bengal also had the Bengal Rhinoceros Preservation Act, legislated in 1932. The West Bengal Wildlife Preservation Act of 1959 replaced the All-India Act of 1912.[139] But from the very beginning these regulations supported both restrictions on land use and the creation of a domain of scientific expertise in land management.[140]

Some of the earliest hunting rules were framed with an eye to preventing forest fires, as the Indian foresters embarked upon ambitious projects of forest regeneration.[141] These rules generally prohibited hunting in a prescribed closed season, poisoning and dynamiting in rivers, and required hunting permits to be issued by the district officer. The restrictions covered deer, antelope, hare, and a variety of jungle fowl that were routinely hunted by tribals like the Kols for food.[142] That such rules undermined the vermin eradication programme was evident to officials in the forest-savannah transition zone, where districts like Midnapore were most under pressure from the extension of cultivation. Separate rules were often framed in these areas, providing another instance and facet of the discourse of frontiers as it was deployed to qualify and thus mitigate the influence of standard administrative procedures.

In such cases the usual sentiments about tribals needing special dispensations were revived in government policy, as zones of

Table 3.4. Death by Snakebite in Southwest Bengal, 1866–9

District	1866-7	1867-8	1868-9
Manbhoom	45	75	86
Singhbhum	-	-	-
Birbhum	40	60	55
Midnapore	341	475	492

Source: WBSA P/150 BJP (Jud), A progs 126–9, 17 Dec 1870, no. 5512, Fort William, 17 Dec 1870, A. Mackenzie, Offg Junior Secy GOB, to Secy GOI Home, p. 115. This table is extracted from a larger one in these proceedings.

anomaly were carved out by the operation of hunting regulations. Tribes like Oraons and Kherwars, keen sportsmen all, might have migrated in the face of rigorous restraint on hunting. Moreover, villages in these regions were 'little more than unappropriated village jungles'.[143] In the jungle mahals, protected forests were mixed up with fields. Villagers who would dam streams and use small nets for fishing invariably operated in the forests. Disallowing these activities was likely to cause hardship.[144] Due to these considerations Chotanagpur and Santhal Parganas got different rules that were less likely to constrain the livelihood of their aboriginal inhabitants.[145]

Despite the special provisions, a few years later, the Conservator of Forests was compelled to note that 'in the Santhal Parganas and Chotanagpur villagers take the law into their own hands and hunt without any licence . . . generally without any firearms.'[146] While such disregard for regulations might have been partly due to the group licensing system (where headmen were given a collective licence for whole villages), and the more cursory control this signified, it also arose in a wider scepticism that many district officers felt for game protection.[147] A view commonly expressed by district officers, and contested by the foresters during the deliberations of the committee constituted by the Bengal government to review hunting and shooting regulations, held foresters responsible for turning reserved forests into sanctuaries for leopards, tigers, and bears, ignoring the consequent hardship to villagers in forest districts like Singhbhum.

The lack of support from district officers for the stringent imposition of hunting regulations on large fauna and carnivora ensured that they continued to be treated as vermin. Since the assistance of local villagers was still actively solicited in destroying dangerous carnivora, it was probably not easy to regulate their bag of small fauna and fowl.[148] Thus the entry of foresters, with their distinct agendas, into the regulation of hunting, made possible an alliance between district officers and private hunters. For a novel set of reasons hunting became a contentious issue and policies to manage it suffered.

There were geographical reasons for ineffective hunting regulations as well. The tributary states on the fringes of British dominion had experienced slower conversion of the forest-savannah complex into agricultural tracts. Even in 1910, a large portion of their total area was reported as forest, if scrubby. Such a landscape naturally harboured dangerous carnivora like tigers, panthers, leopards, hyena, jackal, and wild dogs. *Sambhar, chital,* nilgai and black buck were also found in these tracts.[149] The continuing menace of tigers noted in western Singhbhum in the early 1900s may have been caused by proximity to the tributary states. Like forest fires, timber contractors, and other agents of reserved forest destruction, carnivora had scant respect for administrative boundaries. An average of fifty persons lost their lives to tigers every year while collecting *sabai* grass and clearing forest lines in the reserves of Singhbhum and Porahat. Even camps of coolies, often comprising over two hundred men in one location where road work might be progress, were not immune to raids. One such camp was visited thrice during one night in 1905 by a tiger which carried off a victim on each visit.[150]

All the travails of villagers in forest Bengal and the ambivalence frequently expressed by district officers, especially from the tribal regions, modified but did not prevent the emergence of a Bengal-wide regime of hunting and shooting regulations built on closed seasons and excluded forest tracts. The closed and open shooting-blocks policy adopted by the Indian Forest Departments presaged wildlife park formation from the early twentieth-century. The role of the Society for the Protection of Fauna of the Empire (SPFE), in the history of parks in Africa, has been discussed at length in the recent research of Rod Neumann. The SPFE became inordinately influential in creating Serengeti and other east African parks during the interwar period of the twentieth century.[151] The SPFE got interested in India too in the 1930s, organizing an all-India wildlife conference which recommended the preservation of some animals like rhino, lion, cheetah, musk deer, pangolin and Sind ibex, but did not recommend the creation of parks.[152] Wildlife preserves and parks, the isolation of large forest tracts as animal conservation areas were, with a couple of exceptions, a post-colonial development.[153]

Exploration, settlement, railway and road building were greatly subsidized by hunting, which provided both profits and proteins, even while hunting made these activities safe. In India, the British, through vermin eradication, also appropriated the protective function performed by specialist shikaris and declared them to be poachers.[154] Mackenzie certainly persuades us that the emergence of hunting codes in the colonies did not depend solely on changing European sensibilities. The colonial history of rising conservation ideas, and vermin eradication, which as a special imperial project of the nineteenth century, often obtruded on forest conservancy, clearly shaped hunting management and the allied rise of animal ecology.[155] But even in the limited estate of Bengal it is apparent that regulations did not always succeed eradication, nor were they uniform in design and impact.[156] Thus the emerging overlap and contradiction between the historically transformed spaces of cultivation, hunting, and forest management created a pattern of limitations on the reordering of the landscape.

Forestry Experiments: The Bengal Teak Plantations

Violent turf battles with carnivora were not the only way the wildlands of Bengal were reordered in the nineteenth century. Selective plantation schemes were visualized and implemented in the jungles surrounding civil or military stations and landed estates. Soon after their establishment in 1788, the Calcutta Botanical Gardens became a source of seed, advice, and encouragement for tree planting programmes in Bengal and beyond. From 1794 teak seed and seedlings were distributed from there to every part of the country and plantations were taken up in different parts of Bengal considered favourable for the growth of teak. By 1800 the Superintendent of the Calcutta Botanical Gardens, Dr William Roxburgh, could report with satisfaction that several thousand trees were 'in a forward state'.[157] The greatest success was the Bauleah plantation in Rajshahi where a few plants raised in 1795 under the charge of Assistant Surgeon Henry Barnett had reached thirty feet in height and a foot in diameter in five years.[158]

In the next ten years 715 bighas (about 125 acres) had been planted in Rampur-Bauleah, yielding a count of 36,000 trees including teak. A sample of twenty trees aged ten years, measured for their girth, were all over three feet. Further rapid growth was anticipated with proper thinning and greater spacing of rows planted. But the spread of plantations was constrained by the lack of an establishment to propagate and tend them.[159] These were not easy to obtain since the higher authorities were still sceptical of tree planting efforts. Even the showpiece Bauleah plantation had only been approved at an annual cost of 1,773 rupees due to disruption in the timber trade between the Malabar and Coromandel coasts and Bengal. The survey of Malabar forests and the revival of timber trade was expected to slow down further investments in teak plantations in Bengal.[160]

Where the Board of Control of the East India Company was wary, the colonial scientists were ambitious. They were also more keenly aware of forests as the products of cultivation. After noting the satisfactory growth of the Bauleah plantations, and recommending greater spacing between plants to enhance growth, Roxburgh went on to propose, in what must be considered the earliest farm forestry scheme of modern India, that zamindars in Bengal be encouraged to grow teak with seedlings provided by a gardener appointed under the supervision of the District Collector.[161] He envisioned a scheme of district nurseries receiving technical support from the Botanical Gardens. This was expected to reduce the cost of seeding transport from Calcutta and minimize losses through injury.[162] In Midnapore, Burdwan, and Bihar, the scheme was taken up through several zamindars who did plant teak in their jungly wastes.[163]

In the jungle mahals teak planting was undertaken to introduce both order and industry in wastelands. The Assistant Surgeon there had grown four teak plants received from the Botanical Gardens to a height of twenty-five feet in five years in his garden.[164] Spurred by this success he quickly drew up a wasteland development scheme. Thomas Leake discerned waste tracts of jungle as often having the richest soil throughout the district and diagnosed them to be uncultivated for want of capital and initiative on the part of the local peasantry. Anticipating leases from zamindars lured by the prospect of higher rents, he proposed leasing these lands to raise teak and offered to manage the plantations for a small remuneration to cover palanquin-bearer costs.[165]

These plantations were approved both as a way of extending cultivation and securing a future crop of valuable timber. Planted on the banks of the Dalkishor river, they were expected to profit from the timber trade of native wood merchants who regularly rafted sal down the river to supply Calcutta.[166] The Assistant Surgeon was allowed a hundred rupees a month, the rate prevailing in Bauleah, for maintaining the plantations.[167] Plantations and farm forestry were expected to combine in transforming jungly limits of the arable into productive gardens and groves. By 1820 the Rampur-Bauleah and jungle mahals plantations covered 1,250 bighas and contained 55,000 trees. Yet there had been periodic misgivings about the Company raj creating a forestry establishment.[168] The Superintendent of the Botanical Gardens, Dr Nathaniel Wallich, was ultimately appointed Superintendent General of Plantations and he immediately squelched any doubts about the value of the plantations by declaring them to be 'in so thriving a condition as to hold out a very pleasing prospect . . . of valuable timber.'[169] A plantation committee existed till 1827, but a few years after its discontinuance, in 1831, the plantations were taken out of the control of the Superintendent.[170]

Wallich's enthusiasm was not confined to teak. In fact, being well acquainted with shortages in teak and the expenses involved in procuring it, he became an advocate of mixed plantations where teak and mahogany would be interplanted with sissoo.[171] He found sissoo unrivalled among Indian timbers in its growth and ease of cultivation and a greatly underestimated wood for naval and military purposes. In his view, it could take the place of teak in the manufacture of ordnance carriages. Noting the rapid depletion of supply from traditional sources in Gorakhpur, Rohilkhand, the Ceded and Conquered Provinces, Wallich advocated sissoo plantations where timber trees would be planted at thirty-foot intervals and the intermediate sissoo would be managed on a five-year coppice rotation for firewood and poles. These recommendations were based on thirty years of experimentation in the Botanical Gardens where the annual increment and other aspects of sissoo growth had been studied.[172] Wallich also introduced bamboo plantations for the supply of scaling ladders and tent poles, noticing its enormous consumption in buildings and concluding that native husbandmen were unlikely to plant bamboo.[173] In the next two decades Wallich chalked up an impressive record in distributing several thousand trees of teak, mahogany, sissoo, and other timber species all over India.[174]

The results of this widespread tree plantation effort cannot be evaluated here. In the Bengal teak plantations, however, performance was disappointing in the long run. The jungle mahals plantation was initiated in 1814, on 350 bighas on the banks of the Dalkishor river. Ten years later, floods destroyed a large number of the trees.[175] Forty years after they had been initiated the teak plantations of Bankura contained only 2,000 trees and by 1851 these had been further reduced to 610 growing trees. Tending had been erratic, leading the Superintendent of Tenasserim teak plantations to note on a visit that the cutting had not been systematic.[176] The struggle with these plantations was the earliest Bengal experience of the frustrations involved in raising exotic timber plantations.

Some colonial scientists despaired of artificial propagation of teak on any large scale without proper supervision by a European officer well acquainted with local conditions. While native subordinates might be conversant with particular localities, they were presumed to lack the meticulousness and scientific approach considered necessary.[177] But more than managerial failure, wider ecological forces, as the river Dalkishor changed course, precipitated the premature felling of the jungle mahal plantations.[178] Losing the river as a viable means of timber transport from these plantations cut the final link to plantations already close to being abandoned. Dr Henry Falconer, who had succeeded Wallich at the Botanical Gardens, declared the plantations a failure for the sites being unfavourable for teak, and partly due to the poor management of the plantations.[179]

The Bengal teak plantations were an innovative exercise in extending the civilized arable. They were precursors of other woody and leafy plantation crops that characterized colonial expansion through south Asia. But they denied colonial forest management easy success providing one of the earliest lessons in forestry and its localized, site-specific ecology and politics. As the discussion of silviculture and forest regeneration in Chapters 7 and 8 will show, the lesson was poorly learnt. At this point we must, however, return to the main procedures of colonial forestry in the Company raj, surveying, exploration, classification, inventory, and assessment of trees and landscapes. It was these repeated forays into woodland Bengal and similar regions that gave rise to what Richard Grove and other historians have discussed as desiccationism.[180] Penetrating the jungle revealed the scarcity of trees, to the surgeon-naturalist and shikari-naturalist alike, and provoked speculation on the unpredictability of rainfall and the fragility of food production regimes in the aftermath of deforestation.

Notes

1. M. Jules Clave, writing about Indian forests in January 1879 for the *Revue des Eaux et Forets*, right after the great Paris Exhibition of 1878, quoted in Anon, 'French Writers on Indian Forests', *IF* 5(3): 326-7, 1880.

2. Paul Carter 1989: xxii.

3. For the use of prosopography in history of science, see Thakeray and Shapin 1974. Recent practitioners are Grove (1989a, 1993a); Kuklick (1991); and Rajan (forthcoming).

4. The last named, war, has been covered in the previous chapter, but the rest of the modes will be discussed here.

5. Richard Grove would be sympathetic to the case I am making. But my argument largely runs counter to Ravi Rajan's claims. Rajan does, however, provide a valuable catalogue of the colonial scientists who worked to create a desiccation discourse in India and used it to insist on the urgent need for state forest management.

6. Cohn 1987; Rocher 1993: 241.

7. Nicholas Thomas 1994: 3. See for instance, various essays in Part II of Breckenridge and Van der Veer 1993: 189-340.

8. Stoler and Cooper 1989: 609.

9. Cartmill 1993: 92-117.

10. Prakash 1995: 3.

11. Cohn 1987a, Ranajit Guha 1963, Washbrook 1981, all make this point in different ways.

12. The foundational work has been Said 1979. Other important works are Bhabha 1986: 154; 1984; Spivak 1987. Said, Bhabha and Spivak have recently been called the holy trinity of colonial discourse studies. See Robert Young 1995: 163.

13. Said 1979: 94. Sarah Jewitt (1995) is more sympathetic to Said and finds some utility in his broad categories, while arguing that the value of his approach is lost when his followers take extremely relativistic and anti-modernist positions.

14. In a recent and comprehensive assessment of Edward Said, both his early work on Orientalism and subsequent writing on imperialism, John Mackenzie (1995: 13-39) has convincingly demonstrated Said's lack of historical grounding. He has rightly pointed out that empirically sound research always finds imperialism fragmented by class, gender, regional variations, and penetrated by the ideas of similarly heterogeneous subordinated populations. See in this connection, Said 1993. Another useful study providing a counterpoint to Said is Pagden 1995.

15. See Ahmad 1992, Mohanty 1984: 71-92, Parry 1987, for the most comprehensive versions of the critique. Nicholas Thomas (1994: 171-92) and Robert Young (1995: 160-73) offer creative reformulations by suggesting colonialism be studied at the level of locally mediated expressions rather than de-historicized discourses, where social production of desire and reality are inseparable.

16. I refer here to Inden 1990 and later volumes, especially five through nine, *of Subaltern Studies*. I have dealt with Foucault, colonial discourse, *Subaltern Studies*, and the way they have been combined in modern south Asian social history in other work. See Sivaramakrishnan 1995b. Whether Foucault is the root cause of the problem remains a point of heated debate. Aijaz Ahmad, a trenchant critic of Said and *Subaltern Studies*, tends to argue that Foucault has been misread by Said and some of his followers. The argument in a nutshell is that Foucault shifted attention, especially in *Discipline and Punish*, from the state to technologies of control, but did not recommend historically unspecific analyses of political knowledge. Said, Bhabha and others make precisely that untenable move. See Nicholas Thomas 1994: 42-3; Ahmad 1991. To some other scholars Foucault's disarticulation of power from subjectivity and agency, and his location of it in discourses rather than processes of social production, does in the final analysis enable the unified colonial discourse approach. See Sangren 1995: 5-26, for a careful and clear presentation of this view.

17. Ludden 1993: 250.

18. Ludden 1993 discusses these processes via both Orientalist studies of Indian culture and the pragmatics of colonial administration. Foucault (1973) is a good exposition of the 'Enlightenment episteme', that played into such patterns.

19. Cohn 1987b; Benedict Anderson 1991a; Dirks 1992; Richard Smith 1985; Pant 1987. For the development of statistics in Europe, see Hacking 1975.

20. Appadurai 1993: 324-7; Dirks 1993: 280. For a fine discussion of the optimism and complacency that often guided imperial policy in India after 1860, see Hutchins 1967.

21. Skaria 1995.

22. I have told the story broadly in these terms elsewhere. See Sivaramakrishnan 1995a.

23. For a fascinating account of the aspatial nature of imperial history, see Paul Carter (1989: 64) who says, 'it implies a centre of power around which the boundaries of the unknown are progressively pushed back—a process quite alien to the explorer's experience.'

24. Lefebvre 1991: 48.

25. Lefebvre (1991: 33-9) defines representation of space and spatial practice. I am indebted to Paul Carter (1989: 121-8) for these ideas about surveying as a continuing mode of colonial knowledge, rather than one that precedes other more possessive and controlling or transformative modes. This topic is addressed more fully in the next chapter.

26. Lefebvre 1991: 44–5.

27. Robert Young 1995: 165.

28. For a fine study of this process from Africa, see Neumann 1995.

29. Gregory 1994: 169. Also see Neil Smith 1990. For useful reviews of recent critical writing in geography, see Driver 1992; Watts 1992; Soja 1989; Pundup 1988.

30. Lefebvre 1991; Gregory (1994: 354–401) provides an excellent discussion of these concepts.

31. Bogue 1989; Guattari and Deluze 1977, 1988. In discussing historical geography, Harvey (1989: 238–9) takes up this theme of territorialization.

32. Young 1995: 172–3. See Yang (1989) for a fine recent study framed by a discussion of colonialism and the appropriation of space for local governance in eastern India.

33. Stafford 1990: 86–8. See Cosgrove (1984) for a discussion of the landscape mode of seeing that emerged in western sensibility during the eighteenth and nineteenth centuries.

34. Rangarajan 1994: 158; 1992. For shifting cultivation and its management in colonial India, also see Prasad 1994: 20–50, 78–90, 215–32; Pouchepadass 1995; and Guha and Gadgil 1989.

35. The most important compendious works for Bengal are Dalton 1974 (1872) and Risley 1981 (1891).

36. OIOC MSS EUR F/236, W.G. Archer Papers, note entitled 'Conversion of Tribes into Castes', Risley and Gait (eds), Census of India 1901.

37. See Spurr 1994: 1–67; Suleri 1992: 17–22, 103–10; Appadurai 1993: 328; Cohn 1987b.

38. See Nigam 1990, Radhakrishna 1989, and Skaria 1997 for the British period. Tribes and their relations with Mughal states, sedentarization and its effects, are discussed by Chetan Singh 1988.

39. Beteille 1992: 57–78, 70.

40. This has been discussed in the previous chapter. For a good description of Cleveland's activities in the colonial historiography, see Bradley-Birt 1905. Cleveland as a pioneer of military mobilization in British India has been analysed by Alavi (1991). Paharia agriculture, social organization and its transformation under British rule are exceedingly well treated by Pratap 1987.

41. Sherwill 1851: 545.

42. OIOC W/2126 William Hodges, *Travels in India* (London: J. Edwards, 1793), p. 88.

43. MacPhail 1922: One of the longest serving Deputy Commissioners of Santhal Parganas explicitly compares the creation of the district in 1855 to the carving out of a Paharia enclave in 1783. See Carstairs 1912: 221–2.

44. Sherwill 1851: 546–50.

45. MacPhail 1922: 63.

46. The parganait in the Damin usually had charge of a twelve-village cluster. Sherwill 1851: 552.

47. OIOC P/2800 BRP (LR), Jan–Feb 1886, A progs 20–33, Feb 1886, head I, col 8, no. 3 Camp Dalsinghserai, 14 Nov 1881, J. Boxwell, Magistrate

Darbhanga, to Offg US GOB Rev, p. 69. The persistence of policies of administrative exceptionalism in the Santhal Parganas, in the transformed context of forest conservancy, is dealt with in the next chapter.

48. OIOC L/E/7/185, Dufferin Committee Report on the Condition of the Poorer Classes in India, l/d 6 Apr 1888, R. Carstairs to Commr Bhagalpur, p. 1.

49. Ibid., p. 2.

50. Martin 1833: 153–4.

51. Ibid., 1833: 157–61, 245–9.

52. Cleghorn 1852: et al. 1852: 121–3.

53. W.B. Oldham 1894: 5.

54. Cited in Man 1867: 6. The copy cited was found as a typescript in OIOC MSS EUR, W.G. Archer Papers, F 236 (52).

55. Man 1867: 4–10; Edwardes (1925: 87) also describes the Santhals as squeezed out of their homes in the upper Gangetic valleys by waves of Hindu and Muslim migrants.

56. Dalton 1866a: 194

57. Ibid.: 191; Campbell 1866: 21.

58. Dalton, 1866b: 160.

59. Ibid.: 163–4.

60. WBSA P/107 BJP (Jud), A progs 38-9, Jul 1867, no. 421, Hazarbagh, 18 May 1867, H.M. Boddam, DC Hazaribagh, to Commr Chotanagpur, p. 22.

61. The other colonial ethnographer/historian/tribal officer of note who wrote copiously to contribute to the creation of this narrative, along with Dalton, was of course William Hunter.

62. Beteille 1992: 76.

63. B.B. Chaudhuri 1993: 76.

64. OIOC P/228 BRP (For), May–Aug 1871, A progs 25–40, Jul 1871, no. 1332, Chotanagpur, 16 May 1870, E.T. Dalton Commr Chotanagpu, to Offg US GOB Rev, p. 41.

65. OIOC Tract 334, Anon, *The Kols of Chotanagpur* (undated), p. 9.

66. For a clear discussion of the khuntkatti system of land clearing by Mundas, the emergence of hereditary headmanship and the different tenurial arrangements that smacked of a social stratification not admitted in the egalitarian tribal society pictured in colonial accounts, see Thapar and Siddiqi 1991: 412–25.

67. *Kols of Chotanagpur*, pp. 10–29. In contrast to this narrative of sudden influx and dispossession or oppression, Thapar and Siddiqi (1991: 427) point to a much longer and more gradual involvement of the region with wider economic forces and regional states.

68. Dalton 1865: 4–5.

69. Describing the ambiguity inherent in first naming, Paul Carter asks, 'did names name facts or intentions? Were they mechanically associative or spatially inventive?' See Paul Carter 1989: 60. There was a constant tension between signposts for onward travel and systematics—the abstraction of features of the landscape from their regional geography. The latter did greater violence to local knowledge.

70. For a discussion of sedentarization as de-territorialization, see Paul Carter 1989: 335-6. For an illustrative south Asian case, see Prasad 1994: ch. 5.

71. Mohapatra 1991: 7.

72. B.B. Chaudhuri 1993: 69.

73. Ibid.: 71-2.

74. Mohapatra 1991: 8-9.

75. OIOC MSS EUR F236, W.G. Archer Papers, 'Agrarian Disturbances in Bihar, Etc.', undated, p. 1. Also see Archer's translation and original (Hindi) of Kharia Manifesto in the same collection, one clause of which reads, 'the rajas and landlords have not given us any fields ... we ourselves made them ... and by the command of God himself grew rice.' Petition dated 17 Sep 1936.

76. OIOC MSS EUR F236, W.G. Archer Papers, 'Notes on the Tana Bhagat Movement', undated, p. 5. The petitions of the Sardari agitation, lasting from 1859 to 1880, had demanded that Mundas and Oraons be allowed to re-form eroded village communities, reiterating a long-standing tradition of demands for regional tribal autonomy. See Mohapatra 1991: 32; K.S. Singh 1983: 14-17.

77. B.B. Chaudhuri 1993: 79-110.

78. Under different names this practice has preceded settled cultivation in many parts of the world. Referring to it as protective hunting, John Mackenzie describes the prevalence of vermin eradication in ancient Greece. The extermination of lions over large areas of the Mediterranean and the Middle East was ostensibly for the same purpose. But in England by the late eighteenth century vermin eradication was also restricted to certain people, a prime example being the ritualized fox hunt. See John Mackenzie 1988: 17.

79. Cartmill 1993: 30-62. In the colonial setting, it was not only officials and other colonizers who lived by and purveyed these ideas. Members of the Indian urban elite participated enthusiastically. In a set of letters describing his shikar adventures, a Bengali barrister wrote to his children exhorting them to become good sportsmen as preparation for national service and assured them, 'it has been amply proven . . . that the best sportsman makes the best soldier.' See K.N. Chaudhuri 1918: 4.

80. K.N. Chaudhuri 1918: 18. Elsewhere, elaborating this theme in the context of colonial hunting in Africa, John Mackenzie identifies three phases. First, commercial hunting for ivory and skins; second, hunting as subsidy for conquest and settlement; third, ritualized and ideologically charged hunting. The first two phases destroyed game but in the third conservation came into play. The effect of game regulation was to restrict hunting to the gentleman sportsman, administrator and soldier, This also meant that defensive hunting, to protect African and European lands, became the task of colonial rulers. See John Mackenzie 1989: 42-57. Also see Beinart 1991.

81. Simson 1886: 9.

82. The notion of territoriality and its use in understanding forest management later in the nineteenth century is discussed in detail in Chapter 5.

83. Johnson 1822; Williamson 1808.

84. Johnson 1822: 1-35.

85. OIOC W/3831 Reginald Heber, *Narrative of a Journey Through the*

Upper Provinces of India from Calcutta to Bombay, 1824-25, Vol 1 (London: John Murray, 1828), pp. 190–214.

86. Stebbing 1923. Also see Burton 1952; K.N. Chaudhuri 1918: 6, 14.

87. John Mackenzie 1988: 180.

88. For a pithy description of the rationale for enduring ideas of the tiger as vermin above all, see Edward Braddon *Thirty Years of Shikar* (Edinburgh, 1895), cited by John Mackenzie, 1988: 182.

89 Ball 1880: 229. The great sal forests, extending along the base of the Himalaya also contained many rhinoceroses, according to Forbes-Royle. See Forbes-Royle 1839: lxx. See also Simson 1886: 192; Anon 1892.

90. Hunter 1868: 66.

91. Sherwill 1855.

92. Gupta 1971: 10–11.

93. Dunbar-Brander 1931; W.T. Blanford 1888; Baldwin 1883.

94. Hamilton 1820: 148.

95. Jaundice 1834: 252–4.

96. Sherwill 1851: 553.

97. NAI GOI Home (Public), B progs 62–5, 21 Aug 1869, no. 1467, Chotanagpur, 21 May 1869, E.T. Dalton, Commr Chotanagpur, to Offg Junior Secy GOB, pp. 2–3.

98. Ibid., A progs 86-9, 23 Jul 1870, no. 1321, Chotanagpur, 16 May 1870, Dalton, to Junior Secy GOB, p. 5.

99. Ball 1880: 90–130.

100. WBSA P/140 BJP (Jud), A progs 179–80, 7 Feb 1870, Darjeeling Revenue and Topographic Survey, extract from annual report, 1866-7, p. 116. The same official also noted that the local Mechi people had seen and killed many larger boas.

101. Ibid., Genl Report of the Topographic Surveys of the Bengal Presidency, 1867-8, p. 117.

102. WBSA P/58 BJP (Jud), A progs 340–3, 18 Jul 1863, no. 574, A. Money, Commr Bhagalpur, to US GOB Jud, p. 195.

103. WBSA P/107 BJP (Jud), A progs 38-9, 4 Jul 1867, no. 146, 21 Jun 1867, E.T. Dalton, Commr Chotanagpur, to A, Mackenzie, US GOB Jud; no. 4111, 4 Jul 1867, A Mackenzie, US GOB Jud, to Commr Chotanagpur, pp. 21–2.

104. WBSA P/138 BJP (Jud), A progs 79–80, 10 Dec 1869, no. 2305, Fort Williani, J.P. Wheeler, Asst Secy GOI Foreign Dept, to Secy GOB Jud, p. 65.

105. WBSA P/140 BJP (Jud), A progs 179–80, 7 Feb 1870, extract from Jul 1867 progs of Madras Govt, Public Dept, no. 23 Mangalore, 2 Jul 1867, H.S. Thomas, Acting Magistrate South Canara, to Offg CS Govt of Madras, pp. 123–4.

106. Rangarajan 1992: 132–3.

107. WBSA P/71 BJP (Jud), A progs 12–15, 30 Jul 1864, no. 1166T, Darjeeling, 30 Jun 1864, GOB to all Commrs in Bengal; WBSA P/140 BJP (Jud), A progs 179–80, 7 Feb 1870, extracts from replies to the Jun 1864 circular, pp. 131–44.

108. WBSA P/140 BJP (Jud), A progs 179–80, 7 Feb 1870, no. 23, Mangalore, 2 Jul 1867, H.S. Thomas, Acting Magistrate South Canara, to Offg CS Govt of Madras, p. 124.

109. WBSA P/103 BJP (Jud), A progs 263–4, 28 Mar 1867, no. 507, 14 Mar 1867, E.T. Dalton, Commr Chotanagpur, to US GOB Jud, p. 182.

110. Best 1935: 125.

111. WBSA P/71 BJP (Jud), A progs 12–15, 30 Jul 1864, no. 31, London, 24 Mar 1864, Sir Charles Wood, Secy of State for India to Governor-General of India; no. 1166T, Darjeeling, 30 Jun 1864, Secy Jud GOB, to all Commrs in Bengal, pp. 13–15.

112. WBSA P/138 BJP (Jud), A progs 79-80, 10 Dec 1869, no. 2305, Fort William, J.P. Wheeler, Asst Secy GOI Foreign Department, to Secy GOB Jud, p. 64.

113. WBSA P/140 BJP (Jud), A progs 179-80, 7 Feb 1870, l/d Simla, 9 Aug 1869, Capt. B. Rogers, Bengal Staff Corps, to Major O.T. Burne, Private Secy to Viceroy, pp. 149–50.

114. WBSA P/59 BJP (Jud), A progs 37–8, 7 Aug 1863, no. 241, 25 Jul 1863, G.F. Cockburn, Commr Patna, to Secy GOB Jud, pp. 20-2. The blind well was a hole with a low wall around three sides, covered with brushwood and a kid tied to the fourth side as bait. Strychnine was used only when European doctors were available in the vicinity to supply it.

115. WBSA P/64 BJP (Jud), A progs 91–3, 10 Dec 1863, no. 388, 26 Nov 1863, G.F. Cockburn, Commr Patna, to Secy GOB Jud, p. 51.

116. WBSA P/150 BJP (Jud), A progs 126–9, 17 Dec 1870, no. 5551, Fort William, 17 Dec 1870, A. Mackenzie, Offg Junior Secy, GOB, to Secy GOI Home, p. 112; WBSA P/64 BJP (Jud), A progs 91–3, 10 Dec 1863, no. 8157 10 Dec 1863, J. Geoghagan, US GOB, Commr Patna, p. 52.

117. WBSA P/66 BJP (Jud), A progs 133–5, 26 Feb 1864, no. 6, 1 Jan 1864, H.M. Boddam, DC, Hazaribagh, to Commr Chotanagpur, p. 104.

118. WBSA P/150 BJP (Jud), A progs 126–9, 17 Dec 1870, no. 5512, Fort William 17 Dec 1870, A. Mackenzie, Offg Junior Secy GOB, to Secy GOI Home, p. 112.

119. WBSA BRP (For), B progs 34–5, file lj/l of 1892, Jan 1892, no. 4741J, Cal, 24 Dec 1891, H.J.S. Cotton, CS GOB, to All Commrs, pp. 15. Similar conflicts between district officers and the conservationists intent on forming wildlife parks in Africa are discussed by Neumann (1995: 7–10).

120. John Mackenzie (1988: 170-2) makes a similar point but fails to draw out the implications for understanding colonial statemaking.

121. WBSA P/140 BJP (Jud), A progs 179–80, 7 Feb 1870, l/d Simla, 9 Aug 1869, Capt. B. Rogers, Bengal Staff Corps, to Major O.T. Burne, Private Secy to Viceroy, p. 148.

122. Ibid., 160–71, 21 Feb 1870, no. F/14, Camp Khowang, 7 Feb 1868, Major J. MacDonald, Offg Dy Surveyor Genl and Superintendent of Rev Surveys, Lower Circle, to Junior Secy GOB Jud, p. 96.

123. Reid 1995: 106–7.

124. WBSA P/140 BJP (Jud), A progs 160–71, 21 Feb 1870, pp. 103–5.

125. WBSA BRP (Financial), B progs 17–18 file 14, head II, col 2, 25 Mar 1875, no. 557, Shillong, 19 Feb 1875, H. Luttman-Johnson, Secy to Chief Commr Assam, to Secy GOI Home, p. 1.

126. WBSA P/140 BJP (Jud), A progs 160–71, 21 Feb 1870, p. 107.

127. Ibid., note by IGF, D. Brandis, 10 Apr 1868, p. 100.

128. Sanderson (1879) provides detailed accounts of kheddah and elephant capture. This planter-turned-hunter was stationed at Dacca in Bengal for many years as Superintendent of Kheddah operations. Also see Sanderson 1885; and Hadfield and Bryant 1895.

129. WBSA VRP (For), B progs 1–5 file 38, head II, Col 2, Jan 1879; no. 59, Fort William, 15 Jan 1879: D. Fitzpatrick, Secy GOI Legislative Dept, to Secy GOB Jud; John Mackenie 1988: 185.

130. WBSA BRP (For), B progs 17–21 file 38 head II, col 2, Feb 1879; WBSA P/511 BRP (For), A progs 11–13, 34/3 of 1895, Oct 1895, GOB notification no. 4252, Cal, 14 Sep 1895, p. 127.

131. WBSA P/336 BRP (For), A progs 31–4, Sep 1881, no. 41A camp Mymensinh, 4 Apr 1881, G.P. Sanderson, Superintendent of Dacca Kheddah, to Secy GOB Rev; no. 41, 31 Mar 1881, Sanderson, to Dy Commissary-Genl, Lower Circle, Cal, p. 4.

132. Ibid., no. 41, 31 Mar 1881, Saderson, to Dy Comissary Genl, Lower Circle, Cal pp. 4–5.

133. WBSA P/58 BJP (Jud), A progs 373-5, 25 Jul 1863, no. 2937. 3 Jul 1863, E.H. Lushington, Secy Financial Dept, Secy GOB Jud, pp. 226–7.

134. WBSA P/60 BJP (Jud), A progs 175-7, 8 Sep 1863, no. 208, 22 Aug 1863, C.F. Montressor, Offg Commr Burdwan, to US GOB Jud, p. 88.

135. WBSA P/234 BJP (Jud), A progs 11 file 38, 5 Oct 1876, GOB Res, 5 Oct 1876, pp. 533–5.

136. Guha-Thakurta 1966.

137. John Mackenzie 1988: 283–4.

138. Guha-Thakurta 1966: 276.

139. Ibid.: 278.

140. For a brief but salient discussion of the role of shikari naturalists among the influential foresters in India, see Sivaramakrishnan 1995a: 16–17.

141. WBSA P/379 BRP (For), A progs 18–20, Nov 1884, head I, Col 1, no. 631R, Darjeeling, 31 Oct 1884, Lord Ulicke Brown, Commr Rajshahi, to Secy GOB Rev, p. 23.

142. WBSA P/502 BRP (For), A progs 19–48, file 4A/1 of 1894, Feb 1895, no. 42D-F-G, Darjeeling, 28 May 1894, A.E. Wild, CF Bengal, to Secy GOB Rev, pp. 35–44; no. 819 Ranchi, 25 Aug 1894, W.H. Grimley, Commr Chotanagpur, to Secy GOB Rev, pp. 47–8.

143. Ibid., no. 504 For, Cal, 22 Jan 1895 C.E. Buckland, Secy GOB Rev, to Secy GOI Rev and Agri, p. 53.

144. WBSA P/508 BRP (For), A progs 41–50, file 4A/1, Jul 1895, no. 130LR. Ranchi, 30 Apr 1895, W.H. Grimley, Commr Chotanagpur, to Secy GOB Rev, pp. 45–8.

145. WBSA P/502 BRP (For), A progs 19–48, file 4A/1 of 1894, Feb 1895,

no. 253JG, Burdwan, 4 Sep 1894, R.C. Dutt, Offg Commr Burdwan, to Secy GOB Rev, p. 49; no. 504For, Cal, 22 Jan 1895. C.E. Buckland, Secy GOB Rev, to Secy GOI Rev and Agri, p. 51–3.

146. OIOC P/7032 BRP (For), A progs 12–47, Jan–Apr 1905, file 18R/3, Mar 1905, no. 294, Darjeeling, 2 Feb 1905, A.L. McIntire, C.F. Bengal, to Secy GOB Rev, p. 95.

147. OIOC P/7032 BRP (For), A progs 12–47, Jan–Apr 1905, file 18R/3, Mar 1905, no. 2556 Balasore 23 Dec 1904, W. Egerton, Magistrate Balasore, to Commr Orissa; no. 56J, Cuttack, 13 Jan 1905 E.F. Growse, Commr Orissa, to Secy GOB Rev; no. TR42, 9 Jan 1905, E.A. Gait, DC Singhbhum, to C.F. Bengal, pp. 69–96.

148. OIOC P/7032 BRP (For), A progs 12–47, Jan–Apr 1905, file 18R/3, Mar 1905, no. 2556, Balasore, 23 Dec 1904, W. Egerton, Magistrate Balasore, to Commr Orissa; no. 56J, Cuttack, 13 Jan 1905, E.F. Growse, Commr Orissa, to Secy GOB Rev; no. 234, Dadeeling, 13 Dec 1904, C.F. Bengal, to Secy GOB Rev; no. TR42, 9 Jan 1905, E.A. Gait, DC Singhbhum, C.F. Bengal, pp. 69–96.

149. Cobden-Ramsay 1910: 18–19, 95.

150. O'Malley 1910b: 17–18.

151. Neumann 1995.

152. John Mackenzie 1988: 286–8.

153. Hailey (later Corbett) National Park was designated in 1935 in north India, and Periyar sanctuary was formed in 1938 in south India. I am grateful to Ramachandra Guha for reminding me of these cases.

154. John Mackenzie 1988: 288, 298.

155. See Keith Thomas 1984; John Mackenzie 1988: 300.

156. This theme of regional variations in the regime of restrictions and its diagnostic value in the study of governnance and statemaking will be examined in detail in Chapter 5.

157. OIOC F/4/465, 11269, BC, extract Bengal Rev letter, 5 Sep 1800, pp. 33–4.

158. Ibid., p. 35.

159. OIOC F/4/427, 10478, BC, extract BPC, 23 Oct 1812, prog 6, l/d 4 Sep 1812 W.W. Trant, Secy BOR, to Secy Rev; l/d 28 Aug 1812, E. Barnett, Collr Rajshahi, to President BOR, l/d 21 Aug 1812, George Ballard to Barnett, pp. 6–10.

160. OIOC F/4/501, 11992, BC, extract BRC, 23 Apr 1814, l/d 23 Apr 1814, Govt at Fort William to BOR, enclosing extract l/d 5 Sep 1800, pp. 31–2.

161. OIOC F/427, 10478, BPC, 13 Nov 1812, progs 43, l/d 4 Nov 1811, W. Roxburgh to Secy to Govt at Fort William, pp. 14–16.

162. Ibid., p. 16.

163. OIOC F/4/465, 11269, BC, extract BPC, 8 Oct 1813, progs 11, l/d 21 Sep 1813, R. Rocke and I. Lumsden, Bengal BOR to GG-in-C, pp. 12–16.

164. OIOC F/4/501, 11992, BC, extract Bengal BOR progs, 31 Dec 1813, progs 17, Thomas Leake to Asst Collr Jungle Mahals, p. 8.

165. Ibid., pp. 6–9.

166. Ibid., progs 15, l/d 22 Nov 1813, A. Anderso, Asst Collr Bankura, to Secy BOR, pp. 3–4; extract BRC, 12 Feb 1814, progs 15, l/d 12 Feb 1814, G. Dowdeswell, CS GOB, to BOR, pp. 20–2; extract BRC, 23 Apr 1814, progs 9, l/d Bancoorah, 8 Mar 1814, Thomas Leak Asst Surgeon, to Asst Collr Jungle Mahals, pp. 27–9.

167. OIOC F/4/501, 11992, BC, extract BRC, 23 Apr 1814, progs 8, l/d 5 Apr 1814, BOR to GG-in-C, p. 25.

168. OIOC E/4/687, Bengal Despatches, BPC, 5 Apr 1816, draft 105/1815–16, pp. 437–43.

169. OIOC F/4/655, 18040, BC, extract BPC, l/d 31 Jul 1820, pp. 1–3; extract BPC, 3 Mar 1820, progs 18, l/d 4 Feb 1820, R. Rocke, President Bengal BOR, to GG-in-C, pp. 21–4; extract BPC, 6 Jun 1820, progs 44, l/d 24 May 1820, N. Wallich, Superintendent Botanical Gardens, to C. Lushington, Secy to Govt Genl Dept, p. 61.

170. H. Falconer, *Report on the Teak Plantations of Bengal*, Selections from the Records of the Bengal government, no. 25 (Calcutta: John Gray, 1857), p. 2.

171. OIOC F/4/655, 18040, BC, extract BPC, 6 Jun 1820, progs 44, l/d 24 May 1820, Wallich, to Lushington, pp. 76–9.

172. Ibid., pp. 64–75.

173. Ibid., pp. 81–2.

174. OIOC E/4/760, Bengal Despatches, Bengal Public Dept progs 31 of 21 Aug 1839, draft 473/1839, pp. 244–5.

175. Falconer, *Teak Plantations of Bengal*, p. 3.

176. Quoted by Mactier 1848: 243; Falconer, *Teak Plantations of Bengal*, p. 4.

177. Mactier 1848: 246.

178. OIOC E/4/848, Bengal Despatches, progs 16, 14 Oct 1857, BPC, draft 1201/1857, pp. 173–211.

179. OIOC E/4/839, Bengal Despatches, progs 14, 22 Oct 1856, BPC, Rev letter no. 10, 8 Sep, pp. 1118–20; Falconer, *Teak Plantations of Bengal*, p. 7.

180. Grove 1993a; Skaria 1998; Rangarajan, 1998; Rajan (forthcoming).

4

Visibility, Estimation, and
Laying Down the Ground Rules

Introduction

The Company raj, starting early in the nineteenth century, launched several territorial procedures of statemaking in woodland Bengal. Land surveying, forest exploration and assessment, documenting the lifestyles of forest tribes and engaging with issues of timber conservancy, agricultural expansion, and vermin eradication, were some of the important ones. Francis Buchanan's empirical surveys in Mysore and Bengal were a striking example.[1] This documentation culled the enormous revenue, commercial, and military knowledge of the Indian elite and native administrators over long periods in projects like Rennell's Map of Hindoostan, and Thomas Munro's Administration of the Baramahal. But it also drew on the Orientalism of William Jones, Nathaniel Halhead, and Henry Colebrooke.[2] Colin Mackenzie's survey of the Nizam's territories during 1792–9 combined mapping and description with the collection of oral histories. This is where the negotiation of space occurred. Collective, local memory, the repository of substantialized boundaries, was drawn upon in these surveys, then reinterpreted and effaced.[3]

The historical space of colonial surveyors emerged through what Paul Carter has called the 'language of travelling'. What the surveyor saw, like the explorers of interior Australia in Carter's *The Road to Botany Bay*, was in part what was shown to them and in part the landmarks they designated to mark their way. Carter notes that the aborigines had created a patchwork of meadows and forests, a balance of visibility and invisibility that the explorers named, to possess or transform it, by suppressing the spatiality of the landscape. Buchanan's accounts of a similar patchwork of forest, savannah and fields from his forays into Bihar convey his disappointment about the forests he did not find. His descriptions also provided some of the earliest east Indian evidence of forests that were found to be less dense than assumed by prior travellers. We can sample in Buchanan's work the tone of reproof that later heralded the rise of desiccation discourse.[4]

Dispossession through naming, as Carter discusses in the spatial history of Australia, was made possible by the rise of Linnaean systematics.[5] Colonial expansion created a tension between exploring and classifying, personified in Carter's story by the figures of Cook and Banks: the latter promoting a disregard for history and culture, allowing the subsumption of local knowledge under amateur botany. Banks was, however, often forced to work through the mechanisms of exploration identified with Cook as expeditions to the interior of continents, the *terrae incognitae* of the era of maritime travel, relied on local people and their guidance.[6] Caught in this tension, surveying did encourage new forms of agrarian practice and self-representation and in that sense the space of colonial surveys, in woodland Bengal and elsewhere, remained under constant negotiation.[7]

Surveying, as a mode of knowledge, practised through procedures of exploration, discovery and classification of nature, increasingly piggy-backed on the expansion of international commerce. In return, colonial scientific expeditions generated the knowledge of goods that resulted in the commodification of space. But many sciences, like geology, also gave this process a uniquely territorial dimension with their emphasis on mapping and carefully tracing resources and their routes of extraction.[8] By the 1850s, British geology was institutionalized at home and the first colonial establishment had come up in Bengal. This scientific survey and others collectively imposed order on wilderness, and participated in the procedures through which colonial government imposed law and order on remote territories.[9] At the same time the efforts of surgeon-naturalists and botanists, their relentless surveying and cataloguing of the plants of the Indian subcontinent as they found them, bore fruit in the Great Exhibition in London held in 1851. The Indian exhibit there was the most extensive series of woods from any part of the world. It was assembled mainly from the Wallich and Roxburgh collections.[10]

Throughout the nineteenth century, extension of settled agriculture was a central element of strategies to secure the colonial order in Bengal. Wild animals and their territories in woodland Bengal shaped the spatial patterns of such extension. This was not a problem unique to the British colonial state.[11] But in their attempted solutions, several strands of western thought relating to hunting in general, and the exigencies of dealing with indigenous hunting in particular colonial regions, combined to novel effect. There were tensions in European ideas between Romanticism, which ultimately came out against hunting, and the rise of classificatory science. In the nineteenth century the rise of evolution theory and Darwinian notions of competition and hierarchy assisted in the celebration of hunting as the pursuit of masculinity and powerfulness. Hunting as imperial ritual, celebrating the paternalism of the colonial hunter-administrator, crystallized through ideas that flowered in contact with spreading empire and its demands for symbolic and substantive governance.[12] Through the regulation of hunting the landscape of empire was secured to make, among other things, what we have distinguished as tribal places.

The spatiality of colonial knowledge production took several forms. To distinguish these forms and appreciate their implications for statemaking requires understanding 'the genesis and use of particular forms of knowledge and their immediate historical environment.'[13] The overall impact of colonial knowledge, notwithstanding its specificities, was several dispossessions. The first was creating Others through stereotypes developed to characterize colonized cultures. The second was spatializing, which entailed de-territorialization and re-territorialization of the colonized people and their landscape, a reordering of social and ecological relations. The third was the survey and classification of colonized nature.[14] The first two kinds of dispossession we have already considered through a discussion of tribal placemaking and vermin eradication. We shall now turn to the place of surveying in this ensemble of procedures.

Surveying and Enumerating Natural Resources Beyond Agriculture

Since surveying worked through the imposition of classifications on landscapes and the discovery of landscapes that defied received categories, it promoted an ambiguous spatiality in the formation of colonial geographical knowledge. By the mid-nineteenth century categories like valuable lands, jungly wastes, wild forests, timber yielding species, and the productive areas of empire had undergone significant redefinition. Plant diversity in the uncultivated terrain was both discovered and deprecated, as the vanguard of modern forestry in Bengal searched far and wide for scant hardwoods. By the end of the Company raj, Cleghorn, Forbes Royle, Falconer, Gibson, Dalzell and Beddome had collectively ensured that the forest plants of India were massively documented. They also predicted their imminent disappearance in the absence of a coherent state forest conservation policy.[15] Coming up with this narrative of impending crisis was important to the introduction of forest conservancy in India. This section will trace the experience of forest exploration that constructed the narrative.

Bernier, travelling through Bengal under the Mughals, had compared its natural endowments favourably with those of legendary Egypt.[16] One of the earliest colonial discussions of the potential of forest Bengal was offered by Henry Colebrooke in 1795, who wrote, 'Bengal already possesses many articles which would be brought into notice by a more extended commerce. Redwood is used for dunnage in the present trade, many other woods fit for dyeing are found in abundance.'[17] Bengal was also importing a large quantity of teak from Pegu for shipbuilding, in addition to sal and sissoo from Bihar, Oudh and the forests of the Himalayan foothills. In 1781 Colonel Watson built the first ship in the river Hooghly, and by 1795, in Calcutta, Chittagong, Sylhet and Backergunge, fifty-three ships and ninety-three small vessels had been built with a total tonnage of 39,080. Usually the frame was made of sissoo, beams and inside planks of sal, and the bottoms, sides, decks, keels, stern posts of teak.[18]

Observing the brisk trade in timber down the river systems of north and east India, Colebrooke was inspired to remark, 'the forests of India . . . offer inexhaustible sources for the supply of the finest and most durable . . . timber in the world.'[19] Despite his considerable experience in India, Colebrooke and his cohort of Company officials purveyed a 'wild and untapped bounty' view of the forest landscape they had not travelled in any great measure, let alone observed closely. This view changed rapidly in the next fifty years as the desiccationist opinion gained strength, by challenging the notion of 'inexhaustible forest wealth'. This challenge or revision was made possible by interior exploration and the discovery of long-standing agro-forestry complexes in what was earlier envisioned as a timber mine.[20]

Botanists, army officers, civil surgeons, and administrators of frontier provinces participated collectively in exploring and discovering the geography of empire.[21] In 1814, one Lieutenant Jackson was appointed under the Surveyor-General to survey the limits of Hooghly, Burdwan, Midnapore and the jungle mahaḷs.[22] But the most important surveys of this period were those of Francis Buchanan-Hamilton in Mysore and Bengal. Modelling his approach on the work of John Sinclair in Scotland, he collected statistics in a plethora of detail, tolerated and sponsored by the East India Company only because of Wellesley's penchant for natural history and his conviction that such expeditions were likely to yield profits in the future.[23] Buchanan's Mysore survey reported on timbers as well as wild spices and spice gardens of Malabar. By the time of his Bengal survey, the East India Company was seeking verification of Rennell's maps, but Buchanan was also able to cover much new ground, surveying the less accessible wastes, villages, forests, and jungles that Rennell had avoided because they were 'the refuge of disorderly elements'.[24] In 1823, Nathaniel Wallich, Superintendent of the Calcutta Botanical Gardens, undertook a botanical expedition to Nepal, inspired by the botanical knowledge acquired through the travels of Buchanan-Hamilton.[25]

These travels and their reports were vital to governance more generally, but they also provided the resources for constructing a series of representations of forests and their use. They reproduced the tension 'between botany's concern to reduce the variety of the world to a uniform and universally valid taxonomy and exploration's pursuit of a mode of knowing that was dynamic, concerned with the world as it appeared.'[26] Prospecting for timber in upper Assam, the Commandant of the Assam Light Infantry Battalion reported on thirty-eight species of trees, but noted restrictions on only a few like the Joba Hingoree (a variety of oak, *quercus spp.*) which the rajas of Assam used to build palaces.[27] He also observed that with the decline of Ahom kingdoms, timber usage for making canoes and boats had declined, resulting in the gradual disappearance of general knowledge of timber flotation and boat cutting.[28] So whether timber cutting and rafting in upper Assam had actually declined in the period immediately preceding British entry or not, as abortive attempts to establish saw milling in Jeypore revealed, skilled labour was locally scarce.

Greater involvement in timber trade also taught Company officials to appreciate patterns of deforestation and redefine their portfolio of desired timbers. From Darjeeling at about the same time as Hannay's explorations in Assam came the report that 'wood exists in inexhaustible quantities, of the most enormous size (thirty feet circumference being common), in vast variety and of the greatest strength.'[29] Commenting that exploitation of timber resources in India had been confined to teak, sal and sissoo by the limited skill and technology of Indian carpenters, one surveyor also pleaded for the use of the wider range of valuable timbers, like oak, walnut, chestnut, and others that would yield easily to European tools.[30]

Other values and potential revenue sources were discovered in the forests. A good example is the controversy that erupted over lac in Gorakhpur forests during the 1820s. Initially the disagreements centred on whether lac was cultivated or a spontaneous forest product. Before these could be resolved, perceiving it as a 'new El Dorado in the wilderness', some district officers had slapped a revenue demand on zamindars. Inquiry revealed that zamindars leased out the lac mahal to merchants who organized its cultivation through local villagers adjacent to leased areas.[31] This led the Company raj to contemplate reserving the right to lease out forests for collection of products, by grant of pattas or farming arrangements.[32]

These incipient ideas about establishing state property rights in forests and their products were inspired both by discovering the untapped revenue potential, itself a function of becoming well acquainted with the landscape, and the knowledge that such acquaintance brought of timber resources being consumed fast. This process of exploration, discovery, naming, and tentatively defining forest management through ordering, was complex and variegated. Persisting through the nineteenth century, but transformed by territorialization and intensive forest regeneration efforts, the various procedures we can identify are inadequately captured in any scheme which describes them as timber conservancy and treats that as the precursor of forest conservancy.[33]

Defying accepted periodizations, in the classification and analysis of colonial knowledge, surveying and experimentation remained a significant element of high scientific forestry. In the same way, the segregation of private and public property, sifting of villager and landlord rights, and their containment in the forest estate managed for the larger public good, had already been mooted in the early years of Company raj. Wallich, in pushing for teak and sissoo plantations during the 1820s, had cast his argument in terms of ongoing government deliberations on 'the question relative to securing of both private and public interest in our forests by establishing a set of laws for their management and preservation especially to prevent injudicious felling of their timber . . .'[34]

A close scrutiny of forest rights, particularly in teak regions had yielded much controversy by the 1830s and 1840s. An important outcome of this examination was a sense of unbridled commerce leading to destructive forest exploitation.[35] In south India forests were initially regarded as private property, but in 1800, citing the precedent of Tipu Sultan in Mysore, the Bombay government was authorized to assume the right of timber felling on behalf of the Company. In 1805 a commission was appointed to distinguish public and private forests, and in 1807 a conservator was appointed. The royalties collected by this conservator annihilated the timber trade and imposed great hardship on the peasantry of Malabar.[36] When abolishing the Conservator position in 1822, the Company raj had been influenced by Thomas Munro's powerful plea for a policy of *laissez-faire*:

The people of Malabar and Canara are chiefly agriculturists and merchants, a considerable portion of the ryots are traders, and their country being intersected by rivers and creeks, enables them to bring the produce to the coast in their own boats for sale, and they too are good traders to cultivate Teak or whatever wood is likely to yield a profit. They plant trees for sale, private use, and in order to devote to their temples; and to encourage them no regulation is wanted, but a free market . . . private timber will be increased by good prices, and trade and agriculture freed from vexation.[37]

But local elites like the raja of Nilambur, and principal consumers like the Bombay dockyards, soon raised the question of timber scarcities. In 1834 the Collector of Malabar, Clementson, also reported that Travancore forests were doing better than those in Malabar because the former were under government management. A compilation of reports by the Naval Board, Madras Board of Revenue and district officers soon covered Malabar, Canara, Travancore and Rajamundry forests. One such report noted that proprietors in Malabar allowed their trees, big or small, to be felled at the price of one rupee per tree. So young trees were indiscriminately felled, no sowing was done and regeneration was hampered by annual fires.[38]

Estimated demands for teak in shipbuilding, and the anticipated forest conservancy strategies required to assure a steady supply of teak timber for that purpose, fed this discussion of native forest management. One scholar has gone so far as to claim that

British assertion of control over the forests of the Cape, and then in Malabar and Burma was a direct result of the loss of the North American source of timber supply and the consequent strategic crisis in naval demand for raw materials in the context of wars with the French between 1793–1815.[39]

Given the preoccupation of the times with extending the arable, fiscal measures that would discourage felling young trees were seen as a superior means of safeguarding timber supplies. Both in Malabar and Tenasserim, it was felt that forests should continue to be held by the private sector but taxation was proposed to 'make it in their interest to care for them.'[40]

The failure of this reliance on fiscal measures and market forces to ensure adequate timber supply and the turn to more direct management in Burma has been documented by another scholar.[41] Territorial forest management began to emerge from these Burmese experiences. In that region, even before Falconer's devastating critique of *laissez-faire* forestry in Tenasserim, the government had resolved that the demand of the Royal Navy for shipbuilding timber would not be allowed to supersede arrangements made for teak forest conservancy.[42] It might be argued that the Dangs forest leases of the 1840s exemplified such a move towards more territorial forms of control. Such modes of forest control through state agencies were yet quite disorganized and scattered in the 1850s, being confined to teak-bearing forests in south, central and western India, and Burma.[43]

Elsewhere, as in Bengal, raising plantations was preferred prior to any attempt at forest control. In the North West Provinces (NWP), forming plantations in the Doab region was recommended for both timber and fuelwood. The Magistrate of Bareilly opined that tenurial problems and the practice of cutting tree groves to recover dues from defaulting tenants had combined with the pressure for agricultural expansion to accelerate deforestation. The Military Department concurred that plantations could relieve the problem of timber scarcity.[44] In west and south India, under the first botanist and surgeon naturalists, small plantations of blue gum *eucalyptus globulus*, *acacia spp.* and teak had been commenced. The best known of these were the Nilambur teak plantations started by Connolly in 1844, which by 1860 had 400,000 plants, of which 50,000 were removed that year in thinning.[45]

All these approaches to any kind of 'forest question' that might have occupied the East India Company, reveal the dominance of ideas that treated individual trees as crops—the convergence of agricultural, commodity and botanical perspectives. It was only with the emergence of a more complex understanding of the colonial forest estate as one made up of multiple interrelated products, and the possible source of a variety of environmental services, that the forest came to be represented as a landscape worthy of conservancy.[46] In 1847 the Court of Directors had inquired into the effects of forest destruction on catchments. A few years earlier, Dr Alexander Gibson had reported desiccation in Konkan and South Canara due to deforestation.[47] A new colonial geographical sensibility in relation to forests emerged from the confluence of desiccationism and the imperatives of territorial control.

Following Indian botanists, colonial botanists in Africa too began to argue the connection between drought and deforestation. Over the years they gained the support of important heads of Kew Gardens such as Sir Joseph Hooker, who observed that the flora of British India were more varied than those of any other country of equal area.[48] In other words, the articulation of an environmental vision promoted by surgeon-naturalists, amateur botanists and their ilk, and ideas about efficient regimes of landed property, where forest and agriculture were separated into the public and private domain respectively, began to converge on the issue of colonial forests and their management.

One survey of the forest resources of Bengal, carried out to identify timber for railway sleeper woods, identified sal, *siris*, sissoo, *khair*, *toon*, and *asan*, in the lower provinces.[49] Elsewhere English timber contractors were already extracting sal for sleeper production along the foothills of the Himalayas.[50] By this time the surveys combined economic-botanical interest with explicit assessment of timber resources, listing flowering trees that were used for medicinal and dietary purposes.[51] Sal forests, however, defied neat distinction into plants used for non-timber purposes and trees that were reserved as timber. Sal itself was the best example of the multipurpose tree, being used widely for local house building, also girdled for *dhuna* (resin) production and its associates like *arjun* and asan used for silkworm rearing. Santhals of the Damin planted fruit trees in the clearings around their villages, chironji in particular being grown to exchange its fruit for lowland rice.[52]

As the British became more intimately acquainted with the precise contents of the non-agricultural landscape that lay beyond the cultivated river valleys of their empire, they also gained awareness of its varied and long-standing uses. Some early surveys of the forest wealth in coastal areas like Malabar and Canara, aimed at identifying woods suitable for shipbuilding, noted good stands of teak along river banks but expressed no great distaste for the shifting cultivation practiced alongside.[53] But even by the mid-nineteenth century there was no systematic mapping of forests.[54] This had to await the forest survey initiated in the 1880s. Meanwhile the topographic survey of Chotanagpur, carried out in 1859–61 did not record forests. Gradually, forests became the subject of multiple and conflicting representations, as impediments to agriculture, as the source of many commercial products, and lastly as an influence on climate and health—both human and natural.

This last aspect leading to alarm about desiccation and disease generated a sharp critique of shifting cultivation as something that not only promoted reckless use of the forest but also inhibited the extraction of its multiple values in a controlled fashion. These themes and their combination are apparent in the report by Cleghorn, Baird-Smith, Forbes-Royle and Strachey to the British Association of the Advancement of Science (hereafter called the BAAS report).[55] In contrast to the initial surveys of teak forests in south India and Burma, the mid-century reports of Cleghorn and Brandis criticized shifting cultivation, called *kumri* and *taungya* in their respective provinces, in no uncertain terms. Through the close observation of forest tracts, they stressed the need to exclusively preserve them for timbering and linked forest conservancy to suppression of certain forest-based forms of agriculture.[56] Brandis declared taungya more harmful to teak forests than even jungle fires, for they brought Karen agriculture into direct competition with teak forests for the most fertile lands.[57]

In the next twenty years, foresters took up the surveying that was initiated by amateur botanists, army officers and the administrators of far-flung provinces. They benefited by the plant systematics work done by the botanists. Following the example set by the Dutch physician Koenig, these discoverers and classifiers had sorted through a morass of native tree names, often finding the same name applied to trees of widely differing timber quality.[58] From their reports the blurred categories of timber and non-timber trees were filled in more and distinguished better. All the major trees of lower Bengal were found to have a variety of non-timber uses, both commercial and for local consumption, while many of them were valuable for their excellent timber.[59]

Surveying, hunting management, plantation experiments, and the creation of tribal areas under schemes of exceptional administration may be understood as the spatial history and procedures of stabilizing empire. They served the most important purpose of designating certain spaces in the geography of empire as forest estate. As spatial history they prefigured the arenas in which the struggles over formal structures of forest management would be played out. As procedures they entered the repertoire of such management. Assessing the extent of timber as opposed to non-timber resources in any patch of forest became an important criterion for its classification as a reserved or protected tract. Each of these treatments brought into play a different regime of restrictions. Thus the processes of de-territorialization and re-territorialization, initiated by the Company raj through surveys, plantations, the making of tribal places and vermin eradication, were carried on through the later nineteenth century. In the next part of the book we shall examine more closely the regional variations that emerged in forest management as patterns of territorialization encountered the specificities of history and geography. Before that, in the next section, the story of systematic surveying as a prelude to forest reservation and management is briefly recorded.

Beginning of Forest Conservancy in Bengal

During the period 1795 to 1850, the Company raj chiefly viewed forests as limiting agriculture. In Bengal, forested lands, classified as wastelands, had been included in landlord (zamindari) estates.[60] Colonial administrators of this period also tended to perceive forests as being inexhaustible. Much of the woody vegetation was not, however, timber quality, being the product of a landscape long under shifting cultivation. The East India Company continued the Indian rulers' practice of selling blocks of forests or individual trees to timber merchants for a fixed down payment that encouraged great destruction and waste in their extraction.[61] No attempts to introduce conservancy were made in the North West Provinces or Bengal till after the revolt of 1857, even though the value of NWP sal forests was known from the time of the Gurkha wars in 1814–16, and the reports of Wallich, Superintendent of the Calcutta Botanical Gardens in 1825.[62] In fact large areas of forest lands had been alienated under the Bengal Wasteland Rules, 1853, for a variety of purposes not always connected with the exploitation of forest products. Sometimes these allotments were to encourage particular forms of cultivation. In other cases they farmed out the collection of *bankar* (forest tolls), as had been done in the Sundarbans where bankar blocks had been demarcated and sold on five-year leases.[63]

The introduction of forest conservancy in Bengal as elsewhere in India, Asia and Europe at different times, came out of a wider concern that timber supply was being diminished by reckless felling at the very time that demands were increasing. The sporadic protection of regeneration by major forest contractors like Dear and Company was commended but also feared by the colonial government to be inadequate.[64] Dr T. Anderson, Superintendent of the Royal Botanical Gardens, Calcutta, was appointed as the first Conservator of Forests in Bengal in August 1864, and marked a beginning in formal arrangements responding to this fear. Under his direction, operations for felling, conversion of logs into sleepers and the establishment of plantations began in British Sikkim, Jellapahar hills and the Darjeeling *terai*. A timber slide was installed near Kurseong. In 1868 the Chittagong and western Duars forests were surveyed. Whereas temperate forests of the higher elevations were found to be largely unworked, the lower hill and terai sal showed evidence of serious exploitation. Timber of girth five feet and above was scant in these forests.[65] In 1871, as the Bengal Forest Rules came into force, the province remained a major importer of wood. According to estimates prepared by Leeds, the new Conservator, annually 15,000 tons of teak from Burma, 18,000 tons of sal from Nepal and 10,000 tons of other woods were imported, in addition to 8,300 tons of mixed woods for opium and indigo boxes.[66] So while noting with appreciation the prospect of 3,000 square miles of forest coming under systematic preservation in Bengal within the first decade of conservancy, the Lieutenant-Governor, Sir Richard Temple, remained severely critical of forest contractors and the unnecessary destruction they caused of timber resources.[67]

Against this background, the measures outlined for conservancy wrestled uneasily with the conflict between letting local governments feel their way to a regime of restrictions and insisting on some basic principles like the abolition of all individual proprietorship in forests.[68] Writing to the Government of Bengal in this spirit, Brandis, the Inspector-General of Forests, strongly recommended the establishment of forest administration on cardinal European principles, but encouraged the regional authorities to learn through a survey of divisional commissioners the current state of affairs and possibilities.[69] In the first few years, under the leadership of Anderson and then Leeds, efforts were concentrated in north Bengal.[70] The demarcation of Sikkim and Bhutan forests was carried out. As valuation of their contents followed, in many places timber was found to be scarce, for which both timber contractors and local shifting cultivators were held responsible.[71] Official accounts of deforestation thus drew into their list of culprits all forms of private forest use. They blamed agriculture and commerce, holding farmers, landlords, and merchants responsible for the gaps they came across in the wooded landscape under survey.

Abandoned village sites, redesignated as forest blanks, were planted with exotics. The exotics had all been initially raised in the Botanical Gardens in Calcutta, which served as a seed bank.[72] In temperate forests, oak, magnolia, and walnut plantations were extended by taungya methods already in use for Burma teak.[73] Mahogany plantation, as yet untried in the north, was resumed in Bengal.[74] At lower elevations teak was planted. Here again, the method employed was to settle cultivators with the inducement of free lands for two years and get them to clear the land and grow teak amidst coarse grains. These mixed plantations of toon (*cedrela toona*), teak, and occasionally sal, were expected to supply local demand for tea boxes, charcoal, and construction woods, while also yielding timber.[75]

Operations began to take shape through a pattern of planting exotics and opening accessible areas to timber removal in north Bengal. Anderson acted on the advice of Brandis and initiated the survey of existing resources and their encumbrances. He asked all Commissioners to report what forests of value existed, whether they were government property, if they were in any way conserved, what species of trees were available, what marketing arrangements prevailed and how firewood was locally obtained. The survey concluded with a question on existing rights in forests.[76] The replies received from various Bengal divisions in the late 1860s allow us to construct a picture of the situation in different forested regions at the time.

The Darjeeling terai was found well stocked with sal and some sissoo on river banks. But these valuable trees were in mixed groves with other valueless trees. A good part of what was estimated to be an area in excess of 50,000 acres had been leased to one Wardroper in 1854 to extend cultivation. This lease was cancelled because its conditions were not met.[77] Another 60–80,000 acres of land above 6,500 feet in the district had already been reserved for timber. In Jalpaiguri 25 per cent of the area leased out for cultivation was held by European tea planters while the rest was with cultivators growing rice, tobacco, rapeseed, and mustard. The forest areas later notified from government wastes covered 2,09,180 acres, which included 87,397 acres bearing sal, 42,040 acres of khair (*acacia catechu*) and sissoo (*dalbergia sissoo*), and 68,556 of mixed forest and savannah.[78]

A brisk timber trade, originating in Nepal and the Assam Hills, moved along major rivers of north Bengal. Dear and Company, based in Monghyr, had most of the timber contracts in Darjeeling and from many of the zamindars like the raja of Bykuntapore. In the zamindari forests of north Bihar, villagers were found to enter the forests with carts to cut fuelwood and remove it after paying a toll to the zamindar. In some cases there were rights acquired by prescription, usually pertaining to fuel and small timber for domestic consumption. The fuelwood then entered regional markets through traders.[79] A system of forest leases in *khas* estates of Darjeeling had begun to constrict shifting cultivation, while bringing lessees into conflict with owners of grazing mahals, who fired the forest floor for fresh crops of pasture grasses.[80]

A similar picture emerges from the Santhal Parganas, where the Damin-i-Koh was a government estate, but the rest of the forests were owned by zamindars. Raiyats in the Damin were reported to possess rights to firewood, bamboos, and small trees.[81] Sal was the chief timber tree, and an abundance of bamboo was reported both here and elsewhere in Bihar forests, like those in Champaran and Shahabad.[82]

In Bankura, Birbhoom, and Midnapore, the sal jungles were extensive, entirely owned by zamindars and only occasionally of timber quality. The jungle mahals of Midnapore were about 1,200 square miles in extent, of which 700 square miles were forested. The proprietors were already engaged in timber trade, while also giving annual licences for collecting dhuna, cocoons of silkworms and making charcoal.[83] Subarnarekha and Kangsabati (Kasai) rivers were important to the timber trade, because they carried local timber, and that coming from Chotanagpur in the west and Mayurbhanj in the south, to markets in Hooghly. In the dry season timber also went by carts along tracks running parallel to the rivers. Midnapore town was a major centre for firewood. The firewood was brought by villagers to local markets after it had been cut from the scrub jungle and after zamindari tolls were paid at the rate of two annas per cart or six pice per bullock load. In the Midnapore zamindari and a few others in the jungle mahals, especially estates under the Court of Wards, *patni* leases were held by Watson and Company—a managing agency and predecessor of the Midnapore Zamindari Company. These *patnidars* were more rigorous in collection of fees on fuelwood, because it was a major source of revenue from otherwise scantily cultivated estates.[84]

In the Sundarbans, half the forest area of 6,900 square miles was already under leases for clearing, but only a small portion had been cleared. The rest of the area was brought under government control. Initial management was mainly directed at installing revenue stations at chief points of exit from the forests. Orders issued in 1875 and 1885 prohibiting felling of *sundri* trees under 3¾ feet in girth were largely ignored.[85] In Dacca division, the forests were under a variety of leases, like India rubber extraction in Cachar, tea cultivation and so on.[86] Manbhum, Hazaribagh, and Palamau forests of Chotanagpur were, by the mid-1860s, being steadily broken into for supplying railway sleepers and other public works projects, a trend that had intensified from the 1850s. Here too zamindars collected bankar (forest taxes) from villagers selling fuelwood. Ghatwals had mostly appropriated the timber trade in their *mouzahs* (villages).[87]

These surveys revealed a bewildering variation of arrangements by which local landowners, tenants, and peasantry used forests, acknowledged rights, and adjusted competing claims on them. What also became apparent was that certain areas were more available to large-scale timber removal than others, being less involved in supplying local needs and markets. Focusing therefore on these less problematic areas, the forests at higher elevations in north Bengal were reserved quickly. In the hilly tracts, the regions at medium height (3,000–6,000 feet) were given to tea plantations. The relative lack of political and economic complications was a consideration in selecting areas for quick reservation. The Conservator also favoured areas covered with valuable timber species like sal and sissoo and areas fairly level and close to navigable waterways.[88]

Highlighting the importance of logistical factors, the Government of India recommended the extension of reservation from the north into the southern plateau and deltaic forests, of Chotanagpur and Sundarbans respectively, reasoning that forests in these areas might be scarce but were cheaper to exploit. By 1875, under the leadership of William Schlich, five forest divisions had been created, in Darjeeling, Jalpaiguri, Palamau, Sundarbans and Chittagong, increasing the total area covered to 1,467 square miles.[89] The agenda in the reserved forests was set by the report of Brandis prepared in 1879–80, that recommended fire protection, road building, cultural operations and timber removal. The annual cut was directly linked to the progress of railway and tramway construction.[90]

But such a direct connection between revenue generation by meeting railway needs for sleepers and the purpose of conservancy was not easily made. Differences emerged between different levels of government. The state government argued that forests were preserved at considerable cost and should be exploited for sale of timber and other produce without restriction.[91] The Governor of Bengal, Sir Charles Elliott, initially disagreed, holding that this would interfere with afforestation, which was for him the primary function of the Forest Department.[92] That he had to defer to more pressing definitions of the Forest Department as a 'quasi-commercial agency' is not surprising given the large profits from forest operations in the first two decades.

Over the next twenty years, the questions of fire protection, restriction of grazing, and the efficient removal of sal timber for various uses continued to dominate forest conservancy in Bengal. Direct forest management was spreading through a network of reserved, protected, and unclassed forests. In 1883 the Government of Bengal could report 4,322 square miles of reserved forests and a projection of 6,000 square miles as the ultimate goal.[93] By 1900, the area under reservation had increased to 5,880 square miles, and the total managed area (including protected and unclassed forests) was up to 13,589 square miles in Bengal (Figure 10, p. 152). Tables 4.1–4.4 show the spread of forest reservation by the end of the nineteenth century, the steady increase in revenue generation from government forests, and the tightening of forest administration upon which it depended.[94]

For the last item, the figures on forest crimes are used as an index. Reported forest crimes increased with the increased territory of state forests but they were mostly disposed of at the level of local officials, leaving little in terms of the details of individual cases on the official record. Since seasonal fluctuations in food

Table 4.1. Forest Area in Bengal (in square miles), 1868–1905

Year	Reserved	Protected	Unclassed
1868–9	-	-	-
1875–6	2,585	-	-
1880–1	3,411	2,325	-
1885–6	4,972	2,219	4,034
1890–1	5,189	2,209	4,034
1895–6	5,877	3,437	4,034
1900–1	5,881	3,656	4,033
1904–5	6,049	3,428	3,753

Source: BFAR, relevant years.

Table 4.2. Outturn from Bengal Forests, 1868–1905

Year	Timber (cu. ft. '000)	Fuel (cu. ft. '000)	Bamboos ('000)	MFP (rupee value)
1868–9	-	-	-	-
1875–6	77	-	-	-
1880–1	110	340	-	-
1885–6	6,589	18,853	-	-
1890–1	8,332	16,597	16,525	85,664
1895–6	5,557	32,684	20,700	328,864
1900–1	7,250	33,842	24,033	369,766
1904–5	5,612	33,049	25,117	323,370

Source: BFAR, relevant years.

Table 4.3. Revenue from Bengal Forests (in current rupees), 1868–1905

Year	Revenue	Expenditure	Surplus
1868–9	171,184	126,256	44,928
1875–6	198,274	139,086	59,188
1880–1	561,340	335,381	225,959
1885–6	597,432	370,399	227,033
1890–1	727,392	404,287	394,663
1895–6	935,680	475,601	460,079
1900–1	1,227,077	581,068	646,009
1904–5	1,235,455	713,212	522,743

Source: BFAR, relevant years.

Table 4.4. Forest Crime in Bengal (numbers reported), 1868–1905

Year	Injury by Fire	Unauthorized Felling	Illegal Grazing	Other Offences
1868–9	-	-	-	-
1875–6	-	-	-	-
1880–1	5	37	6	14
1885–6	15	252	112	245
1890–1	20	632	354	1164
1895–6	153	847	431	61
1900–1	114	1781	711	268
1904–5	68	1984	860	192

Source: BFAR, relevant years.

production and the vulnerability of forests to fire damage were both related to rainfall, not surprisingly, higher incidence of forest crimes may be seen in years of drought or scarcity.

The descriptive statistics of the tables tell their own story. By the end of the nineteenth century the colonial forest estate had emerged as a chequer-board of reserved, protected, and unclassed forests. In each of these distinct forest spaces distinct patterns of forest management emerged. But we cannot easily perceive a transition from the relatively tentative period of survey, classification, exploration, and discovery, to one of confident management or forest regeneration. Each regional forest space—where regions were defined administratively by land classification or socially and geographically by patterns of settlement and accommodation with bio-physical features of the environment—evinced a different temporal rhythm in its internally varied transitions into intensive forest management. In many places such transitions never occurred or they were significantly reversed.

The second part of the book describes and explains these variations, reversals, and contentions that marked forest management in Bengal in the last ninety years of colonial rule.

Notes

1. Cohn 1987a; Vicziany 1986.
2. Ludden 1993: 255–8. For Munro see Stein 1989: 139–74; Mukherjee 1968; Rocher 1983.

3. Dirks 1993: 284. In the seventeenth century, the rise of surveying in England similarly imbricated the turn to instruments and measurement in traces of memory and experience. See Bartolovich 1994: 19; Stigloe 1976.

4. Paul Carter 1989: 1–56, 67, 342–50; *Journal of Francis Buchanan Kept During the Survey of the District of Bhagalpur in 1810–11* (Patna: GOBO, 1930).

5. In a wider rhetorical analysis of colonialism, it has been pointed out that objectification was important to surveillance, appropriation by naming and description was important to possession and control. 'Given the expansion in the knowledge of nature that accompanied the exploration and colonization of new territories in the nineteenth century . . . the principles of natural history inevitably provided paradigms for the rhetoric of colonial rule.' See Spurr 1994: 24–6, 29–33, 67.

6. Paul Carter 1989: 19–22; Pratt 1992: 23–34.

7. Appadurai 1993: 315–18. For the more general argument about constant negotiation of space, see de Certeau 1988: 117.

8. Pratt 1992: 33–4; Lefebvre 1991: 48–50, 234–40; Mackay 1985; Stafford 1990: 69–79; Emery 1984; Barron 1987.

9. Stafford 1990: 71; Anon 1859: 124. The key role of the Great Trigonometrical Survey cannot be underestimated here. For its records, see Phillimore 1945–50.

10. Balfour 1862: 9.

11. Rangarajan (1994: 149–52) describes the various ways in which pre-British states had used, managed and appropriated forests for hunting, ecological warfare, protection of forts, complicating land revenue assessments, extending cultivation and so on.

12. Cartmill 1993: 119–40; John Mackenzie 1988; Ritvo 1987.

13. Rocher 1993: 215. This idea creatively informs several recent studies of the early years of Company raj, especially the statemaking aspects of law, military tradition and forest polities. See Singha 1993a, 1993b; Alavi 1993; Skaria 1992.

14. For examples of 'othering' studies in Indian historiography, see Inden (1990) for a comprehensive survey of all major categories of Orientalist colonial knowledge; Nigam (1990); and Skaria (1997), for a discussion of tribes; Pant (1987), Dirks (1989 and 1990) for a discussion of caste. 'Naming' has been best discussed by Paul Carter (1989). For 'spatializing', see Mitchell 1988; and Gregory 1994: 175–82.

15. See Rajan (forthcoming: Appendix) for a useful compendium of some of the important writings of these colonial scientists.

16. Bernier 1968 (1891): 437.

17. Henry Thomas Colebrooke and A. Lambert, *Remarks on the Present State of the Husbandry and Commerce of Bengal* (Calcutta, 1795), p. 143.

18. Ibid., pp. 165–206.

19. Ibid., p. 213.

20. Forbes Royle 1839: 12.

21. Kumar 1990: 52–61; Burkhill 1965.

22. Toynbee 1888: 67.

23. Sinclair 1794; Vicziany 1986: 626–9.

24. James Rennell, *Memoir of a Map of Hindustan or the Mughal Empire* (Calcutta: Editions Indian, reprinted from 3rd edn, 1793; Vicziany 1986: 647; C.E.A.W. Oldham 1930.

25. OIOC F/4/655, 18040, extract BPC, 14 Apr 1820, progs 68, l/d 8 Apr 1820, Wallich to Secy Genl Dept, pp. 45–52; OIOC E/4/707, Bengal Despatches, BPC, 27 Dec 1822, draft 751/1822-3, p. 1044. Buchanan-Hamilton had visited Nepal twenty years before on a mission that had to be abandoned a few months after he reached there, and nobody had been since. Vicziany 1986: 636.

26. Paul Carter 1989: 18.

27. Hannay 1845: 119–28, 131.

28. Ibid., pp. 132.

29. Irvine 1846: 184.

30. A similar search for a wider range of 'valuable woods' was going on all over India, see for instance, W. Griffith 1842.

31. OIOC F/4/1410, 55691, BC, extract BRC, 9 Jun 1829, progs 27, l/d 14 May 1829, R.M. Bird Commr Gorakhpur to Sudder Board of Revenue, pp. 12–13.

32. Ibid., progs 28, l/d 9 Jun 1829, E. Molony, Dy Secy Rev, to Sudder BOR, p. 26.

33. This mode of analysis is best presented by Skaria (1998).

34. OIOC F/4/655, 18040, BC, 'extract BPC, 6 Jun 1820, progs 44, l/d 24 May 1820, Wallich, Superintendent Botanical Gardens, Calcutta, to C. Lushington, Secy to Govt, Genl Dept, pp. 78–9.

35. Bryant (1994a: 161-3) describes these trends for Burma. He also argues that the lack of forest management was due to the influence of *laissez-faire* ideas on Company administrators. Such an explanation does not deal with the colonial process of knowledge gathering and changing representations of forests and their possible management.

36. H. Falconer, *On the Teak Forests of the Tenasserim Provinces, with other Papers on the Teak Forests of India and Summary of Papers Relating to Madras and Bombay Forests*, Selections from the Records of Bengal Government, no. IX (Calcutta: Military Orphan Press, 1852), hereafter SRBG 9, pp. 177–9.

37. Ibid., p. 206.

38. Ibid., pp. 180–98; OIOC E/4/772, Bengal Despatches, progs 5, 30 Nov 1842, India Marine Dept, draft 765/1842, pp. 713–14.

39. Grove 1990: 28.

40. OIOC E/4/772, Bengal Despatches, progs 5, 30 Nov 1842, India Marine Dept, draft 765/1842, pp. 725–52, 743; NAI GOI Foreign Dept, Political Consultations, 66–9, 26 Aug 1831, l/d Moulmein, 6 Aug 1831, Commr Tenasserim to Chief Secy at Fort William.

41. Bryant 1994a.

42. NAI GOI Foreign Dept, Secret Despatch no. 1150, Court of Directors, 6 Feb 1846.

43. For Dangs, see Skaria 1992: ch. 4. This west Indian region, home to the Bhils, had been involved with timber and other forest product trades from well before the colonial period. See Hardiman 1994: 98–100. By the 1850s restrictions on sal and sissoo were also emerging. See Rangarajan 1994: 161–2.

44. OIOC E/4/783, Bengal Despatches, progs 3, 16 Apr 1845, India Rev Dept, draft 252/1845, pp. 362–8; OIOC E/4/814, Bengal Despatches, progs 32, 17 Mar 1852, Bengal Military Consultations, draft 177/1852, pp. 856–7.

45. Balfour 1862: 17, 234.

46. For a compact summary of the work of Connally, Cleghorn, Gibson and others that generated this vision in the 1840s and 1850s, see Ribbentrop 1900: 63–72.

47. OIOC E/4/792, Bengal Despatches, progs 21, 7 Jul 1847, India Financial Consultations, draft 509/1847, pp. 1051–7, reports ongoing research on the relationship of forests to ambient humidity and soil moisture. Also see S. Eardley-Wilmot, *Notes on the Influence of Forests on the Storage and Regulation of Water Supply.* Forest Bulletin no. 9 (Calcutta: Superintendent of Government Printing, 1926), pp. 4–26; Balfour 1849.

48. Hooker 1904. A detailed discussion of the rise of desiccationist ideas in India may be found in Grove 1993a: 339–50. For parallel trends in Africa, see Grove 1989b: 23–32.

49. Anon 1852: 117–19.

50. OIOC E/4/819, Bengal Despatches, progs 5, 16 Mar 1853, India Railway (financial) Dept, draft 181/1853, pp. 697–700.

51. Long 1854–6.

52. Sherwill 1851: 569–88.

53. See OIOC MSS EUR E-309, Charles Greville Papers, box 1, file 2, l/d 20 Feb 1808, from P.W. Briscoe to C.F. Greville, enclosing extracts from the reports of the CF in Malabar and Canara, pp. 2–10.

54. WBSA P/550 BRP (For), A progs 1–5, May 1898, file 12-5/1-1, no. 19, Camp Jalpaiguri, 15 Apr 1898, E.G. Chester, CF Bengal, to Secy GOB Rev, pp. 67–8.

55. Cleghorn et al. 1852: 120, 135.

56. Cleghorn 1861; NAI GOI, Foreign Dept, Foreign Consultations, 108–11, 29 Aug 1858, Report by Brandis on teak forests of Pegu, pp. 262–337.

57. NAI GOI, ibid., pp. 324–7.

58. Balfour 1862.

59. OIOC P/232 BRP (For), May–Aug 1872, A progs 2–11, May 1872, no. 857, Cal, Mar 1872, H. Leeds, CF Bengal, to Secy GOB Rev, pp. 7–8, lists eighteen main trees, all having non-timber uses and local timber uses in the lower provinces of Bengal.

60. Ribbentrop 1900: 60.

61. Stebbing 1922: 61, 35.

62. Ibid., pp. 66–7. In 1827 Wallich had reported that the sal forests of upper India were beginning to fail, Stebbing 1922: 201.

63. OIOC P/432/74 BRP(For), Sep–Dec 1868, A progs 19, Nov 1868, l/d Cal, 8 Oct 1868, T.H. Lloyd, Provisional Secy to the Port Canning Land Investment Reclamation and Dock Co., to Secy GOB, p. 11.

64. WBSA P/42 BRP (LR) May 1862, progs 52–3, no. 5, London, 27 Mar 1862, Sir Charles Wood, Secy of State for India, to GC-in-C, p. 66.

65. WBSA P/482 BRP (For) Jun 1893, A progs 1–4, file 5B/3, no. 40M-G, Darjeeling, 26 Apr 1893, E.G. Chester, Offg CF Bengal, to Secy GOB Rev, pp. 151–4.

66. WBSA P/232 BRP (For) May–Aug 1872, A progs 2, May 1872, Leeds to Secy GOB Rev, no. 857, Calcutta, March 1872, p. 6.

67. OIOC MSS EUR F86, Richard Temple Papers, Handlist 4, l/d Darjeeling, 27 Jun 1875, from Temple to the Marquis of Salisbury.

68. OIOC P/66/57 BRC (For) Aug–Dec 1864, A progs 1–19, no. 75, 1 Nov 1862, GOI to Sir Charles Wood, Secretary of State, pp. 5–7.

69. Ibid., memo dated 18 Dec 1862 by Brandis, IGF, GOI; memo by Brandis dated 6 Jun 1863.

70. OIOC P/432/62 BRP (For), Jan–Apr 1866, A progs 5–8, Feb 1866, T. Anderson, CF (LP), to Secy GOB, no. 126, 7 Feb 1866.

71. BFAR, 1866–70, particularly, report, 27 Jul 1870 by H. Leeds, p. 3; NAI, PWD (Rev-For), Mar 1871, A progs 121–4, no. 222F, 24 Mar 1871 from Secy PWD, GOI, to Secy GOB Rev; OIOC P/432/65 BRP (For), Nov–Dec 1866, A progs 4, T. Anderson, CF Bengal, to Secy GOB, no. 62, 5 Dec 1866, pp. 16–18.

72. Planting of exotics remained a limited enterprise in Bengal, as less than ten species were tried. In other provinces like Madras (59), Punjab (32) and Burma (31) the planting of more exotics, according to one account, reveals the greater emphasis on plantations as opposed to natural regeneration. See Hole 1913: 10–12.

73. There appears to have been a short lived controversy about the relative merits of pines and hardwoods as planted species, but the foresters prevailed over the Lieutenant-Governor of Bengal in the matter. Carving out a domain of expertise increasingly inured to lay administrative opinion was facilitated by parallel developments in the federal government. NAI, Rev, Agri and Commerce (For), Oct 1871, A progs 13–15, Res by GOB on CF's report on Chotanagpur forests; OIOC P/66/57 BRC (For) Aug–Dec 1864, A progs 1–19, note dated 14 Aug 1863 by Dr Hugh Cleghorn and D. Brandis on the forest rules proposed for Madras, pp. 23–5.

74. OIOC P/432/66 BRP (For), Jan–Feb 1867, A progs 11, T. Anderson, Supt Royal Botanical Gardens, to Secy GOB, no. 106, 27 Dec 1866, pp. 15–18.

75. BFAR, 1870, p. 6; Champion (1933: 23) notes, however, that sal plantations were extremely rare before 1910.

76. OIOC P/66/57 BRC (For), Aug–Dec 1864, A progs 1–19, Circular no. 3478, dated 19 Oct 1864 from GOB, Dept of Rev (For) to all Divisional Commrs.

77. OIOC P/432/62 BRP (For), Jan-Apr 1866, A progs 5–8, Feb 1866, A. Money, Commr Bhagalpore, to CF Bengal, no. 439, 10 Feb 1865.

78. Hatt 1905: 1–2.

79. OIOC P/432/62 BRP (For), Jan-Apr 1866, A progs 5–8, Feb 1866,

A. Money, Commr Bhagalpore, to CF Bengal, no. 439, 10 Feb 1865, p. 3; J.P.H. Ward, Collr Shahabad, to Commr Patna, no. 840, 13 Feb 1865; C.H. Campbell, Comr Rajshahi, to CF Bengal, no. 21, 17 Feb 1865.

80. Stebbing 1924: 379–403.

81. OIOC P/432/62 BRP (For), Jan–Apr 1866, A progs 5–8, Feb 1866, A. Money, Commr Santhal Parganas, to CF Bengal, no. 150, 25 Apr 1865; across the state the question of village rights produced a wide variety of responses, but most divisions did report regulation of rights by zamindari levies. How these rights were apportioned between arbitrary categories of rights and privileges, I have discussed elsewhere. See Sivaramakrishnan (1995a: 13–15).

82. OIOC P/432/62 BRP (For), Jan–Apr 1866, A progs 5–8, Feb 1866, G.F. Cockburn, Commr Patna, to CF Bengal, no. 68 CT, 15 Feb 1865.

83. Ibid., C.F. Montressor, Commr Burdwan, to CF Bengal, no. 188, Mar 1865.

84. Ibid., pp. 10–14.

85. Trafford 1911: 8.

86. OIOC P/432/62 BRP (For), Jan–Apr 1866, A progs 5–8, Feb 1866, C.P. Casperz, Commr Sundarbans, to Commr Nuddea, no. 169, 24 Nov 1864; C.T. Buckland, Commr Dacca, to CF Bengal, no. 456, 24 Feb 1865, pp. 15–20.

87. Ibid., E.T. Dalton, Commr Chotanagpur, to CF Bengal, no. 2050, 29 Nov 1864; Major G. Hunter Thompson, Supt. Rev Survey Chotanagpur, to DC Hazaribagh, no. 898, 6 Sept 1864; Lieut R.C. Money, DC Manbhoom, to Commr Chotanagpur, no. 1664, 28 Sep 1864, and no. 1914, 16 Nov 1864.

88. BFAR 1871-2, to 1877-8. These annual reports written by William Schlich as Bengal Conservator are detailed and richly descriptive in a way that is unsurpassed by subsequent reports.

89. BFAR 1877-8; Stebbing 1926.

90. D. Brandis, *Memo on the Forests of Jalpaiguri and Darjeeling* (Calcutta: Government Press, 1879); Brandis 1897.

91. WBSA P/518 BRP (For) May 1896, A progs 27, file 8-B/1, no. 40T-R, 24 Apr 1896, C.W. Bolton, Secy Rev GOB, to Gillander, Arbuthnot and Co., Managing Agents, H. Dear and Co., pp. 135–6.

92. WBSA P/518 BRP May 1896, A progs 16–17, file 8-B/1, no. 191, Cal, 10 Jan 1896, C.E. Buckland, Secy Rev GOB, to Secy GOI, Rev and Agri.

93. See 'Forest administration report for Bengal, 1882-1883', abstracted in *IF* 10(5), 1884, p. 223.

94. The figures are given till 1905 because Bengal was partitioned that year, and soon after reunification, Bihar and Orissa were made a separate state in 1912. So the geographic area of Bengal and consequently of its forests underwent frequent changes during that period. After 1912 and till 1947, the area treated as Bengal prior to 1905 was comprised in the separate states of Bengal, and Bihar and Orissa, furnishing separate reports on forest administration.

Part Two

Formal Structures of Forest Management

The next four chapters deal with the last ninety years of the colonial period when through laws, administrative machinery, and scientific planning and execution of operations, forestry became a formal, directed endeavour of statemaking in Bengal. Chapter 5 explores regional variations in statemaking through territorialization and access restriction regimes. Chapter 6 locates aspects of forest management in the agrarian environments, and debates on agricultural development, from which they were inseparable.[1] Chapters 7 and 8 examine the regionally diverse experiences of forest regeneration in Bengal and show how they shaped the emergence of scientific forestry as development discourse.

Unlike previous studies of colonial forestry in India, or other studies that treat modern forestry as territorialization writ large, Chapter 5 carefully distinguishes between the import and consequences of procedures like reservation and protection. By focusing on instances when both these procedures failed, were revoked, or were only partially realized, this chapter suggests that variations, when documented, serve not only to qualify more generalized accounts of colonial forest management. They indicate that regions like the jungle mahals became zones of anomaly to forest reservation in ways historically prefigured by their encounter with the Permanent Settlement.

Highlighting both regional variations in forest management, and the gap between plans and outcomes, Chapters 7 and 8 describe the conflicted construction of scientific forestry in Bengal and its emergence as a development regime. Together these chapters track the vicissitudes of forest conservancy in Bengal, to demonstrate that it was implicated in the elaboration of a development and conservation administration that was the product more of specific historical experiences and less the result of imported European models. We may thus unpack the notion of scientific forestry, and question the simple divide often instituted between universal western modernity and its indigenous Other.

When forest management geared up under scientific principles, revenue and markets were not the sole concerns. Legitimacy was a constant worry since that affected the degree to which plans deviated from concept in reality. The way forest regulations became operational, the nature of rights curtailed and denied, and the groups affected, were matters shaped by the kind of credence given to local practice in official surveys. These outcomes were also affected by the attitude of villagers towards new laws or policies.[2]

While surveys to gather detailed information on the forest resources of the region were still under way, the Government of Bengal passed blanket orders to reserve all kinds of land bearing valuable timber, 'that is to say land in which such trees [teak or sal] abound.'[3] Similarly the question of forest demarcation was taken up and resolved in favour of the Forest Department, using a Central Provinces ruling that enabled demarcating forests in a way that 'such lands not covered with sal be included as may appear necessary to facilitate the protection, management and improvement of forests.'[4] The Forest Department was to be consulted prior to transfer of any land under wasteland rules. This decision, and the one relating to demarcation overruled the civil administration's concern that present grazing and future agricultural expansion would be jeopardized by this policy, and reinforced the ideas promoted by foresters that efficient forest management required large contiguous tracts.

The first thirty years of forest conservancy in Bengal were a period of conservative lumbering. Physical control of forested land, assured supply of timber, and profits in forest operations were driving concerns of forest management. In 1899, the Lieutenant-Governor of Bengal deprecated the decline in revenues from forestry saying, 'forest officers must realize more than they do that they are the agents of a great commercial undertaking ... and as such they are not merely the scientific protectors of an important property.' Active exploitation of forests to the limit of possible annual yield was the professed goal.[5]

Within the next decade the priorities of the government were being radically reformulated and primacy was being given to regeneration. This concern with regeneration voiced an older anxiety about desiccation, namely the relationship of forest to rainfall, surface hydrology, drought and floods; but it also arose from the newer discoveries about difficulties in sal reproduction in reserved forests.[6] Scientific forestry as a regionally specific set of practices emerged from this policy transition. Silviculture was introduced as a major departmental activity by 1908, requiring detailed annual reporting. But hopes remained pinned on natural regeneration of sal in all provinces of Bengal.[7] So the movement from restrictions to regeneration flowed along existing channels.

Notes

1. The consequences of delinking rural environmental issues from agrarian relations have been discussed in Agrawal and Sivaramakrishnan (forthcoming).

2. The response of villagers to restrictions arising from colonial forest management cannot simply be discussed under the rubric of resistance and acts of incendiarism. Over the long term, sporadic acts of violence or repudiation of government-imposed controls by villagers fit into wider patterns of accommodation through which forest management incorporated such response as local knowledge. In Chapter Five I have illustrated such an instance by discussing the shifting arrangements through which grazing regulation was adjusted to silvicultural planning in Darjeeling district.

3. OIOC P/66/57 BRC (For), 1864, progs 6–12, memo by Brandis dated 6 Jun 1863; note dated 14 Aug 1863 by Cleghorn and Brandis, pp. 18–25.

4. OIOC P/432/75 BRP (For), Jan–Apr 1869, A progs 2, Feb 1869, no. 409F, Fort William, 28 Nov 1868, Col C.H. Dickens, Secy GOI PWD, to Secy GOB Rev.

5. NAI Rev and Agri (For), A progs GOI 32&33, file 352 of 1904, Mar 1905, no. 4969, Cal, 24 Dec 1904, A. Earle, Secy GOB, to Secy GOI, p. 409; A progs GOI 12–16, file 45 of 1901, Jul 1901, no. 680-F, Simla, 18 Jun 1901, J.B. Fuller, Offg Secy GOI, to Secy GOB Rev, p. 1031.

6. NAI Rev and Agri (For), A progs GOI 41–3, file 11 of 1908, Feb 1909, GOB Res no. 5110, Cal, 11 Dec 1908, p. 597. The discussion here of forests and climate must be seen in the context of a larger discussion generated from the Famine Commission reports of the preceding twenty years. Its relation to an older conservation discourse cannot also be missed. The nature of these

older ideas of colonial surgeons and botanists are discussed in detail by Grove (1993a, 1995a). Their role in late-nineteenth-century forest management has been assessed by Skaria (1998a) and Rangarajan (1998).

7. NAI Rev and Agri (For), A progs GOI 76&77, file 336 of 1909, Jan 1910, GOB Res no. 3558 For, Cal, 20 Nov 1909, p. 132.

5

Varied Regimes of Restriction and Lumbering

The difference between forest officers and civil officers . . . [is] . . . that the former are scientific men mainly concerned with wood and trees and their production, and the latter . . . with the people, their interests and doings.[1]

The history of the forest in all countries shows the same period of development. First hardly recognized as of value or even personal property, the attitude of settlers to the forest is of necessity inimical—the need for farm and pasture leads to forest destruction. The next stage is that of restriction in forest use and protection against cattle and fire, the stage of conservative lumbering. . . .[2]

Introduction

Colonial involvement in forest exploration, timber trade, and promoting timber plantations had all occurred during the first half of the nineteenth century within a political-economic regime where productive land was private property.[3] Earlier, territoriality under the Company raj was largely confined to demarcating external boundaries of empire. Growing knowledge of the landscapes and natural resources of the empire, as the Company raj gave way to direct rule, made possible an internal or managerial territorialization that forest conservancy demanded.[4] This process culminated in forest reservation, where land—irrespective of current land use—was designated for management as forests, thus creating a class of state-owned land dedicated to specific types of tree production.[5] Making state property of natural resources thus brought 'a wide range of inter-related problems under a single relatively manipulable legal abstraction.'[6]

The road to reservation was often long and tortuous. For instance in Assam, when woodcutters from Bengal appeared in quest of sal in the 1850s, forests, then classified as government wastelands, were farmed out on five-year leases with no felling restrictions by species or girth. In 1868, *mouzadars* were appointed who paid a nominal tax of four and a half rupees per axe per year. Since each axe yielded fifty logs per year and each log sold for at least ten rupees in markets along the Brahmaputra river, the mouzadars hardly felt restrained by the impost.[7] Similarly, in Burma during 1830–52 timber merchants had a field day under a system of licensing that was incapable of preventing over-harvesting. Lumbering regulation through territorial control was a long time coming.[8] It was preceded by a phase of early conservancy which did not extend to 'forests as a defined organism in the household of nature, but merely to a few marketable species.'[9] Forest management through territorialization, when it happened in India, seemed to follow the European experience. In France, for instance, the failure of private enterprise and limited regulation preceded national forest codes that reserved areas for forestry in the face of violent peasant protest.[10] Ideas and practices imported from Europe, however, constituted but one ingredient in a recipe produced in different colonial locations.

In Bengal the category of managed state forest territory was further classified by different regimes of restrictions. Reserved and protected forests differed significantly in the extent and nature of internal territorialization. This put forest regions in Bengal at different places on a continuum of forest management the end points of which were defined as local determination and central direction. Reserves were areas where the designated land was entirely brought under government management. Protected forests contained specific tree species that were not to be felled without government sanction. Other forms of land management in these areas respected local practices and needs to a great extent.

From the very beginning there were differences—between standardized forest management and locally varied approaches—in the way central government visualized forest conservancy. It also came to be practised in diverse social, political, and ecological settings. These factors combined to generate a variety of regimes. While detailing these regimes, this chapter will not only indicate the manner of their production. Organized around the theme of territorialization, this chapter will delineate the limits to internal territorialization as a distinct mode of modern forest conservancy in colonial Bengal. What these limits were, how they were set and their consequences for forest management are best understood by a close examination of governance as it was debated and then manifested in forest reservation, protected forest rules, and the variety of procedures developed to control land, revenue, and trade in forest products.

Bureaucratic proliferation characterized colonial government in India by the late nineteenth century. But in woodland Bengal it was accompanied by fission and conflict in the departments, and their hierarchies, that made cohesive statemaking difficult. Foresters and district officials saw their mandates in sharp contrast, especially in regions of southwest Bengal that we have already discussed as zones of anomaly. The character of vegetation and the range of its commercial and subsistence use in this region also posed special problems for neat systems of internal territorialization. The Bengal Forest Rules had to distinguish between reserved and open forests. In doing so they re-created the distinction between imperial reserves and village reserves that the Bombay Forest Committee had devised in 1863.[11] Later this difference was institutionalized through reserved and protected forests by the Indian Forest Act of 1878 (Figure 10). William Schlich, the Bengal Conservator at the time, was quick to note that most protected forests (excepting the Sundarbans) were in the non-regulation provinces and remained particularly encumbered with questions of local rights.[12]

By carefully considering the shifting locus of governance in forest Bengal, using indices of revenue, rights, local welfare, national interest and forest composition I shall argue that the social and ecological diversity of actual contexts in which forest conservancy was introduced forced the modification and, at times, the abandonment of carefully laid plans. This chapter will demonstrate that regionally varied modes of forest management built on prior forms of administrative exceptionalism. Forest reservation, and hence

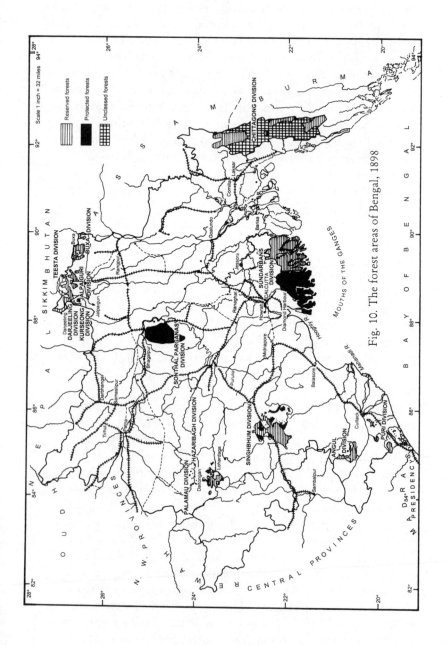

Fig. 10. The forest areas of Bengal, 1898

Reserved forests

Protected forests

Unclassed forests

Scale 1 inch = 32 miles

centralized management, was least successful in the jungle mahals and other such areas that have been characterized as zones of anomaly. In these areas protected forests were created as an alternative system of forest management sensitive to historical particularities and local demands.

Territorialization Refined: A Fractured State

Modern statemaking has been characterized by the drawing of boundaries around territories to an extent probably not witnessed by prior political history.[13] Cartography rapidly became the authoritative mode of representing the meaning of territory.[14] Recently, arguing that 'territory as a key practical aspect of state control has been relatively neglected by theorists of the sources of state power,' Peter Vandergeest and Nancy Peluso have developed a concept of territorialization and applied it to the study of forest management in Thailand.[15] They extend the concept beyond the more usual application to boundaries and political identity and suggest that it can usefully be applied to spatial organization of state administration, especially the creation and control of state property under agencies endowed with territorial and functional jurisdictions.[16] Qualifying these subdivisions as internal territorialization, the process is unravelled through the reservation, demarcation, mapping and management of forests in Thailand.[17] In the sense that control of resources is predicated on control of land, the idea of internal territorialization is useful to distinguish a special form of state-enforced resource control.[18] This chapter utilizes the concept in that fashion. Such territorialization can imply both inclusion and exclusion for social groups using the resource in question. But it does differ from other forms of inclusion and exclusion.[19]

While territorialization as discussed by Vandergeest and Peluso is a valuable starting point to understand the spread of conservative lumbering in Bengal, the complex and contradictory processes that we notice there require further analysis. Neil Smith's observation that 'rediscovery and rewriting of the imperial past . . . should be done in explicit connection with a sense of the lived geographies of empire' draws our attention to the ecological diversity of sites where natural resource control was territorialized.[20] His words also remind us that diversity is not only a product of natural processes but of cultural construction of those processes as well.[21] Territorialization itself produces many structurally different forms of control and management that relate to differences in landscape. These pluralities and contingent indeterminacies are missing from any unilinear account of modern forestry that treats it as territoriality refined. As Totman reports for Japan, and the Bengal case will demonstrate, woodland management often worked through the regulation of use rights rather than land ownership.[22] Such alternate forms of statemaking in the forests were not always proto-territoriality. They coexisted with more directly territorial forms, and sometimes superseded them.

Many of these variations derive from what William Cronon and other environmental historians have sought to reserve as nature's autonomous place in history. While the degree of nature's freedom or primacy in determining human history remains a topic of inconclusive debate, we could bracket that argument and discuss nature as a 'hybrid . . . lively if socially constructed actor.'[23] Such natural agency, combining bio-physical processes and human interaction with them, also exhibits regional patterns. From the perspective of Bengal and territorialization, these patterns are significantly different. These differences are marked by the distinction between reserved and protected forests. Several recent studies have told the story of forest reservation and recognized its territorial basis without using the precise terminology applied here. They have focused on demarcation of forests, the extinction of traditional usage, rights and privileges, and the curbing of public access to forests, reserved or protected.[24] The work of these scholars, however, sometimes blurs the distinction between reserved and protected forests as they emerged in the aftermath of forest conservancy. The consequence for woodland Bengal is evident in the work of Stuart Corbridge on Chotanagpur and Santhal Parganas, which gives an overdrawn and homogenized account of colonial transformations in the region, summed up by his remark that, 'in place of open access to forests there now appeared fences and forest guards.'[25]

Generality of this sort obscures the insights that variations provide into processes of statemaking. Seeking those insights, this chapter proposes a careful study of where reserves were established and where protected forests were designated. My argument is that the making of reserved and protected areas displays a regional pattern in Bengal that is insufficiently explained by merely accepting the expressed legislative intent of the distinction, or worse, ignoring it altogether. We have something to learn about colonial statemaking by attending to the spatiality of its regime of restrictions.[26] The following discussion draws out these lessons by relating forest management to the diverse political and ecological conditions in which it was practised in woodland Bengal.

During the latter half of the nineteenth century valuable trees and estates were listed and concurrently the possible modes of efficiently extracting this value were devised by newly appointed forest conservators in different provinces. In many instances, land settlements being made in khas (government) villages served as an opportune moment for surveying tracts of forest and wastelands as separate blocks and reserving them for government use.[27] This was then a phase of asserting exclusive government rights over the commercial value of forests and inventing a regime of restrictions on other rights in these forests. These included the restriction of forest firing, grazing, cutting sal and other valuable timbers, girdling, shifting cultivation, and private lumbering.[28]

While installing the regime of restrictions the creation of government forests came in for debate. Arguments within the colonial state were about what were the valid restrictions and also who would enforce them. Ramachandra Guha vividly portrays the struggle that went on across provinces and within the echelons of the colonial state before the Indian Forest Act of 1878 was legislated.[29] The Madras Presidency repudiated the Act for doing violence to the principles of rule of law.[30] But the matter did not end there. The Act, and all its subsequent amendments, remained ambiguous on systems of management, leaving regional administrations enough latitude to fashion interpretations of its provisions that best suited particular contexts. As a result sharp conflicts emerged between district and forest officers, despite the Government of India's exhortation that foresters should work in concert with civil officers and acquaint themselves with local details of timber trade.[31]

Bengal forest rules were framed in debates within the government that reveal how technical-managerial aspects of forest conservancy were decoupled from wider issues of rural welfare and order even while the success of scientific forestry depended on the existence of a politically and socially stable agrarian order.[32] If district officers spoke as if the peasantry were their constituency, it was only because they needed the daily cooperation of various sections in the rural population. The foresters, in contrast, seemed to assume that they did not have to nurture a constituency in the rural populace. They apparently envisioned an assignment that lifted them from the realm of mere officialdom.[33] So while working toward achieving the tricky balance between revenue and more efficient use of resources, different branches of government developed discrete, often opposed departmental visions of how these goals might be achieved.[34]

Reservation, followed by the preparation and execution of working plans leading to separate jurisdictions for forest and civil officers, became the measure of territorialization fulfilled. But these procedures that would take territorialization to its full extent were thwarted by a variety of factors. Even the forest reserves faced problems of conservancy that were often really problems of forest working—the feasible modes of harvesting and removing timber. By examining the range of arrangements through which forestry was organized we shall see how forest management through territorialization was limited. We shall learn about the strategic modifications imposed by the landscape and the local people on forest conservancy in Bengal. This chapter will discuss forest working, the elaboration of forest control through rules and forest settlement, and the interdepartmental disputes over control of revenue and land. The next chapter will focus on the agrarian contexts of forest management. Differences in forest composition, socio-political history and government interests are revealed as both outcomes and influences in those regions where protected forests were established in southwest Bengal.

Forest Working Arrangements

When the Bengal Forest Department was created, timber extraction was already a widespread activity and in the first few years, this activity was organized in ways involving the department to different degrees. Most forest districts were under long leases to British contractors who extracted timber using logger's choice methods. Since these were leases intended to convert waste into arable lands they could be annulled for failure to do so.[35] In the permit system the department simply marked trees to be felled. The permit holder would extract the timber after paying a fixed royalty per log and remove it on pack ponies and bullock cart.[36] Otherwise the government employed petty contractors, often different ones for felling, sawing, dragging, and floating. Sometimes seasonal labour was recruited to carry out felling, conversion, and transport of timber. In addition to hill tribes, *daffadars* (labour contractors) would bring up labour from the plains for forest operations in the Duars.[37] Thus in Jalpaiguri Noonia coolies were brought from Bihar for road work, while Nepalis were encouraged to settle so that they might provide labour for creeper cutting, line clearing and fire protection. Such imported labour was often paid less than locally hired labour. These outsiders were offered land for tillage to encourage them to settle and become a permanent labour force.[38] In these two types of situations the costs were borne by the department. Given the unpredictability of demand, and the intensive supervision that direct operations called for, the permit system remained largely in vogue. It seemed to possess the added merit of securing Forest Department revenues before timber removal and released forest officers, still scarce in the 1870s, to discharge other 'developmental' duties.[39]

Until 1876 the forests were mostly worked by private enterprise, with the Forest Department selling selected trees at certain rates and firewood by weight. The purchasers undertook cutting and carriage. Surveying the arrangements for forest working in Darjeeling in 1881 Schlich, then the Inspector-General of Forests, recommended that cutting operations be confined to specified blocks to remove firewood, scantlings and other wood together and thus prepare the area for regeneration in blocks.[40] This had led to auction sale of coupés. A new method of selling timber and fuel, called the monopoly or bonus method, was introduced in 1905–6. The new method required close supervision of wood conversion by the purchaser, depriving him of the incentive of making the best of all parts of the tree; but it did help inexperienced buyers who might fail to accurately estimate the value of outturn.[41] The older method, auction sale of coupés at wholesale prices, also continued alongside. The consolidation of forest working by private

enterprise steadily increased with a corresponding rise in departmental efforts for securing forest regeneration.

Though departmental working always remained the vigorously
pursued ideal, departmental operations never came to occupy a
major place in forest working. In some places like Jalpaiguri they
were started and given up by the 1880s.[42] Most frequently, they
ran into problems of labour supply. Solutions to these problems
remained elusive, given the deep aversion of foresters to shifting
cultivation. In their view such cultivation not only cleared away
miles and miles of valuable forest but also brought agricultural
competition to the very lands on which the best forests grew.[43]
Leeds did note that the existence of habitation in the forest reserves
secured labour for forest working and facilitated inspection of tracts
by watchers and subordinate officials, but could only conceive of
allowing their continuance where these settlements and the area
they farmed were clearly demarcated. He also insisted that in any
such arrangement jhum must be excluded.[44] Consequently the
secular upward trend in the expansion of the arable through the
late nineteenth century, and the concurrent spread of forest conservancy, meant a severe constriction of the areas historically occupied
by the tribes dependent on an agro-forestry-based economy in Bihar
and Bengal.[45] The attack on shifting cultivation and forest
reservation displaced available labour. Other trends in the agrarian
economy, notably the spread of plantation crops like tea and
cinchona in north Bengal and Assam; the commercialization and
intensification of agriculture in the south for rice, oilseeds and
jute; the increase in public works and mining, all combined to
tighten and restructure labour markets.[46]

During the period 1865–82, the Assam tea industry grew steadily
in response to rising prices and thereafter it continued to grow
until 1900 despite falling prices. This generated a massive demand
for labour and much of it was derived through the indenture
system. Long-distance recruiting took place principally from Chotanagpur, Santhal Parganas, Bihar and eastern UP. During 1880–
1900, of the 710,000 coolies recruited for the Assam tea gardens,
46 per cent were from Chotanagpur.[47] According to another
estimate, during 1878–94, half the total import of labour into
Assam was from Chotanagpur and Santhal Parganas, and during
1895–1930 this region still contributed 35 per cent of the labour

supply to Assam.[48] In north Bengal too, adivasis (autochthones) of southwest Bengal were the chief source of plantation labour, and women and children went in large numbers. Thus between 1890 and 1920 more than half the workers in Darjeeling and Jalpaiguri tea gardens were adivasi women. A spatial pattern later emerged, with Nepalis in the higher elevations of Darjeeling and Santhals and other Chotanagpur tribes in the lower elevations of Darjeeling and Jalpaiguri.[49]

Another aspect of this movement of labour is what has come to be known in the labour history literature as ethnic segmentation. This meant that permanent migration to tea plantations was preponderantly from particular tribes of the Kolarian group of southwest Bengal.[50] Foresters often expressed a preference for the same aboriginality that Kaushik Ghosh has evocatively discussed in connection with tea garden recruitment. They could not compete, however, with the range of coercive and deceptive methods used by an elaborate hierarchy of sardars, *arkattis* (contractors), and others engaged in capturing these desired tribal populations for labour markets on plantations. Forcible recruitment of plantation labour has generated a vigorous debate on push-and-pull factors operating in colonial migration from the labour catch-ment areas of India, details of which need not concern us here.[51] Expanding forest management encountered a tribal peasantry burdened with debt, insecure tenures, extortionate rents, and declining food availability due to famine and the ecological deterio-ration of already impoverished, and now more inelastic, margins of the arable. Coupled with these processes, the forcible recruit-ment of these tribal groups as plantation workers meant that these people were not easily available to the Forest Department.

The problem was compounded by the expansion of cash crop cultivation in east Bengal and riverine Bihar. Even before the massive out-migrations caused by recruitment for coal mines and tea cultivation in north Bengal and Assam, regular seasonal migration brought hill tribes to plains Bengal in search of employ-ment in winter cropping.[52] Such migration increased with the increase in jute and cotton cultivation in eastern, central and western India and also contributed to a form of segmentation in the labour market that kept a certain proportion of labourers in localized networks.[53] In north Bihar, for instance, the tightening of the labour markets encouraged local zamindars to arrange mutually stable seasonal migration patterns between neighbouring labour-surplus and deficit regions. This combined with the preference of poor peasants for seasonal movement, since that gave them an opportunity to augment family income without losing land and ties in their home villages. As a result, when the government sponsored migration to north Bengal districts like Jalpaiguri and Cooch Behar, in the aftermath of famine in the 1870s, it faced great difficulty implementing such resettlement from what were considered overcrowded districts.[54]

The preceding discussion of rural labour in colonial Bengal after 1860 indicates both the displacement and absorption of labour around woodland Bengal. These processes created grave difficulties in securing timely and sufficient labour supply for forest working. To stabilize and ensure labour supply in reserved forests, a system of forest villages was, therefore, introduced. But two obstacles had to be overcome. First, there was no provision whereby land included in a demarcated forest could be leased for other purposes if they did not include commercial activities like tea cultivation. Second, ensuring labour for forest working throughout the year required a captive labour supply that had to be permitted a living in the forests. Such a solution had to also avoid conferring permanent rights on the forest villagers for that might impair forest reservation.[55] So forest villages were formed on a service tenure arrangement that resembled the privileged tenures prevalent in zamindari estates. The idea was to attract Garo and Mechi woodcutters to settle in the Duars and other north Bengal reserves. They were allotted land to till for subsistence on condition of being on call as forest labourers.[56]

Similar practices emerged in Singhbhum reserves. In some cases forest villages were formed by taking existing tribal villages in the reserves off the revenue roll and converting them into service tenures under the Forest Department.[57] With the intensification of forest regeneration operations labour needs rapidly increased. Every felling series necessitated eighty houses of forest villagers, since each household could regenerate one acre under taungya. To retain these villagers they were assigned an additional three acres per house for permanent wet rice cultivation, thereby setting aside about 250 acres in each felling series to support the forest villages.[58]

Once created, the forest villages imposed their own constraints on silviculture.

Thus the mobilization of forest labour, responding to local constraints and governmental apprehensions, determined one set of restraints on the territorial management of forest reserves. This had to do with the way departmental control created through forest reservation was dissipated in the modes of forest operation or working. Territorialization also developed through divergent and partial strategies in different regions of forest Bengal. We shall turn to that next by considering the making of rules, the creation of reserves through forest settlement, and the contentious emergence of protected forests.

Illustrative Forest Settlements: Porahat and Dhalbhum

Forest rules 'were a mechanism to transform the state's nominal ownership . . . into an actively exercised proprietorial right.' By also creating classes of forests the same rules modified this exercise in significant ways.[59] In the reserved category the entire custody, administration, and control of forests and their products were given to the Conservator. In the open forests (another term for protected forests) the control of the Forest Department was to extend only to specific reserved trees.[60] By creating two classes of state forests the Government of India recognized both the impracticality of detailed demarcations and the potential for admitting, and thus to some extent reconciling, a plurality of interests in these forests.

The rules, as framed, had already departed from the scheme suggested in the Forest Act of 1865. In their application these rules brought other sources of disagreement and modification into play. These transformations, as part of wider uneven processes of change, reflect the impact of both conflict and collaboration between broadly constructed entities of state and society. This allowed social actors as humble as graziers and woodcutters to leave their impress on political discourse through the structures of bureaucracy and its internal contradictions.[61] These pressures created divergences between policy and law, practice and code, that draw our attention to the silences in codified law that have traceable motives. This means legal structures cannot be understood without the events and processes that produce them.[62] In amplification and illustration of this argument, I shall now take the case of two forest settlements arising from attempts to reserve forests in southwest Bengal.

Forest reservation entailed forest settlement, a procedure that defined legal title of the state in reserved forests and proscribed accrual of rights other than those recognized at the time of settlement. Forest settlements were to determine the extent to which state proprietorship of reserved forests was circumscribed by adverse rights and propose a scheme for exercise of rights that facilitated forest working for revenue and its preservation for the future.[63] The wise men of Indian forestry described settlement as the procedure that fixed the matrix of rights, while management was a fluid procedure that was expected to adapt to changing circumstances. But when instituting forest settlement the Bengal government found that rights determined only from preferred claims were likely to provide a poor depiction of actual patterns of forest use, and in an unusual instruction, the Government of Bengal recommended that the forest settlement officer investigate practices of forest use to assess what customs should be recorded as rights.[64] Thus the forest settlement returned to a more complex and troublesome conception of rights enmeshed in custom that had always puzzled British legal codification and standardization in Bengal.[65]

The reservation of Porahat forests in Singhhum gives us a rare instance where forest settlement was carried out, as in north Bengal this was mostly avoided.[66] The estate, populated mostly by Kols, was 791 square miles and the forest reserve constituted about 26 per cent of it. These forests comprising largely sal, lay in the hill country, catchment to the Sanjai river, the valley of which was soon to be traversed by the Bengal-Nagpur railway.[67] The settlement operations stirred up considerable agitation on the part of the Kol peasantry. Inquiring into the settlement proceedings and consequent aggravation of the affected villages, the Deputy Commissioner, Singhbhum, found that the settlement was made peremptorily and was incomplete.[68] The complaints were that insufficient forests had been left for grazing and collection of firewood, thatch grass had become scarce, boundary marking had deviated from the demarcation line. In places the line had passed so close to habitation as to make villages constantly liable to charges of cattle trespass.[69]

Irate headmen encouraged their Kol villagers to confound and obstruct the settlement operations. They would not provide supplies to Maulvi Wahidullah, the revenue official, or Manson, the forest officer, deputed to carry out demarcations and record claims. Fowl would be running all over the village in which these officials had struck camp and yet the answer to any purchase requests to buy fowl for provisioning the official mess would be 'none are available.'[70] Some of the resistance to forest settlement was sponsored by Christian leaders of the agitation against rents, who campaigned among their Kol brethren to decline compensation for land taken for reservation.[71] But by the end of the operations the raiyats were reported to be satisfied overall, both with the compensation for rice fields and dwelling, and the alternate sites to which some of them had to move and resettle in consultation with the Mankis of their Pirs.[72]

The Porahat reserves were created in 1890, just in time for the opening of the Bengal-Nagpur railway to goods traffic in 1891. The timing was significant, as the Conservator put an immediate end to the system of purchasers selecting the trees they would cut at ten rupees apiece, introducing systematic logging that by 1898 had yielded 8,07,627 broad gauge sleepers from 20,921 trees in the estate.[73] The completion of the forest settlement in Porahat produced varied responses from the affected peasantry. In the north of the estate, there had been a recent influx of settlers, from whom the Mankis and Mundas had made considerable income as rent from new lands brought under cultivation. These settlers, who had left Lohardugga to avoid a recent settlement which had assessed uplands for the first time, had prepared terraced paddy fields and were cultivating them without paying revenue. Thus both they and the Mankis were resistant to forest settlement. But in the south, the Kols were more amenable to forest conservancy, seeing in it a way to restrict the entry of strangers and the invasion of native Christians.[74]

The Bengal government assumed direct management of Dhalbhum estate in 1887 under the Chotanagpur Encumbered Estates Act, 1876, displaying another facet of the legal and institutional arrangements through which the former jungle mahals and adjoining tribal-dominated regions were being administered to create and preserve a tribal place or homeland in Bengal.[75] As these forests were valuable both for timber and trade in other forest products, both district officials and foresters tended to advocate management for 'environmental' and 'stewardship' reasons.[76] This would regulate the trade in sal, timber and a wide range of products including tasar silk cocoons, myrobalans, wax, honey, vegetable dyes and tans, lac, sabai grass, horns of buffalo and deer, ivory, mahua, and *chandani* seeds.[77]

The initial proposals were confined, however, to government leasing the forests, an arrangement that would give the encumbered estate due profits from forest products, without making conservancy conflict with private profits. Dear and Company, and other Calcutta firms, who were already enjoying forest leases from various Singhbhum landlords, were willing to work in Dhalbhum as well.[78] In the absence of any intervention by the estate management, permanent tenure holders in the vicinity of khas jungle were including these forests in areas being leased by them to private firms. The Forest Department was leased 161 square miles out of 438, and the rest was leased to private traders. In doing so the government as manager of the estate was acting to secure the revenues of Dhalbhum and hasten its financial recovery.[79] But this elicited an irate response from the Bengal Conservator who pointed out that any system of forest leases left the proprietor free to give further leases of minerals that would interfere with forest working.[80] Considering the likely abridgement of peasant and villager rights to grazing, small timber, fruits, and flowers, the government was persuaded by the Chotanagpur administration and the Conservator, working in concert but for different ends, to reserve the Dhalbhum forests.[81]

The work of forest settlement in Dhalbhum was taken up in 1893 and completed in 1896. A buffer zone of jungle was left half a mile on each side of forest blocks to permit extension of raiyati cultivation and meet their needs of firewood, small timber, and grazing. Boundary lines were drawn straight to facilitate management. They consisted, as in reserves elsewhere, of ten-foot wide cleared lines, sticks planted in heaps of stones and natural features of sufficient prominence.[82]

Forest settlement was particularly difficult as it involved fixing rights constantly negotiated in local usage. For instance, many holders of privileged tenures, who held them free or on quit rents, had in long periods of possession made considerable encroachments on proprietor's khas, wastelands and forest lands.[83] A large number of claims were filed about these special tenures, but of 119 such claims, only 14 were appealed to the Deputy Commissioner of Singhbhum. As few as two appeals went further and reached the government.[84] In one of these, Birain Majhi and other *pradhans* and raiyats of Dhalbhum petitioned to claim all kinds of customary rights that were being endangered by reservation. The Commissioner, otherwise prone to be a tribal paternalist, found this offensive to the idea of zamindari ownership of wastes and forests. Thus the principle of landed property, that had informed the Permanent Settlement a hundred years prior to this forest settlement, returned to quash tenant and raiyat claims.[85]

Forest settlement worked to restrict the rights of tenants and raiyats so that proprietors could benefit fully from the growing opportunities for timber trade provided by railway expansion.[86] But the railways had also earned the estate large compensation for acquired lands that made possible a plea for release of the estate to its proprietor. By the time the government responded to the petition of Shatrughan Dhal, and restored the estate, 235 square miles of territory in the estate had been reserved as forests.[87] But the Commissioner of Chotanagpur pointed out, deprecating the tone of the Forest Settlement report, that reservation in this case was not aimed at securing revenue as much as it was intended to prevent reckless waste of forest wealth by landlords.[88] In southwest Bengal, forest settlements were caught in the contradictory influences of identifying areas for unfettered forest management generating major produce and strengthening state institutions guarding the rights of peasantry.

Biodiversity, Political Diversity, and Regional Variations

As a regime of restrictions, forest reservation was mainly intended to secure to the colonial government a monopoly over the commercial value of the forests. These restrictions were always being rearranged so that they would make the Forest Department the chief beneficiary of timber trade without paralysing the trade. Thus the private trade of forest tribes in valuable timber was prohibited but supply of forest products for their domestic consumption was free.[89] While petty commerce could be regulated, or generate profits through royalties and tolls, the control of *jhumiyas* (shifting cultivators) was more complicated. They were useful pioneers and the only labour pool for forestry works. The fear was that excessive restriction might lead the frontier tribes people to migrate or resort to resistance through raiding, robbing and murder.[90] Depopulation of forest reserves was seen as a potential calamity because it would deplete the labour pool. This meant that curbing shifting cultivation was not a simple matter.[91] In the Chotanagpur region, the contiguous block of Saranda forests, well located for timber flotation down several rivers, was considered worth the effort to control shifting cultivation. Elsewhere, especially in the Damin-i-Koh, forests were left alone. The scattered distribution of the forests and the potential for conflict with Santhals and Paharias were important considerations influencing this decision.[92]

Similarly restrictions remained tentative in Chittagong government forests, where headmen of timber gangs were permitted to go in and cut what they liked as long as they brought away no timber under a certain girth.[93] In such cases the discourse of frontiers is much in evidence. Lord Ulicke Brown, Commissioner of Chittagong, advocated little forest conservancy in Kuki territory, saying

these people are wild and easily moved one way or another and we are obliged to treat them exceptionally . . . an attempt can be made to induce them to jhum elsewhere, but if they are decidedly opposed . . . I recommend they be allowed to remain where they are.[94]

He saw this as a strategy that would consolidate Kuki settlement and bring peace to the frontier. Government reticence reflected the concern of civil officers like the Assistant Commissioner Daltonganj, who felt that in the face of jhumming restrictions villagers would move to neighbouring tributary mahals like Sirgoojah and Jushpur.[95]

District officers with long years of experience in Chotanagpur, and similar tribal regions, also reminded the government that shifting cultivators not only did the commendable task of extending cultivation, but they rarely did so by denuding forests. They left small blocks of forest in the midst of their clearings, and planted groves, so 'the best cultivated tracts in Chotanagpur are still finely wooded.'[96] Thus these officers generated representations of shifting cultivators as careful husbandmen that opposed strong views expressed by the Conservator, Leeds, that jhumming induced idleness and poverty and its abolition alone would make the wild tribes settled, industrious and useful members of the community.[97] Commending restraint and caution in such areas, the Government of India remarked,

the explorations of Capt. Losack and Mr. Davis have doubtless been of much use in accustoming the wild tribes to the sight of forest officers moving about amongst them which, in an entirely new country, is the first and most important step in the direction of forest conservancy.[98]

The discourse of frontiers thus took on a new valence by slowing and qualifying forest conservancy in the regions where it had first been developed as colonial rule was stabilized.

Bengal foresters toured extensively during the early years of conservative lumbering, performing reconnaissance surveys of forested areas. They found that the definition of valuable species proved difficult. In the jungle mahals, sal, asan, and mahua were used for construction, mango and bombax wood for boxes to transport indigo and shellac, and tamarind wood for carts.[99] From his tour of Bihar districts, Schlich reported that valued species (hardwoods) were widely used for a range of domestic purposes and in the local economy of lac production, silkworm rearing, cutch extraction, dhamar collection and so on.[100] Sal alone had a variety of uses in the tribal agrarian economy. The timber was used for making agricultural implements and putting up fences; sal resin was a medicine, burned as incense, and used for caulking boats; sal butter obtained from the seed was used for cooking and lighting; the seed was also roasted and eaten with the mahua flower (which after the sal was the most widely used food tree); while the sal leaf was both fodder and excellent material for making plates and containers. But these subsistence uses were only one type of impediment to the regime of restrictions imposed by the state on various village-level forest uses.

Another problem was the complex interlocking of forest and subsistence food supply in the jungle mahals and Chotanagpur region, a condition made possible by the considerable biological diversity of the forests and their scattered intermingling with poly-cropped farms.[101] The coexistence of a multiplicity of cropping systems and different ecological zones in a small geographical area characterized eastern India in varying degrees. This biodiversity correlates with diversity in material culture and subsistence strategies that marked most sal forest region of India.[102] Therefore, as in the case of California fisheries during the same historical period, 'as many regulatory efforts emerged . . . as there were strategies for production.'[103] Such biodiversity posed problems for the rationalizing and centralizing tendencies, via reservation and standard forest operations, that were important to conservative lumbering. The diversity made the landscape less legible and hence less amenable to the standardization and control regime that grew out of the regime of restrictions.[104] To combat this problem scientific forestry sought not only to restrict access to lands that were designated as reserved forests. Even where some access or privileges were admitted the degree to which these rights or acknowledged customs could serve the needs of peasant communities involved was gradually diminished by forest degradation or the reduction of biodiversity.[105]

If the confusing demands of a diversified local forest-based economy defied docketing and exclusion in certain parts of Bengal, in others the landscape itself stood in the way. The Sundarbans posed unique problems. These littoral mangrove forests were completely different from forests in the rest of Bengal, thus challenging taxonomic comprehension. They were also infested with man-eating tigers. As a result woodcutters were wary of entering the forests without *fakeers* (ritual specialist mendicants) who would propitiate the spirits of tigers and other wild animals. Naturally forestry operations in this area were dependent on the fakeers, and initially the Bengal Forest Department was satisfied with realizing some revenue through a system of boat tolls and licences.[106] In Assam, Schlich made an initial assessment that noted the scattered nature of sal forests. Since the main revenue potential seemed to lie in India rubber (*Ficus elastica, f. laccifera, f. obtusifolia*), a non-timber forest product, the state government was reluctant to take matters out of the hand of district officers or introduce strict conservancy.[107]

In the Chittagong Hill Tracts, Schlich proposed control of

jhumming and prohibition of boat preparation from *jarul* timber. The district administration went along, because these two restrictions were expected to promote plough cultivation, sedentarization and progressive habits. But they demurred on the important issue of river tolls on timber transport. At various toll stations, on the banks of rivers, hill chiefs were collecting a 5 per cent *ad valorem* duty upon the market value of produce in Chittagong.[108] The district officials felt this was better left with hill chiefs as they could not be materially compensated for loss of dignity involved in being divested of these collections.[109] Thus from the very inception of conservancy, regional variations began to emerge in the regime of restrictions. These variations were marked by the local practices of fakeers in the Sundarbans, hill chiefs in Chittagong, jungle zamindars in southwest Bengal. Partially modified local arrangements created a political space where forest use was negotiated through routines familiar to peasantry, landlords, and traders in forest products; but under novel pressures of state forestry.

Control of Revenue, Control of Land

With forests becoming a state resource, management focused on control over their disposition, and control over all manner of revenues derived from them. Some of the earliest conflicts between foresters and district administration etched out these issues. In Cooch Behar Division, the Assistant Conservator, Darjeeling and the Deputy Commissioner, Mynagorie clashed over cutting sal trees for building roads and bridges.[110] *Joomai* (shifting cultivation) was the dominant mode of farming in Darjeeling, and the Conservator was strongly opposed to the Deputy Commissioner making allotments on patta to extend this form of agriculture. His argument was not only that jhumiyas destroyed virgin forests, but also that pattas given to them precluded the settlement of farmers other than Lepchas and Paharias, who could 'obtain abundant crops by the simplest rotation.'[111]

One of the first things that Anderson, the Bengal Conservator of Forests, had proposed after the creation of the Forest Department was the collection of a whole range of forest revenues by the fledgeling department. These had included royalties on the sale of firewood, bamboo and charcoal; duty on foreign timber; permit fees and rent on jungle privileges from villagers; and grazing and fishing tolls. The argument was that the Forest Department must have unbridled powers to curtail a whole range of subsistence activities that impeded timber production and removal.[112] Many of these revenue demands were not new impositions. In the Duars buffaloes from the Rungpore and adjacent regions would graze in the cold season only after their owners had paid tax based on the number of buffalo grazed. Tibetan and Bhootea graziers would leave their cattle in the lower hills during the dry season after paying tribute in the form of butter to the local landlord.

In opposing the regulation of these forms of forest use by the Forest Department the Commissioner of Cooch Behar, Haughton, raised two important issues. First, that much of the area was devoid of trees. Second, that in the absence of a local officer capable of resolving disputes, frontier populations would be subject to the tyranny of petty officialdom.[113] Haughton also pointed out that much of the timber for house and boat building in east Bengal came from the Duars and Forest Department interference in this trade had doubled prices in two years. District officers were also aware that grazier communities used forests often because they were denied access to village grazing lands. Some of these exclusions had gained validity through the demarcations carried out during survey and settlement operations. Not surprisingly, Haughton insisted that reserved forests be clearly demarcated and the Forest Department be strictly confined to them.[114] Demarcation, as a technology of forest conservancy, was available not only to assert Forest Department control. It equally served as a means of restricting the Forest Department while participating in the wider processes of sorting out the rural landscape into the domains of distinct social groups in agrarian society.[115]

The Bengal Conservator, Leeds, could not carry out the demarcations and insisted that sal and sissoo forests of the Duars be placed under the Forest Department, based on the survey by Assistant Conservator, Gustav Mann. In 1869, he could only remonstrate that four years of attempted conservancy had not regulated private fellings and those organized under civil authority.[116] Seeking to accommodate the district officers, Leeds suggested that all local needs of sal and sissoo timber would be met free or according to past rule by the local forest officer in consultation with the civil officer.[117] Haughton responded by disputing the ability of any agency to determine the existence of sissoo forests. Sissoo, he noted, grew naturally on the banks of rivers, and in the Duars existed mainly as seedlings and saplings that he had seen in his marches through various forest areas where regeneration was prevalent. His explanation for lack of timber was that periodic flooding of the rivers swept away the seedlings, clearly a situation where what appeared to be a promising young grove one year could be prostrated the next.[118] As Haughton put it in his letter to the Conservator, using a colourful turn of phrase

your departmental officers would no doubt on seeing these patches of healthy thriving trees, desire that they might be made over to the department, but this would be useless. They could as easily conserve a flight of swallows.[119]

Astutely discerning the forester's claim to the landscape as one based on the fixity of certain defining features—the predominance of sal and sissoo regeneration—Haughton emphasized in contrast the impermanence and transience of these very features. He wrote:

it must be borne in mind that every acre of ground (in the Duars) left out of cultivation for a few months becomes overgrown with jungle . . . the Mech tribe from time immemorial have occupied the forests at the foot of the hills cutting the jungle, cultivating the soil with cotton for two or three years and abandoning it for nine to thirty years.[120]

Haughton noted further that villagers burned the grass in the sal forests for fresh pasturage and to assist sal regeneration.[121] Both by precipitating initial dieback and later scorching multiple shoots, fire was believed to promote strong single-stem sal regeneration. It was acknowledged to eliminate forest pests.[122] He thus reiterated the case for demarcating reserves but also stressed the intimate local knowledge that such an operation called for, asserting that the foresters did not possess it.

The exchanges between Colonel Haughton, Commissioner of Cooch Behar, and the Conservator of Forests stemmed from the politics of forest conservancy in the Duars, but the issues highlighted there gained wider currency as the amendments to the Indian Forest Act of 1865 were formulated.[123] In Bengal Haughton resumed his vigorous critique of arbitrary powers in the hands of

foresters by objecting to provisions of the draft legislation that
would 'give the Conservator a claim to the disposal of every tree
. . . the ownership of which was doubtful.' He also pointed out
that many villagers on the banks of the Teesta made their living
by the collection and sale of driftwood. This livelihood was
jeopardized by a provision that required them to establish title to
every individual piece of wood they picked up.[124] Hardship for
villagers gathering fuelwood, thatch grass, and fodder was anticipated by other Bengal divisions responding to the draft legislation.[125]
The added apprehension of civil officers was that conservancy
would not only infringe rights and customary uses of the forests
by villagers, but the manner of doing so would be determined and
implemented without their supervision while they remained
responsible for managing the consequent unrest.[126]

When placing the consolidated view of the Bengal government
before the Government of India, Rivers Thompson, the Revenue
Secretary, emphasized the need for recognizing and providing for
prescriptive rights and customs. He also pointed out that district
officers possessed detailed knowledge of land tenures and customs
which they used to balance the demands of revenue and welfare.[127]
Two decades later, the Commissioner struck the same chord in
his protracted opposition to de-localized forest management in
Chotanagpur:

much has been done by the exercise of personal influence, and much
can still be done, but the Bengal Nagpur railway will naturally tend to
lessen the force of this influence by gradually opening out this country
to outsiders, who having no interest in the country, for the sake of
present gain . . . will lay waste the jungles . . . .[128]

He listed tasks like preservation of forests in political states, leasing
of forests in encumbered estates, mining concessions and regulation
of timber markets in the vicinity of private forests as best left to
the locally knowledgeable civil officer suitably advised on technical
matters by a locally stationed forester.[129]

These controversies created patterned variations in the regime
of restrictions throughout the phase of conservative lumbering,
highlighting the 'conflicting and irreconcilable opinions . . . held
by forest and civil departments'.[130] At one point the civil officers
seemed to gain the upper hand as the government framed rules

that placed forest officials under the district magistrate, stating, 'all distinctions and practices which are likely to encourage the impression that forest work lies outside the ordinary duties of land revenue officials should be gradually eliminated.'[131] None the less, the civil officers mostly fought a losing battle against the creation of an 'abstract space' of forest territories in Bengal.[132]

My purpose is not to affirm the sharp, overdrawn contrast between free forest use and conservancy that has spilled from superficial readings of such colonial intra-bureaucratic conflicts into the polemics of anti-state, environmental, and tribal rights movements in the last two decades. I want, on the contrary, to point out a long-standing tension between localized forest management arrangements and central control, and to examine the conditions that privileged one or the other form of governance in any place and time.[133] For a protected forest administration that elevated the interests of tenants over the landlord worked by stifling the voice of the forestry expert, by blending its universal tunes to a local rhythm. In practical terms this meant disagreements over the control of subordinate forest officials and the frustration of forest working schemes.[134] Collectively these processes generated local variations in the regime of restrictions that favoured particularized arrangements, and promoted the idea of autonomous places—zones of anomaly—in the standard territory of forest conservancy.

Protected Forests and Limited Conservancy in Southwest Bengal

In a fine study of local agrarian relations and the extent to which British rule was dependent in a Bihar zamindari on local collaborators in the landed elite, Anand Yang has proposed the notion of a 'limited Raj'. He thus parallels and particularizes for eastern India, similar arguments made for north India by Tom Metcalf and for south India by David Ludden.[135] In some ways colonial forest management, as it developed through the later part of the nineteenth century, was an ambitious experiment in direct administration that went beyond anything attempted in the agrarian sector. Protected forests in Bengal signified, in that context, a 'limited Raj'.

In 1894 all wastelands which were the property of the government in Chotanagpur Division were declared protected forests.[136]

As the Porahat Estate was then under government management, all wastelands there were constituted into protected forests. The Mundas of the northern part of the estate objected strongly as they felt that this move was an abrogation of the agreements made when the Porahat reserves were created. So the civil administration hastened to clarify that the Protected Forest Rules would apply only to certain marked plots of wastelands, and were not based on territorial exclusion in the same way that reserves were.[137] In other wastelands the nominal influence of the Forest Department was limited to sale of dead wood certified by the headman as not needed by villagers. But when settlement operations began the proprietor of the estate requested a demarcation of the protected forests and release of other wastes into his charge. He filed an application under section 38 of the Indian Forest Act, volunteering certain forest areas of the estate for Forest Department management.[138]

Initially twenty-five large blocks, 246 small blocks and residual waste were categorized. But later the government confined its interest to the large blocks, an area of 34 square miles, leaving the rest, an area of 125 square miles to the proprietor, but subject to record of raiyati rights.[139] Unconvinced Mundas and raiyats still opposed demarcation and obstructed it in a variety of ways. Settlement pillars, used to orient demarcation, were often concealed or rendered irrelevant by the extension of cultivation disregarding them. The opposition was sharp because the protected forests were being carved out of what after the reservation process had been regarded as bahar jungle (outside forests), and the Mundas found this to be both a breach of earlier understandings, and a repudiation of their role in forest protection.[140] Many old jhums were taken as protected forests on the principle that they had not been cultivated since the last settlement. Mundas actively opposed this by intimidating amins (land surveyors) who were non-local since the local ones had not been willing to work. Police camps had to fortify the operations in many places.

The Protected Forest Rules therefore had to recognize the legitimacy of clearing wastelands for bona fide cultivation or face opposition. The Forest Department was gradually confined to preventing forest destruction by 'outsiders'. These rules thus became another instrument in the hands of the local administration to devise the paternalist protection of tribals. In other words, the rules were subordinated in Chotanagpur to the project of defining and administering zones of anomaly, only now it was done in the context of forest conservancy.[141] Another feature of the protected forests was the division into areas managed by the Forest Department and those managed by the district officer.

As protected forests came up during the same period that cadastral survey and settlement operations took place in the Chotanagpur region, the settlements became another procedural opportunity for raiyats to leave their impress on the management of protected forests. In the Porahat protected forests, villagers had effectively permitted only trees ringed by them to be cut as timber, preventing the felling of green trees. This was duly recorded by the Settlement Officer.[142] In the Santhal Parganas protected forests the Settlement Officer permitted villagers to lop reserved trees in certain blocks, recognizing customary usage that reduced Forest Department control to a token proscription.[143] Such attenuation of Forest Department control in the protected forests led the Bengal Conservator to conclude:

other than Sundarbans, none of the protected forests in Bengal can be said to belong wholly to the government. All of them are subject to exercise of numerous privileges or concessions under such conditions that privilege holders must for all practical purposes be regarded as part owners.[144]

Thus the protected forests limited the processes of internal territorialization through which forest management was introduced in Bengal. But through the local variations permitted in protected forest rules this partial territorialization was also regionally diversified. If protected forests defined a general zone of anomaly in late-nineteenth-century patterns of forest conservancy, this was further particularized in specific tracts like Chotanagpur, where extra privileges were given to raiyats to cut down certain reserved trees. Such specially licensed areas in southwest Bengal emerged in the regions already subject to administrative exceptionalism there. They were produced by the combination of sustained resistance from the local populace and the discourse of frontiers.[145]

Notes

1. WBSA P/555 BRP (For) Aug 1898, A progs 20–48, file 4-F/1, no. 1812R, Bhagalpur, 17 Aug 1897, W.B. Oldham, Commr Bhagalpur, to Secy GOB Rev, p. 1978.

2. Fernow 1907: 6.

3. On plantations, Stebbing (1922: 202) cites only the report of R. Baird Smith in *Calcutta Review*, 1847, no. 23, regarding plantations on western and eastern Yamuna canals started in 1821 and 1831. I have discussed, in the previous chapter, Bengal teak plantations of even earlier dates.

4. I refer here to the botanical and other landscape surveys carried out by Nathaniel Wallich, Francis Buchanan-Hamilton, J.D. Hooker, W. Jackson, among others; the work of the fledgeling Geological Survey; the Great Trigonometrical Survey, and so on.

5. Rangarajan (1998) shows how, in this pre-territorial stage of colonial forestry in Central Provinces, the protection of timber trees was more decisive than half measures for establishing government control of timber-bearing land.

6. McEvoy 1986: 118.

7. Stebbing 1926: 213–16.

8. Bryant 1994a: 161–3.

9. Ribbentrop 1900: 62.

10. Freeman 1994: 173–8; Sahlins 1994.

11. Amery 1876: 31.

12. Ibid.: 30.

13. Sahlins (1989) provides an elegant study of the process of territorial nation-state formation. Benedict Anderson (1991a: ch. 10) discusses the role of cartography in this process. Jessop 1990; Mann 1986; Sack 1986; Soja 1989; Tilly 1992.

14. Baigent and Kain (1984) provide a worldwide appraisal of the rise of cartography and cadastral surveys, and their becoming, by the nineteenth century, axiomatic to state control of land. Harley (1992: 231–48) has discussed mapping from the perspective of cultural critique. Peluso (1995) provides a survey of recent writing on the politics of mapping.

15. Vandergeest and Peluso 1995.

16. Ibid.: 387–90, 401.

17. Ibid.: 402–14.

18. Peluso 1993 and 1992 are successful analyses in these terms.

19. Describing the development of nineteenth-century national forest policies in China one scholar says, 'the focus shifted from land to trees . . . inclusion gave way to exclusion with the introduction of legislation establishing forest reserves', putting things in terms of inclusion and exclusion, but fails to note that forest reservation is a territorial form of exclusion. Menzies 1992: 730.

20. Smith 1994: 491. This also recognizes what environmental historians have called nature's autonomous place in history, without resorting to ecological determinism. See Cronon 1990, 1991, 1992; Crosby 1986; White 1990; Worster 1985, 1990; Merchant 1989.

21. There is a fast burgeoning literature in this genre. Some notable samples would include, Barnes and Duncan 1992; Cosgrove 1993; Daniels 1993; Cosgrove and Daniels 1988; Duncan 1990; Duncan and Ley 1993.

22. Totman 1989: 88–92.

23. Demeritt 1994: 165. For nature as a motor of history, see Cronon 1990; Worster 1988: 289–307; and Merchant 1989; among others. For a pithy critique stressing the power/knowledge relations enabled by the material and discursive preservation of nature's essential reality, see Latour 1993.

24. Ramachandra Guha 1990, 1989; Rangarajan 1992, 1994; Skaria 1998; Haeuber 1993; Richards and Flint 1990: 25–7.

25. Corbridge 1993: 131.

26. Some interesting work in this direction is emerging from African studies. See Fairhead and Leach 1995; Millington 1989: 229–48; and most recently the excellent collection on Zimbabwe: Grove and McGregor 1995.

27. OIOC P/432/75 BRP (For), Jan–Apr 1869, A progs 11, Feb 1869, Col E.T. Dalton, Commr Chotanagpur, to H.L. Harrison, Junior Secy GOB, no. 69T, Camp Keonjhar, 27 Jul 1868, p. 3.

28. Shifting cultivation and the hesitation in dealing with it in Bengal are discussed later in this chapter. Girdling was engaged in a protracted struggle where village headmen were unsuccessfully recruited to the cause of the administration. See OIOC P/432/76 BRP (For) May–Aug 1869, A progs 6, Jun 1869, Col E.T. Dalton, Commr Chotanagpur Div, to US GOB, no. 1426, Chotanagpur, 19 May 1869. In different regions, controversy arose about girdling as it was found useful to season teak before felling and flotation, while girdled sal became vulnerable to pest attack. See Stebbing 1924: 26–30. Control of fire and grazing, where similar methods were adopted, is taken up in the next chapter because they were incorporated into debates about the regeneration of sal.

29. Ramachandra Guha 1990.

30. Stebbing 1924: 14–16.

31. OIOC P/232 BRP (For) May–Aug 1872, A progs 10–11, Jul 1872, A.O. Hume, Secy GOI Agri, Rev and Commerce, to Secy GOB Rev, no. 8/742, Simla 28 Jun 1872, pp. 13–14; C. Bernard, Secy GOB Rev, to All Commrs and CF Bengal, nos. 39 and 2971, Cal, 22 Jul 1872. For a discussion of similar conflicts elsewhere in India during this period, see Rangarajan 1992; Skaria 1992: chs 8 and 9.

32. Similar disputes in Africa and their influence on the policy process are documented by Millington 1989: 234–5; David Anderson 1989: 250–62.

33. See Stebbing (1924: 13–14) for a lengthy complaint, about the misperception of foresters by civil officers, that conveys this sense clearly.

34. Vandergeest and Peluso (1995: 409–11) report similar conflicts between the Ministries of Interior and Forestry in Thailand.

35. BFAR 1866-7 to 1868-9.

36. Grieve 1912: 15; Ribbentrop 1900: 214.

37. OIOC P/432/76 BRP (For) May–Aug 1869, A progs 9, Jul 1869, H. Leeds, CF LP, to Secy GOB Rev no. 57A, Cal, 5 Feb 1869. p. 15.

38. Trafford 1905: 8; Hatt 1905: 8.

39. BFAR, 1870, pp. 9–10.

40. OIOC W3022, William Schlich, *Memorandum on the Management of the Forests in the Darjeeling Division, Bengal* (Simla: GOI Press, 1882), pp. 7–10.

41. WBSRR Rev (For), A progs 1–4, Feb 1908, file 9-R/1, no. 141, Darjeeling 27 Sep 1907, CF Bengal to Chief Secy GOB, pp. 91–2; Singhbhum *DG*, p. 103.

42. Hatt 1905: 5.

43. OIOC P/432/75 BRP (For) Jan–Apr 1869, A progs 1, Apr 1869, H. Leeds, CF Bengal, to Secy GOB Rev, no. 33A, Darjeeling, 13 Oct 1868, p. 12.

44. OIOC P/228 BRP (For) May–Aug 1870, A progs 25, Jun 1870, H. Leeds, CF Bengal, to Secy GOB Rev, no. 840, Cal, 28 Feb 1871, p. 19.

45. The rate of arable expansion and its impact on forest and forest fringe areas has been estimated recently in large statistical studies. See Richards et al. 1985: 705–11; and Richards and Flint 1990. Agrarian historians further distinguish between arable expansion at the extensive and intensive margins, both of which continued in southwest Bengal into the first decades of the twentieth century. See Prabhu Mohapatra 1985: 272–9; Bose 1993: 27–8. Though the restrictions placed on the extensive margin by forest conservancy meant that agricultural land was prepared mostly at the expense of the forest-savannah transition zone, having the most adverse effect on Santhals, Hos, Mundas, Oraons and Kharias, the chief occupants of that zone.

46. Behal and Mohapatra 1992; Mohapatra 1985; Lalitha Chakravarty 1978; Bose 1993; Prakash 1992a; Dasgupta 1986; Bates and Carter 1992; Kerr 1983; Robb 1993.

47. Behal and Mohapatra 1992: 146–53.

48. Computed from figures provided by Corbridge (1993: 140–1) based on *Annual Reports on Labour Immigration into Assam* for relevant years.

49. Engels 1993: 225–7.

50. Mohapatra 1985: 259–63; Prakash 1992a: 31; Ghosh 1994; Dasgupta 1981.

51. Push factors related to land tenure, ecological fragility and commercial agriculture are discussed by Schwerin (1978: 26–35); Lalitha Chakravarty (1978); Prakash (1992: 28–9); Mohapatra (1985: 271–98). The pull of demand and coercion, especially, are discussed by Behal and Mohapatra (1992); Bates and Carter (1992, 1993: 159–85); Robb (1993: 17–21); Bhowmick (1981); Dasgupta (1981).

52. Mohapatra 1985: 248–50.

53. Bose 1993; Bates and Carter 1992: 221–3; Richards and McAlpin 1983.

54. Hill 1991: 269; Yang 1989: 181–205.

55. WBSRR Rev (For), A progs 6–10, Apr 1905, file 10R/1 (1–3), no. 43, Darjeeling, 4 May 1904, A.L. McIntire, CF Bengal, to Secy GOB Rev, p. 1. OIOC P/9798 BORP (For), Apr–Jun 1915, A progs 1–3, Jun 1915, file IIIF/79 of 1915, no. 610-74, Ranchi, 17 May 1915, H.H. Haines, CF B&O, to Secy GOBO Rev, p. 1.

56. WBSRR Rev (For), A progs 6–10, Apr 1905, file 10R/1 (1-3), no. 320, Darjeeling 26 Mar 1904, A.L. McIntire, CF Bengal, to DC Jalpaiguri; no. 74G, Jalpaiguri 11 Apr 1904, R.G. Kilby, DC Jalpaiguri, to CF Bengal; no. 1276, Cal 8 Mar 1905, A. Earle, Secy GOB Rev, to CF Bengal, pp. 2–5; Grieve 1912: 16.

57. OIOC P/8414 BRP (For), Apr–Sep 1910, A progs 6–8, Aug 1910, file 10R/3, no. 126, Chaibassa 2 May 1910, J.C. Baker, DCF Singhbhum, to DC Singhbhum; no. 2301, Cal, 17 Aug 1910, W.R. Gourley, Offg Secy GOB Rev, to Commr Chotanagpur, pp. 59–62.

58. WBSRR Rev (For), B progs 9–10, Mar 1917, file AT/2, no. 17C-2WP, 28 Feb 1917, G.S. Hart, IGF, to Secy GOB Rev, pp. 3–4.

59. Bryant 1994a: 164; Bryant (1994b) discusses territorialization more directly. In Burma teak forests, regeneration through taungya whittled control in another way.

60. NAI, PWD (Rev-For), Aug 1866, A progs 46–50; Jul 1867, A progs 25–33; Dec 1868, A progs 24-9; Oct 1869, A progs 2–4, 18–22; May 1870, A progs 1–4, 41–4; Feb 1871, A progs 62–8, notification no. 13F, 16 Feb 1871, pp. 2–5.

61. On a wider comparative scale, Skocpol (1985) re-emphasized the importance of studying the structures and activities of the state closely. This has initiated a spate of re-examinations, some of the most creative of which may be sampled in Migdal et al. (1994). A comprehensive review may be found in Nugent (1994). I counterpose this literature here to suggest that statemaking is a process that bears the impress of social forces precisely because the state is a heterogeneous entity that is not clearly differentiated or always distinguished from society.

62. Humphreys (1985: 251-9) makes this case well at a theoretical level.

63. WBSA P/411 BRP (For) Jul 1887, A progs 129_32, head IV, col 1, IGF's memo, 27 Apr 1887, p. 11.

64. Ibid., no. 1T-R, Darjeeling, 15 Jun 1887, P. Nolan, Secy GOB Rev, to all Commrs, p. 21.

65. See Singha (1990) for a similar discussion of the process through which the Indian Penal Code was developed.

66. This was done by recording that no rights existed in these forests at the time of reservation. A standard technique of dispossession that may be seen at work in all the early north Bengal working plans.

67. OIOC P/3178 BRP (For) Jun–Aug 1888, A progs 147–54, Aug 1888, head I, col 2, no. 77S, Cheybassa, 6 May 1887, Jt FSOs Porahat, to DC Singhbhoom, p. 41.

68. OIOC P/2803 BRP (For) Aug–Oct 1886, A progs 142-3, Oct 1886, head I, col 2, no. 263, Cheybassa, 3 May 1886, Lieut Col W.L. Samuels, DC Singhbhoom, to Commr Chotanagpur, p. 42.

69. Ibid., p. 42.

70. OIOC P/2937 BRP (For) Aug–Sep 1887, A progs 132, Sep 1887, head I, col 2, no. 273C, Ranchi, 10 Aug 1887, F.B. Manson, DCF, to CF Bengal, p. 80.

71. OIOC P/3178 BRP (For) Jun–Aug 1888, A progs 147–54, Aug 1888, head I, col 2, no. 246R, Chyebassa, 5 Aug 1887, Major H. Boileau, DC Singhbhoom, to Commr, Chotanagpur, p. 35.

72. Ibid., no. 1099R, Ranchi, 18 Nov 1887, C.C. Stevens, Commr Chotanagpur, to Secy GOB Rev, p. 35.

73. O'Malley 1910b: 102.

74. OIOC P/3178 BRP (For) Jun–Aug 1888, A progs 147–54, Aug 1888, head I, col 2, no. 77S, Cheybassa, 6 May 1887, Jt FSOs Porahat, to DC Singhbhoom, pp. 39–40.

75. WBSA P/564 BRP (LR) April 1899, progs 61–73. file 4E/1, no. 192TW, Camp Purulia, 16 Aug 1898, A. Forbes, Commr Chotanagpur, to Secy BOR, LP, pp. 887–900. The making of tribal places in southwest Bengal has been discussed in the preceding chapter.

76. WBSA P/473 BRP (For) Sep 1892, A progs 7–26, file 4F/1(1-19), no. 905W, Chotanagpur, 6 Nov 1888, C.C. Stevens, Commr Chotanagpur, to Secy GOB Rev, pp. 79–80.

77. Ibid., p. 81.

78. Ibid., no. 1008W, Ranchi, 28 Oct 1890, W.H. Grimley, Commr Chotanagpur, to Secy GOB Rev, p. 95.

79. Ibid., p. 96.

80. Ibid., no. 190DF-CN, Darjeeling, 2 Jan 1891, E.F. Dansey, Offg CF Bengal, to GOB, p. 98.

81. Ibid., no. 373T-W, Camp Hazaribagh, 17 Mar 1892, W.H. Grimley, Commr Chotanagpur, to Secy GOB Rev, pp. 105–6.

82. Brandis had stressed the importance of straight lines in his first inspections of Bengal forests. See D. Brandis, *Suggestions Regarding the Management of the Forests in the Jalpaiguri and Darjeeling Districts* (Calcutta: Home, Rev and Agri Dept Press, 1881), p. 5. This was faithfully executed in many north Bengal forest demarcations. See Trafford 1905: 3; Tinne 1907: 1–2.

83. WBSA P/482 BRP (For) Jun 1893, A progs 23–30, file 4F/1-1-8, no. 308TR, Camp Pathargaon, 7 Feb 1893, W.H. Grimley, Commr Chotanagpur, to Secy GOB Rev, pp. 185–6.

84. WBSA P/571 BRP (For) Nov 1899, A progs 4–5, file 4F/2-1, res no. 4502 For, 6 Nov 1893 GOB Rev; no. 1386W, Ranchi, 30 Sep 1899, J.G. Ritchie, Offg Commr Chotanagpur, to Secy GOB Rev, pp. 225–9.

85. WBSA P/518 BRP (For) May 1896, A progs 1–5 file 8P/4, no. 2074W, Ranchi, 31 Mar 1896, W.H. Grimley, Commr Chotanagpur, to Secy GOB Rev, pp. 115–17; note of Nazimuddeen, FSO, p. 120. For interesting south Indian parallels presented in terms of the conflict between state and village spaces, see Murali 1995: 102–18.

86. WBSA P/571 BRP (For) Nov 1899, A progs 4–5, file 4F/2-1, no. 1386W, Ranchi, 30 Sep 1899, J.G. Ritchie, Offg Commr Chotanagpur, to Secy GOB Rev, p. 231.

87. WBSA P/551 BRP (LR) Jun 1898, progs. 1–5, file 4-E/5-1-3, no. 32W, Cal, 2 Apr 1898, E.W. Collin, Offg Secy BOR LP, to Secy GOB Rev, pp. 1299–1306; WBSA P/564 BRP (LR) Apr 1899, progs 61–73, file 4E/1,

no. 1028, Cal, 14 Mar 1899, M. Finucane, Secy GOB Rev, to Secy BOR, LP, p. 905.

88. WBSA P/487 BRP (For) Nov 1893, A progs 3–7, file 4-F/1-9, no. 597W, Ranchi, 31 Jul 1893, W.H. Grimley, Commr Chotanagpur, to Secy GOB Rev, p. 129–36.

89. NAI, Rev, Agri and Commerce (For), Sep 1817, A progs 39–43, circular no. 316–22, 19 Sep 1871, GOI to GOB, NWP, Punjab and CP.

90. NAI, Rev, Agri and Commerce (For), Oct 1871, A progs 13–15, Res by GOB on CF's report on Chotanagpur forests; district officers in Bengal and elsewhere were acutely aware that forest tribes intimately acquainted with the forests they were being demarcated out of could easily enter and use these forests in defiance of rules. OIOC P/432/68 BRP (For), Apr–May 1867, A progs 17, May 1867, W. Mcleery, Asst Secy GOI PWD, to Secy GOB, no. 7, 22 Apr 1867, pp. 15–17.

91. NAI, Rev, Agri and Commerce (For), Sep 1871, A progs 39–43, circular no. 316-22, 19 Sep 1871, GOI to GOB, NWP, Punjab and CP. For a similar point regarding south India, see Pouchepadass 1995: 139–45.

92. NAI, Rev, Agri and Commerce (For), Oct 1871, A progs 13–15, no. 234, 31 Jul 1871, H. Leeds, CF Bengal, to Secy GOB Rev, p. 9; OIOC P/229 BRP (For) Sep–Dec 1871, A progs 18, Sep 1871, GOB res dated 26 Sep 1871, pp. 59–64.

93. Ibid., res by GOB on CF's report on Chotanagpur forests.

94. OIOC P/227 BRP (For) Jan-Apr 1871, A progs 10 Jan 1871, Lord Ulicke Brown, Commr Chittagong, to Secy GOB Rev, no. 290, Chittagong, 28 Nov 1870, p. 11.

95. NAI, Rev, Agri and Commerce (For), Oct 1871, A progs 13–15, no. 234, 31 Jul 1871, H. Leeds, CF Bengal, to Secy GOB Rev; OIOC P/243 BRC (For), 1873–75, A progs Jul 1873, file 23-1/2, no. 485, Jalpaiguri, 29 Mar 1873, Capt. C.W. Losack, DCF Cooch Behar, to CF Bengal, p. 15.

96. OIOC P/243 BRC (For), 1873-5, A progs 18, Mar 1873, no. 215, 20 Feb 1873, Col E.T. Dalton, Commr Chotanagpur, to Secy GOB Rev, pp. 113–14.

97. Ibid., p. 5; of course even foresters could not help noticing, after living close to these tribes in the forests, that jhum supported a frugal and modest lifestyle that they often had to imitate, living in bamboo huts and hunting for food. See, Stebbing (1930) for a more personal and sympathetic account of jhumiyas in the Chittagong Hill Tracts, than we get from his official four volume history of colonial forestry in India.

98. NAI, Rev, Agri and Commerce (For), Oct 1871, A progs 13–15, no. 462, Simla, 11 Oct 1871, A.O. Hume, Secy GOI, to Secy GOB Rev.

99. OIOC P/243 BRC (For), 1873-5, A progs, Finance Dept, Mar 1875, col 3-3, no. 111, Burdwan, 15 Jun 1874, C.T. Buckland, Commr Burdwan, to Secy GOB Finance, pp. 1–10.

100. OIOC P/243 BRC (For), 1873-5, file 23-2/3, no. 70c, Gowhatty, 7 Jun 1873, W. Schlich, CF Bengal, to Secy GOB Rev, pp. 22–40.

101. For details see Sivaramakrishnan 1997.

102. Similar findings in Central Provinces are reported by Rangarajan (1992) and Prasad (1994). The central and eastern Indian mixed deciduous forest area being quite different from the Himalaya or the rain forests of the Western Ghats in south India gives us an ecological basis for findings that differ from those of Ramachandra Guha (1989).

103. McEvoy 1986: 101.

104. I use the expression legibility following Holston (1989) and his discussion of the way the planning for the modernist city eliminates private spaces and other irregular patterns of social organization and spatial arrangement. This is done with a view to promote and facilitate central control and intervention in the daily life of the citizenry being accompanied by a rhetoric of public health, community, enhancing light, welfare and other collective goods. The parallels with modern state forestry and its vision of the managed forest as the source of natural and physical health, pleasant views, climate moderation and so on, are striking. For a wider comparative discussion of the pursuit of legibility as a basic aspect of the modernist state, see Scott 1998.

105. The programme of scientific forestry to raise pure stands of commercially desired hardwoods is discussed in detail when the next chapter takes up the struggles over sal regeneration in Bengal.

106. OIOC P/243 BRC (For) 1873–5, A progs, Sept 1873, file 30/2, no. 15c, A.L. Home, DCF, to CF Bengal, pp. 83–99.

107. Ibid., file 3-2/3, memo no. 94c, Dibrugarh, 10 Jul 1873, W. Schlich, CF Bengal, to GOB, pp. 103–30; no. 2819, Cal, 19 Sep 1873, Secy GOB Rev, to CF Bengal, p. 132.

108. WBSA P/482 BRP (For) Jun 1893, A progs 1–4, file 5B/3, no. 40M-G, Darjeeling, 26 Apr 1893, E.G. Chester, Offg CF Bengal, to Secy GOB Rev, p. 155.

109. OIOC P/243 BRC (For) 1873–5, A progs Jun 1875, Finance Dept, col. 2-26/27, no. 265c, 15 Feb 1875, W. Schlich, CF Bengal, to GOB; no. 81f, Chittagong, 25 Feb 1875, R.C. Mangles, Commr Chittagong Div, to Secy GOB Finance, pp. 39–47.

110. OIOC P/432/67 BRP (For), Mar 1867, A progs 6, Lieut Col J.C. Houghton, Commr Cooch Behar, to Secy GOB, no. 208T, 2 Mar 1867.

111. OIOC P/432/68 BRP (For), Apr–May 1867, A progs 14–16, Apr 1867, T. Anderson, CF LP, to Secy GOB, no. 77, 17 Dec 1866; ACF Sikkim to CF LP, no. 98, 10 Dec 1866, pp. 24–5.

112. OIOC P/432/63 BRP (For), Aug–Oct 1866, A progs 1, Oct 1866, R.B. Chapman, Secy BOR LP, to Secy GOB, no. 762A, 18 Jul 1866, T. Anderson, CF, to A. Money, Commr Bhagalpur, no. 16, 25 Jun 1866, pp. 1–3; OIOC P/432/70 BRP (For), Sep–Oct 1867, A progs 14, Lieut Col J.C. Haughton, Commr Cooch Behar, to Secy BOR, LP no. 541, 19 Mar 1867, pp. 9–10.

113. OIOC P/432/70 BRP (For), Sep–Oct 1867, A progs 14, Haughton to Secy BOR, LP no. 541, 19 Mar 1867, pp. 10–11.

114. OIOC P/432/68 BRP (For), Apr–May 1867, A Progs 17, May 1867, W. Mcleery, Asst Secy GOI PWD, to Secy GOB, no. 7, 22 Apr 1867, enclosing

remarks of Major Waddington, Supt revenue survey and assessment, Thana and Rutnagherry, on demarcation of forests in the Thana Collectorate; OIOC P/432/70 BRP (For), Sep–Oct 1867, A progs 14, Haughton to Secy BOR, LP no. 541, 19 Mar 1867, pp. 10–11.

115. See Skaria (1998) for a limited discussion of demarcation as a tool of scientific forestry in western India. I am suggesting that demarcation had a more complex and changing relationship to forest conservancy.

116. OIOC P/228 BRP (For) May–Aug 1871, A progs 1–12, Jun 1870, H. Leeds, CF LP, to Commr Cooch Behar, no. 197B, Gowhatty 14 Jan 1869, p. 5.

117. Ibid.

118. Ibid., Col J.C. Haughton, Commr Cooch Behar, to CF Bengal LP, no. 22, Julpigoree 5 Feb 1871, p. 7.

119. Ibid.

120. Ibid., Haughton to Offg US GOB, no. 657, Julpigoree, 15 Mar 1869, p. 9.

121. Ibid., A progs 8, Jul 1871, Haughton to US GOB Rev, no. 174, Julpigoree, 19 Jan 1871, p. 8. Foresters were most dismissive of civil officers and their allusions to local farmer knowledge relating to things like fire and sal regeneration. The story of their view of fire and its consequences is taken up in the next chapter.

122. Ibid.

123. See Ramachandra Guha (1990); and Rangarajan (1994) for a discussion of the national debates about the making of Act VII of 1878, Indian Forest Act.

124. OIOC P/228 BRP (For) May–Aug 1870, A progs 8, Jul 1871, Col J.C. Haughton, Commr Cooch Behar, to US GOB Rev, no. 174, Julpigoree, 19 Jan 1871, pp. 7–8.

125. Ibid., A progs 12–13, Jul 1871, Col Henry Hopkinson, Commr Assam, to Secy GOB Rev, no. 46T, 14 Apr 1871.

126. Ibid., A progs 14, Jul 1871, Col E.T. Dalton, Commr Chotanagpur, to US GOB Rev, no. 902, Chotanagpur, 24 Apr 1871, p. 19.

127. Ibid., A progs 16, Jul 1871, Rivers Thompson, Offg Secy GOB Rev, to Offg Secy GOI, Agri, Rev and Commerce, no. 2640, Fort William, 17 Jul 1871, pp. 21–3.

128. OIOC P/3871 BRP (For) May–Jul 1891, A progs 98–121, Jul 1891, head IV, col 2, no. 74R, Ranchi, 15 Apr 1891, W.H. Grimley, Commr Chotanagpur, to Secy GOB Rev, pp. 23–4.

129. Ibid., p. 24.

130. Ibid., no. 453(2) For, Cal, 6 Aug 1891, W. Maude, US GOB Rev, to all Commrs and CF Bengal, p. 14.

131. OIOC P/3873 BRP (For) Nov–Dec 1891, A progs 66–78, Nov 1891, file 18R/21, no. 1074-152 For, Cal 21 Mar 1890, H.W.C. Carnduff, US GOB Rev, to Commr Presidency Div, p. 47.

132. Ibid., no. 5RL, Cal, 14 May 1890, A. Smith, Commr Presidency Div, to Secy GOB Rev, p. 49. The Santhal Parganas, where forests were reserved

and de-reserved within two years, marked an exception to this trend. This case is fully discussed in Sivaramakrishnan 1997: 90–6.

133. In a similar vein, discussing the inability of states or peasantry to impose their will, Totman (1989: 41) shows how various shared forestry arrangements emerged in Japan during the fourteenth through sixteenth century.

134. As happened in Singhbhum, Manbhum and Ranchi. OIOC P/10102 BORP (For), 1917, A progs 3–32, Feb 1917, file IIIF/2 of 1917, no. 541R-XIV/4, Ranchi, 21 Feb 1916, E.H.C. Walsh, Commr Chotanagpur, to Secy GOBO Rev; no. C-114-XXIV/2, Ranchi, 20 Mar 1916, H.H. Haines, CF B&O, to Secy GOBO Rev, pp. 5–9; A progs 1–9, Apr 1917, file IIIF/3 of 1917, no. C-106, Camp, 17 Mar 1916, H.H. Haines, CF B&O, to Secy GOBO Rev; no. 1753-R, Chaibassa, 26 Jul 1916, M.G. Hallett, DC Singhbhum, to Commr Chotanagpur; no. 4366-XXIV-P-1, Camp, 30 Dec 1916, H.H. Haines, CF B&O, to Commr Chotanagpur, pp. 2–8.

135. Yang 1989; Ludden 1989; Metcalf 1979.

136. OIOC P/7573 BRP (For), May–Jul 1907, A progs 1–12, May 1907, file 19-P/3-4 of 1906, no. 3126T-R, Darjeeling 26 Oct 1906, Chief Secy GOB, to Secy GOI Rev and Agri, p. 139.

137. Ibid., no. 579LR, Ranchi, 25 Jun 1906, E.A. Gait, Commr Chotanagpur, to Secy BOR, p. 131.

138. OIOC P/6561 BRP (For), Apr–Jul 1903, A progs 25–30, Jun 1903, file 10R/2, no. 32LR, Ranchi, 13 Apr 1903, F.A. Slacke, Commr Chotanagpur, to Secy GOB Rev; no. 896T-R, Darjeeling, 3 Jun 1903, GOB Rev notification, pp. 117–18.

139. OIOC P/7573 BRP (For), May–Jul 1907, A progs 1–12, May 1907, file 19-P/3-4 of 1906, no. 3126T-R, Darjeeling, 26 Oct 1906, Chief Secy GOB, to Secy GOI Rev and Agri, p. 140.

140. Ibid., no. 105S, Chaibassa, 30 Jul 1906, Settlement Officer Porahat to Commr Chotanagpur, pp. 134–5.

141. OIOC P/8134 BRP (For), May–Jun 1909, A progs 30–2, May 1909, file 19P/1, no. 1052R, Chaibassa, 14 Sep 1908, H.D. DeM Carey, DC Singhbhum, to Commr Chotanagpur; no. 572T-R, Darjeeling, 13 May 1909, F.W. Duke, Offg Chief Secy GOB, to Secy GOI Rev and Agri, pp. 225–7.

142. O'Malley 1910b: 108.

143. OIOC P/9548 BORP (For), Jan–May 1914, A progs 4–11, Mar 1914, file IIIF/1, no. 1818-254, Ranchi, 17 Sep 1913, H. Carter, CF B&O, to Secy GOBO Rev, p. 50.

144. OIOC P/8135 BRP (For), Jul–Aug 1909, A progs 33, Jul 1909, file 19P/6, no. 139, Darjeeling, 31 Oct 1908, A.L. McIntire, CF Bengal to Chief Secy GOB, pp. 41–2.

145. Ibid., p. 43.

6

Forests in a Regional
Agrarian Economy

Introduction

Moving from the procedures of forest reservation to the struggles over protected forests we can see how forest management was really about disembedding wooded tracts from an agrarian landscape for a variety of economic and political reasons. Strategies of territorialization could only create forests where a range of rights and land uses were identified and erased. To the extent that forest demarcation remained incomplete the management of forests continued to be caught up in a regional web of agrarian relations.

The official wisdom of district administrators that recommended abjuring demarcation also promoted caution in preparing any detailed record of rights. Discussing the issue in relation to Kolhans protected forests, the Chotanagpur Commissioner remarked, 'it is better not to raise the question of forest rights as the aboriginals never understand . . . and always think their rights are going to be interfered with.'[1] Before 1892 the unreserved forests of the Kolhans had been fully enjoyed by the tenants subject only to the restrictions imposed by headmen. When they were declared protected, only a few blocks had been demarcated and the rest were released under the care of headmen, under rules *not framed under law but by the government as estate proprietor*. When the government took up preparation of the record of rights the exercise threatened to widen permissible forest usage considerably. Hunting, shooting, quarrying, sale of sabai grass, sale of wooden handles for agricultural implements, sale of fruit and flowers were discovered among the practices routinely carried on, often with government approval, but in direct contravention of the rules.[2]

Thus the record of rights observed more closely the everyday forms of forest use, and noted the 'heterogeneous traverses' that de Certeau talks about, as 'guileful ruses of different interests and desires . . . (which) . . . overflow and drift over an imposed terrain . . . theoretically governed by the institutional frameworks that it in fact erodes and displaces.'[3] The gap between rule and practice that the record of rights illuminated was not closed. By dropping the operation to reconcile the record and the rule, the government allowed these discrepancies in protected forest management in southwest Bengal to remain as 'technical infringements' and become the historical provenance of the region as a zone of anomaly.[4]

To better understand the nature of agrarian society in which woodland Bengal took shape we would have to discuss the formation and change of agrarian regimes in the colonial period. This major task, if attempted here, would take us too far from the focus on forest management that is organizing this book.[5] This chapter takes on the more modest project of highlighting the manner in which the agrarian world directly obtruded into the realm of forest management in woodland Bengal. It begins with a discussion of agrarian livelihoods that were subordinated to forestry by the distinction between major and minor forest products. The examples are taken from the striking hardships imposed on Paharia tribes in the Rajmahal hills of southwest Bengal. Moving then to north Bengal, this chapter examines the issues of grazing regulation and hill-side conservation. These are different but complementary examples of the contradictions that emerged when forest management became entangled in wider issues of hill areas development. All the cases in this chapter bring home the basic point that forest landscapes took shape in the eyes of the colonial foresters only when they were disengaged from regional agrarian economies. But this very partitioning and re-formation of landscapes made forest management more difficult and contentious in practice.

Conservancy Looking Between the Stems: Minor Forest Products

Forest reservation became one means to regulate the thriving trade in minor forest products (MFP), usually by way of permits, but occasionally, as in the Santhal Parganas, the trade was controlled by requiring sale of MFP in specified markets.[6] These arrangements were largely unaltered except where the MFP traded became more valuable than the major products of timber and fuel. This happened with sabai grass in the early twentieth century, so that will be considered in detail. But in the first decade following the Bengal Forest Rules, only a limited range of MFP had been noticed. Bamboo, cane and grass were not counted as MFP. In the next fifty years the category of MFP came to include bamboo, though it was often separately accounted for and sold in coupés like wood.[7]

A number of resinous products were among the exports from the forests of lower Bengal. Shellac, *cutch*, and lac dye were the main ones. Dhuna, the sap extracted from the sal tree, was a staple of the trade from the jungle mahals and the forests of Singhbhum.[8] Locally the trade marked the exchange of forest products for food, usually rice, between forest tribes and neighbouring farmers. But dhuna from lower Bengal was also exported to places like Mirzapore in United Provinces, indicating a larger commerce in a product used as incense, gum and insecticide, among other things.[9] Since the extraction of dhuna was done by girdling the tree, thus killing it, girdling came under criticism from the foresters.[10] The Bhumij people, who principally engaged in the dhuna trade in the jungle mahals and adjoining Chotanagpur estates, were initially not restrained as the resulting forest clearing was found desirable. But the growing importance of forests as timber source made the absolute control of conversion—by girdling or felling—more lucrative, facilitating an alliance between village elites and foresters to restrict girdling for dhuna.[11] By the 1870s the vigorous trade in gums, dhuna, cutch (prepared from acacia catechu) and other MFP produced at the expense of timber had been seriously regulated.[12] By the first decades of the twentieth century, the rising contribution of MFP to forest revenue encouraged the government to cast its net wider for these products. The Forest

Act of 1927 expanded the definition of forest produce to include timber, charcoal, cutch, wood oil, resin, natural varnish, bark, lac, mahua flowers and seeds, and myrobalans.[13]

The regime of restrictions operated two ways upon MFP. First, all potential revenue from MFP was controlled through permits, licences and tolls. Second, the extraction of MFP was curbed where it interfered with the exercise of government monopoly over major produce, as in the case of attempts to end sal girdling for dhuna (dhamar). In this latter case MFP trade was also adversely affected in some cases by more general restrictions on forest use. Thus the tasar silk industry declined as cocoon gathering became difficult after forest protection.[14] While these two procedures prevailed in most of Bengal in the odd case we notice a third mode of dealing with MFP. This was characterized by high government effort in managing the production and trade of a particular MFP, and seemed to occur in zones of anomaly—places where paternalist tribal administration converged with forest management that was not exclusively interested in timber.

In southwest Bengal the third, rare, mode of dealing with MFP is demonstrated by the case of sabai (*ischaemum angustifolium*), a grass that grew abundantly in the dry deciduous sal forests of Singhbhum and Santhal Parganas. It was an important trade good which the Paharias of Rajmahal exchanged for grains, salt and cloth from the Santhals and other plains peoples, and later became a valuable raw material in the manufacture of paper. Fire protection in forests sharply reduced sabai grass production in Singhbhum reserves. This had its own adverse consequences for the Paharias.[15] Islands of *kurao* (shifting) cultivation that had been preserved in the Rajmahal Hills became the principal place for commercial production of sabai grass as it was excluded from reserved forests in Singhbhum.[16]

In the period 1899 to 1907, production of sabai grass in Rajmahal went up sharply, from 2,54,740 maunds to 5,23,879 maunds and this increased government royalties from 15,358 rupees to 32,742 rupees per annum.[17] The production covered a total area of 40,000 acres, of which less than half was cultivated by Paharias themselves.[18] *Mahajans* (traders) firmly controlled production through a system of advances to the Paharia cultivators. Initial advances covered costs of clearing and sowing and subsequent advances were paid to organize weeding and cutting of the grass but were subject to high interest rates. The traders thus obtained the grass bundles for a pittance and sold them to contractors for baling and further sale to paper mills.[19] Finding the Paharias captive to the mahajans under this system the government initiated proposals to regulate sabai grass production and sale. In 1895 a system of passes had been introduced that required mahajans to seek permission from the Santhal Parganas administration before entering into any agreement with Paharias for sabai grass. In 1899 restrictions had been imposed on import of labour for cultivation.[20] But this could not break the stifling monopoly established by a small group of mahajans from Sahebganj and their employment of upcountry labour, as they converted the annual contract, in practice, into an annual usufructuary mortgage.

In October 1907 Keso Sardar of Simaria and other hill manjhis submitted a memorial complaining about the mahajan combine and the suppression of sabai grass prices by the cartel resulting in the sale of grass worth rupees hundred for rupees five.[21] As the subsequent inquiry revealed, the *chukti* (pact) between mahajan and Paharia landholder virtually resulted in the latter signing away all rights in the produce of the land. The mahajans not only employed coolies from outside, they debarred the entry of the Paharia landowner onto the fields when they were being harvested, thus concealing the actual yield.[22] While Patterson, Subdivisional Officer Rajmahal, felt that the problem owed its origin to the improvidence of Paharias, Stark, the inquiry officer, described the gradual process by which the Paharias began conversion of their kuraos into sabai fields only to have them usurped by mahajans through advances. While Paharias always sowed sabai by broadcast methods interspersed with their Indian corn, the mahajans desired and secured field clearing and sabai mono-cropping through the leverage of debt bondage.[23] Stark gave a moving account of the impoverishment of the Paharias, whom he had known from the 1880s, in Teliagarhi and Madhuban. The clearing of jhums for sabai had diminished their income from wood sales, and directly affected their self-provisioning. He wrote:

I was struck by the great change for the worse in their condition—there was scarcely a family with sufficient grain in store to last more than a month; whereas in the past there were large bamboo chacki (stores of Indian corn and bajra), cobs of corn suspended from the rafters and the simri inside the house.[24]

Working through agricultural advances to Paharias and direct sale to a single large contractor, sabai grass cultivation was reorganized by the government to secure remunerative prices for the Paharia growers and limit their displacement by outsider hired labour.[25] These measures temporarily ameliorated the commercial exploitation of Paharias and their lands. But they also strengthened the processes by which all sabai grass production was tied to external markets and industrial supply, thereby restricting its availability in local markets and for domestic consumption.

Similar things happened in neighbouring Singhbhum. Over the years, Balmer Lawrie and Company of Calcutta had become monopoly purchasers of Singhbhum sabai grass, making it scarce for local consumption where it was used in the manufacture of tethering ropes, grain morhas (storage), house building, harnesses, cots, thatching, and so on. Sabai theft from forests dedicated to industrial raw material production became, as the Commissioner of Chotanagpur observed, the only way Hos (the local peasantry) could get any for their bona fide needs.[26] The Commissioner of Chotanagpur went on to recommend a return to the permit system that was followed for all other MFP, pointing out the potential for increase in revenue while serving all markets. To the Forest Department, the monopoly was an easier system to manage and hence preferable. The Conservator argued against taking a local view of the problem, suggesting that local needs could be met through regulated distribution by Balmer Lawrie, since theft was also a manifestation of competition from Marwari traders operating in the depleted proximate forests of the Tributary States.[27]

Ultimately the question boiled down to there being insufficient sabai available for Paharia cottage industry, rope and string making, that supplemented their family income and provided employment to older men who were no longer able to cultivate or hunt. Since long-staple sabai was needed for all village uses and paper mills were unconcerned with staple length, seasonal surpluses from Balmer Lawrie were actually useless for local distribution as proposed by the Conservator since the grass had been cut in variable lengths. Permits were opposed by foresters mainly to facilitate monopoly administration by eliminating other sources of sabai in the regions covered by the monopoly. However, wage payment in sabai grass often did put some of the product into local circulation.[28] If the promoters of the 'narrow local view' were concerned about Paharia livelihood, the foresters challenged both the view and its definition of traditional livelihood. To them the local trade in ropes made from sabai grass was itself a recent development. Thus the case of sabai, and its playing out in what we have described as zones of anomaly, brought under the spotlight an organizing opposition between 'local welfare' and 'national interest' in forest management, that had begun to crystallize through various controversies surrounding grazing, timber and environmental services from woodland Bengal.

In Singhbhum the leases for sabai extraction had passed into the hands of Balmer Lawrie by 1895, and despite a higher lease rate paid by other lessees in 1903–9, Balmer Lawrie regained the leases thereafter at a lower rate and kept it till 1930. The rate they paid did, however, rise from 3,000 rupees for a three-year period starting 1895, to 1,40,000 rupees for a six-year period starting 1921. From then on the leases were renewed annually till the decision in 1929 to return the leases to Mankis in their circles. While this resolution of the question illustrates the persistence and influence of exceptional administration in southwest Bengal, it was also made possible by the increase in supply of sabai from Saranda and Angul forests that reduced the pressure on Singhbhum and Rajmahal from paper manufacturers. A concurrent and rapid rise in timber revenue from the Singhbhum forests also made feasible some sacrifices in sabai revenues.[29] The important point is that the switch to 'traditional modes of control' in sabai production and trade was not smooth or very successful in optimizing profits. Its justification lay in promoting local responsibility for forest management.[30]

As we have seen, such arguments and the degree to which they translated into forest management practices varied sharply across the reserved forest/protected forest divide. Another source of variation was the proportion of MFP to timber and poles in different forests. A forest yielding a diversified range of products, and where the revenue was significantly derived from products other than timber, was more likely to be under localized management.

Evidently the extent to which foresters fought to retain their stranglehold on forest management varied with the type of forest. Among all the sources of revenue timber held a special place, for a forest that appeared rich in timber, or held promise of it, was where scientific forestry combined with regimes of restriction to make a separate domain that foresters were most loath to share. This point is further illustrated by the way the twin objectives of forest conservation and the development of Darjeeling as a hill station came to be pursued in the latter part of the nineteenth century.

Hill Station Development, Forests, and Grazing

From the beginning of forest conservancy in Bengal one thorny question was how to distinguish government forests from those of the Municipality of Darjeeling.[31] After the Bhutan war, when the British took over the area, there was rapid Nepali migration and settlement into Kalimpong, Rangpo and other localities around Darjeeling. In contrast to Lepcha and Bhutiya residents who had jhummed gentler slopes and practised transhumant grazing, these Nepali settlers prepared terraced rice fields and grew wheat and mustard as winter crops.[32] This constricted pastures to some extent. Later, with the growth of tea plantations in the Darjeeling area and Duars, forest grazing increased as the cattle kept by coolies working in the gardens entered the forests adjoining the tea estates.[33] Brandis, the Inspector-General of Forests, had suggested a fee to reinforce the notion that this was a privilege that should not become a recorded right in a future settlement. Transhumant Bhutiyas, moving seasonally from the Chel river valley to its head waters, were already accustomed to paying such a levy.[34] In 1878 the Government of Bengal decided to grant to cultivators east of the Tista, grazing privileges in certain forest blocks free of charge.[35] But as recommended by Brandis this largesse was not extended to professional graziers. Village headmen, like mandals, were made responsible for the protection of such blocks opened to grazing. They had to prevent the damage of these forest blocks by fire, lopping, and other forms of sapling destruction.[36] A certain amount of grazing had to be provided within forests to pasture cattle of neighbouring villages and milch cattle supplying milk to civil stations, garrisons and towns. Conflict arose between fuel and timber demands for local community and ensuring milk supply. Darjeeling cattle owners vehemently opposed rules which necessitated periodic change of blocks grazed and the total closure of others.[37]

By the late 1870s the construction of *bathans* (cattle enclosures) was being taken up in north Bengal. But the provision of milk and the supply of fodder for the purpose of feeding milch cattle were demands on the forest areas that were subordinated to the organizational demands of the Forest Department which had timber-related management as its top priority. This meant that the Forest Department was not involved initially in the retail trade of fodder or fuelwood. Such an approach invited the criticism of the Commissioner of Cooch Behar who noted that the graziers and timber contractors had made a spate of complaints against the Forest Department while milk and butter supply to Darjeeling and other civil stations had deteriorated.[38] A few years later the grazing rules for Darjeeling were drafted. They laid down a system of permits, prohibited construction of camps, restricted grazing to those supplying the needs of Darjeeling, and banned buffaloes, goats, and sheep from the forests. This again invited an irate response from the Commissioner who cited a report in the *Indian Planter's Gazette* that linked recent murders of forest guards to the ever-growing restriction on forest use by graziers.[39] But these policies were justified by the state government, speaking exclusively with the voice of the Forest Department, as necessary to prevent wasteful use of forests.[40]

Temporal and spatial restrictions on supply of grass from forest lands interfered with the practice of big, locally powerful cattle merchants who sent large herds of cattle to graze in forests remote from their headquarters and competed for the grass available to villages in the locality.[41] Another problem was the negotiation of an acceptable level of restriction on grazing. For goalas who had the monopoly over supply of fresh milk to civil stations, grazing restrictions, which were varied by permits based on occupation, meant that their monopoly was strengthened. The leverage provided by this monopoly allowed goalas to constantly nibble at the edges of state-enforced grazing restrictions. In 1884, for instance,

Darjeeling goalas petitioned the townspeople that grazing grounds were insufficient and bad, milch cattle were producing less, and they were harassed by incessant petty fines on violating restrictions. They threatened to leave the district.

The Secretary to the Government of Bengal responded

the fines referred to are judicial fines, and with the very small staff we have, it may be safely assumed that for one offence in which the offender is caught red-handed (we cannot prosecute in any other case), fifty are committed of which we know nothing except the results to the trees in the forests.[42]

Here we have an example of regulations observed mostly in the breach, but a very public process by which the informal accord was being reiterated. This case is comparable to what E.P. Thompson has to say about forest law and statute law in Windsor Forest in England, where 'the keepers were more likely to get cooperation from the foresters if they affected an easy going toleration of small offences.'[43]

With the growth and development of Darjeeling civil station, and military cantonments around it, the problem of milk supply became more acute. By the turn of the century the Forest Department was finding it much more difficult to persist in a high-handed policy of excluding graziers from the forests as the problems in milk supply for Darjeeling were noticed by the state government.[44] But all this was being done under the Darjeeling Forest Working Plan of 1892. This plan, assuming a rotation for valuable timber tree of 160 years, had divided the total area of 26,624 acres into eight working circles. In one of these, mature trees were being removed by annual selection while the other seven were further divided into five blocks for regeneration over thirty-two years. In blocks not under regeneration, improvement fellings were prescribed at four year intervals.

In 1885, the grazing rules issued under Act VII of 1878 had required the forest officer to open five to ten acres to grazing per head of milch cattle permitted. By 1900 an average of 630 cattle per month were grazing to supply the milk needs of civil stations and cantonments. The figure had gone up from 440 in 1892 to 928 in 1902, though it had been stable at around 450 in the previous ten years.[45] The problem was accentuated by the decline in availability of grass in well-stocked forests while the increase, or constant yield of grass, retarded the dense stocking of the forest with valuable timbers. The separation of grazing and timber production was thus visualized as essential to scientific forestry. All proposals for amelioration of the grazing problem, thereafter, took the form of arrangements that would physically segregate these two demands on forest management into distinct enclosures in a move towards internal territorialization.

The government, responding predictably to the growing pressure of the question which was being manifested as a disagreement between different branches, appointed a committee to inquire into grazing regulations in the reserved forests of Darjeeling.[46] This committee was headed by Commissioner Rajshahi, with Conservator of Forests, Bengal, Deputy Conservator of Forests, Darjeeling, Deputy Commissioner, Darjeeling, and Superintendent, Veterinary Department as members. The committee held that the demands of silviculture and grazing were incompatible, suggesting that some areas be set aside for grazing. Disagreeing with the recommendation of the Inspector-General of Forests for stall feeding (tried since 1896 without success), this committee proposed the deforestation of certain areas to produce fodder grasses. They thus wished to discontinue the prevailing arrangement of grazing permits.[47] Necessary revisions of the Darjeeling Working Plan to accommodate the idea of a grazing reserve were taken up.

The Government of Bengal had introduced a system of partial stall feeding but only three bathans were built by 1911 and only one of them was in use. Free grants of cut fodder also could not induce villagers away from grazing in wastes and jungles.[48] However, pressure to build more remained. On the other hand the new grazing regulation had the curious effect of reducing the incidence of grazing in the grazing reserves. This fact assisted the Forest Department in concluding that grazing was of diminishing importance to cattle keepers in the Darjeeling area. Grieve's Working Plan for Darjeeling noted in 1912 that only 11 per cent of Darjeeling's milk supply was coming from forest-grazed cattle.[49] It is instructive to see how this happened. In 1909, the grazing fee had been raised to one rupee per head of cattle from eight annas (half a rupee). A limit of 918 cattle grazing per month was also placed. The permitted grazing density was one head of cattle for fifteen acres, thus creating a grazing reserve of 13,770 acres.[50] In the next seven years the area was reduced to 7,738 acres and the permitted incidence of grazing to 515 heads of cattle per month. This was done ostensibly to bring policy in conformity with reality when it was found that Bhutia graziers had consistently grazed only 50 to 60 per cent of the permitted cattle during this period. In the ten years from 1905 to 1915, grazing incidence had come down from 748 head of cattle per month to 279 head of cattle. The origin of Darjeeling milk supply had also changed, as Table 6.1 indicates.

Yet, in 1917, the government received from one Noden Bhutia and others a petition complaining against the arbitrary curtailment of their grazing privileges in Darjeeling forests. The case is revealing in many respects. Earlier, the graziers had kept as many cattle as they needed and obtained permits for grazing. With the new regulations they were held to fixed number of cattle for each grazier. Individual graziers were accustomed to reducing their cattle in winter and increasing them in summer. In addition to this seasonal variation, their cattle population also fluctuated with lactation cycles, as temporarily dry animals would be sent to Sikkim. Any decline in the number of cattle grazed by one Bhutia then meant that his permit was reduced by that number. Over time this approach led to a net decline in cattle grazed.[51] That this was in fact the intent of the foresters becomes clear. Recommending reduction in the size of the grazing reserve, H.S. Gibson, the Deputy Conservator Forests, Kurseong wrote:

Table 6.1. Origins of Darjeeling Milk Supply (percentage)

Source	1905	1911
Forests	50	11.25
Khasmahal	14.3	33.75
Tea Gardens	28.4	25.25
Bastis	nil	17.25
Europeans	7.3	8.50
Nepal	nil	4.00

Source: OIOC P/10122 BRP (For), 1917, A progs 4–6 Mar 1917, file 3I/1, G.S. Hart, IGF note on Darjeeling forests, 28 Dec 1916; no. 85F/54–1, Simla, 14 Feb 1917, A.E. Gilliat, US GOI Rev and Agri, to Secy GOB Rev, p. 13.

In the first place the grazing being concentrated in a smaller area will be easier to supervise and control. In the second place forest which is used for grazing cannot be used for any other purpose and is unable to yield timber and fuel, which are the main legitimate forest produce . . . it is therefore a sheer waste to allow grazing in a larger area of forest than is sufficient to meet the requirements of the number of cattle being grazed.[52]

In 1913 the Board of Forestry had discussed the question of grazing and fodder supply and the Inspector-General of Forests had followed thereafter to review the position in different provinces. He noted that while grazing could be injurious to natural regeneration in certain instances, it was beneficial in others. Regular removal of grass in closed and plantation areas released regeneration and helped avoid fires.[53]

Grazing regulation in the north Bengal region had largely proved unnecessary, fodder being plentifully available. The only regulation was for Darjeeling, which was a special case. But in other regions, as in highland Burma, the considerable uncultivated arable and scrub forest that was interspersed between wet paddy fields and the few pockets of high forests, was an inadequate grazing ground for cattle kept in large numbers to supplement rainfed agriculture. Here under the pressure of an overall policy of grazing regulation, a policy of cheap grass and dear grazing was followed without much success.[54] The limitation of grazing and control of numbers of cattle admitted to graze in any forest area remained key considerations for the silvicultural conference in 1939, though by then this silvicultural requirement had become dependent on what grazing management was administratively feasible.[55] Where grazing was a more strongly contested issue, such as the Central Provinces, working plans were prepared by making a grazing settlement. In such places different rates of grazing fees were fixed and the distinction between domestic cattle of forest villagers and those of *malguzars* and professional graziers continued, although there were many ingenious ways in which it was subverted.[56]

Commerce and Conservation in North Bengal

The travails of grazing regulation in the reserved forests around the Darjeeling area had started as the hill township became a major administrative and military station in Bengal. Conflicts between conservation and diversified agrarian development increased in north Bengal as tea and cinchona plantations began to spread across the region. Tea planters started an agitation about deforestation in Darjeeling. They complained that the rapid denudation of the Darjeeling hillsides was diminishing rainfall in the area to the detriment of tea production.[57] There was some truth in this complaint as the Deputy Commissioner Darjeeling could confirm that the advent of British rule had impelled the clearing of large forest areas. In 1830, Captain Herbert, the Deputy Surveyor General had written of these Himalayan mountains as 'completely clothed with forest from very top to bottom.'[58] Ten years later other travellers and the first superintendent of Darjeeling reiterated this position and added that the population of the district, which had an area of 138 square miles, was no more than a hundred souls. The sparse population of the area, mainly made up of Lepchas, was partly due to the oppression of the raja of Sikkim, who had caused the flight of 1,200 Lepchas in the 1830s.[59]

By 1905, the population density of Darjeeling had risen to 214 per square mile, cultivation had extended to about 21 per cent of the district area, tea covered 50,000 acres (33 per cent of net cropped area), and reserved forests accounted for 38 percent of the district. The local Lepchas numbered only 10,000, clearly a minority of the district populace, having been confined to the well-wooded tracts of Kalimpong and moved into Bhutan, where they could still jhum without government interference.

The Divisional Forest Officer of Tista was asked to prepare a report on the deforestation, especially its effects on rainfall and soil erosion.[60] The findings were predictable enough. Denudation was caused, so foresters held, more by native-owned *jotes* (fields) in *khas mahal* (government estates) where steep slopes were cultivated, terracing was poor, and heavy grazing and indiscriminate clearing of wastes was practised. Tea, as it was on gentler slopes, was less damaging. The recommendations were stern. All wastelands on steep slopes were to be planted with forest trees so being closed to grazing and irrigated crops. Dry stone revetments were to be built for fields terraced for dry crops (like tea).[61] While tea planters wanted curbs on native agriculture they were not prepared to increase their capital costs by undertaking intensive soil conservation measures. They opposed terracing and engineering works, arguing that exclusion of grazing and limiting cultivation could amply regenerate the forest areas and prevent further erosion. A committee made up of the Deputy Conservator Forests, Darjeeling and two representatives of the tea industry was set up to report on feasible measures.[62]

Their report said that tea estates were well managed and concentrated hostility on the khas mahals. In these areas the holdings, unlike the tea estates, were small. Most cultivators were recent Nepali immigrants, who had created terraced fields on the hillsides earlier jhummed by the Lepchas. Most of these terraces had been formed below 4,000 feet by cutting into the hills and banking the edges to a height of one foot. A channel from the nearest stream was led to the topmost terrace and provided irrigation.[63] The committee observed, in a rare appreciation of jhum, that since the Lepchas did not clear the land of all tree roots and stumps, coppice regeneration usually occurred in their abandoned fields, which acted to preserve hillsides. The Lepchas grew Indian corn and kodo (millets) by hoeing and dibbling crops according to elevation. The Bhutiyas, mainly graziers, were the other ethnic group in Darjeeling district. They invited the ire of the committee for being as destructive as their cattle by lopping valuable species like *tun* and *champ* for fodder and hacking down saplings and small trees. Cardamom, again grown mainly by Nepalis, needed running water and therefore occupied banks and beds of streams.[64]

This committee proposed a special legislation that would allow the Bengal government to reforest, and close to grazing, areas prone to landslip. It also proposed prohibiting rice cultivation on steep slopes; prohibiting cardamom cultivation altogether; reserving lands on river banks to raise protective forests; and further extending forest reservation to catchment areas. The committee also suggested that the *mandals* (headmen) of respective villages be made responsible for maintaining the reserves. They were to be compensated for performing this task by receiving a monopoly right to collect fodder in the reserves through a system of monthly passes.[65] This last proposal did not find support from the district administration, who felt that the mandal did not enjoy the respect of the villagers to the extent that would make possible on his part the execution of such a policing role. But cardamom cultivation, carried on by a system of advances to villagers by local moneylenders called kayas, was opposed by all local officials.[66]

The district official, as may now be anticipated from some of our discussion in the previous chapter, spoke forcefully for raiyati interests. He offered three distinct lines of argument. First, he noted (relying on the authority of Joseph Hooker) that landslips had always occurred in the region even when it was well forested. Second, he predicted that placing raiyats under mandals and amins in the matter of managing their cattle and choice of crops cultivated would lead to oppression by petty officialdom.[67] Third, he pointed out that jhumming in the khas mahals by Lepchas was a relic of primitivity that with the efflux of time, and increased contact with more sophisticated forms of plains cultivation, was bound to be transformed into 'scientific methods'. A corollary to this observation was the fact that tea and forests already occupied more than 60 per cent of the district.[68]

Having articulated the hardships any regulation of land use would impose on raiyats, Forrest, the Deputy Commissioner of Darjeeling, went on to propose a Darjeeling Hillside and River Conservancy Act that would prohibit cultivation of wet rice and cardamom on environmental grounds, but 'would give the proprietor every opportunity to become his own conservator.'[69] The proposal faced one practical stumbling-block created by the complex hierarchy of sub-tenancies in the privately held estates. Regional environmental conservation could only work through proprietors (government or private persons) when their control of the land and its uses was not vitiated by subordinate rights. A balanced disposition of rights between proprietors and tenants, as increasingly favoured after the Bengal Tenancy Act of 1885, was inimical to the simplification and concentration of those rights. This meant that sweeping environmental regulation became more difficult because it relied on such concentration of rights.[70]

These contradictions emerged in Forrest's recommendations. He was irritated by the constant complaints by foresters and tea planters against Lepcha, Nepali and Bhutiya cultivators, who if subject to all the recommendations of the committee, were in danger of 'being improved off the face of the earth.' Forrest was clearly opposed to restrictions on wet rice cultivation, cardamom cultivation and grazing that would place the average hillman in such danger. Yet, where forests had to be preserved proprietors and mandals were ill-equipped to deliver the goods. The Settlement

Officer for Kalimpong had reported that mandals had been regular in paying revenues, maintaining roads, providing coolies and supplies, but failed in protecting wastelands and trees.[71] On balance, Forrest came out in favour of the proposed legislation that would not extinguish private rights but 'take power to intervene when a man's action is detrimental to his neighbour or to the public interest.'[72]

The proposal ultimately got bogged down on issues of compensation. The Government of India ruled that since compensation would have to be paid in all lands taken over for conservation, from both revenue paying and freehold tenures, this would invite proceedings under the Land Acquisition Act. They recommended against the new law for Darjeeling conservation. Faced with the prospect of tiresome litigation and uncertain financial liabilities, the proposal was quietly shelved by the Government of Bengal, after eight years of active consideration.[73] In dealing with hillside cultivation and grazing, a limited regulation and modified forest management practices were used to acknowledge their role in the regional agrarian economy. The cases of minor forest products in southwest Bengal as well as grazing and plantation crops in north Bengal reveal that forest management, at local and regional scales of operation, remained imbricated in agrarian economies. Conservancy was thus always tempered by the demands of rural livelihood that had to remain viable if scientific forest management was to expand its domain.

Forestry, Agriculture, and Silviculture

One of the themes of this chapter, and the last one, has been the ambivalence in forest policy between favouring commercial or national-interest forestry and welfare or local-benefits forestry. At the turn of the century, the last great German Inspector General of Indian Forests, who also had the distinction of holding that office for the long period of eighteen years, wrote:

a reserved forest has not necessarily the object, as is frequently believed, of producing large timber for export or public works; but more often that of supplying the local demands of small timber, fuel, grass or any other forest produce.[74]

These words expressed a growing recognition in the senior echelons of the Indian forestry establishment that there was considerable forest dependence among tribal peasantry living in dryland agriculture areas. The nature of the landscape in such areas also compelled this recognition to some extent. Resembling northwestern China, southwestern Bengal presented a mosaic of vegetation, sparse deciduous woodlands on plains opening into savannah grasslands with scattered trees and brush.[75] This patchwork landscape stood in the way of forest conservancy through reservation which aspired to work like a system of land-use zoning.[76] Reserved forests were managed through working plans, and forests burdened with rights were not amenable to working plans.[77] Forest rights that eluded limitation or extinction were a measure of the complex and tight relationship between wooded and farmed lands. This had to be accepted and dealt with ultimately in the policy process as forest management moved from restriction for logging to regeneration for timber production, from lumbering to silviculture.[78]

The control of vast areas of land meant that forest policy had significant implications for agriculture, in particular for cultivation and stock raising. The creation of a manageable arable landscape had long been an imperial ideal.[79] By the middle of the 1890s the progress of demarcation, reservation and systematic timber exploitation had made forestry a valuable revenue source. The process of timber extraction, for instance, had to serve both the goals of revenue generation and supply of sleeper woods and other industrial goods, for the expanding commercial economy and growing political integration of the late nineteenth century.[80] In the early twentieth century, wartime needs had to be met too. There seemed to have been a constant tension between managing forests as part of a larger project to maximize agricultural production and productivity, on the one hand, and the effort to delink forests from annual crop production. This had implications for management and control systems, rights of villagers, and so on. It was often a conflict between local versus state priorities, or annual versus perennial crops, or multipurpose trees versus timber trees.

In Bengal, as in other provinces, state control over wastes and forests, the contraction in available culturable land and pastures, and the increasing market value of land were strengthening proprietary groups like zamindars and rural elites. This power and resource control came at the expense of poor tenants, nomadic graziers, agricultural labourers and artisans, sharpening intra-village contests over forests. David Ludden points out that 'among the interventions into village society that nurtured the Anglo-Indian empire, dividing public from private land stands out as the most important.'[81] I would add that the separation of forestry and agriculture into distinct domains of production was no less important as it created further exclusions and conflicts in both public and private lands.[82]

These struggles took distinct trajectories in different parts of Bengal with implications for forms of property and territoriality in forests. This history is important for at least two reasons. First, because 'forms of property and other legal and social institutions are . . . creatures of history.'[83] Second, because it illuminates the mounds and hollows of the geographies of empire, while reminding us that material and imaginary landscapes are mutually constitutive in a way that positivist political sociology, teleological environmental history or relativist cultural geography alone cannot adequately comprehend.[84] Regional variations then become important, not as versions of a theme, but for the different temporalities and diverse accommodations we can detect in what is often subsumed as a single trend in statemaking processes.[85]

Fractured by the multiple tensions generated by regional variations in conservative lumbering, internal territorialization, as manifest in the separation of the domains of agriculture and forestry, became a question on which the national forestry establishment was divided in at least three ways. First, there were the emerging multiple-use forestry ideas sponsored by Brandis. Whether it be an argument for investing in extending rubber plantations in Assam, or promoting forest reservation so that the supply of less valuable woods increased, and in turn supported charcoal furnaces in the iron-rich southern districts, Brandis was arguing for the enlargement and diversification of forestry operations.[86] This did prove a departure from the earlier conservationist concerns of Cleghorn or Gibson, which have been alluded to earlier, and constituted the second strand.[87] Third was the view that forestry should subserve agriculture and its development in India.[88] The influence of this position may be seen on the forest policy resolution of 1894, which made a fourfold classification of forests in the following terms:

- forests to be preserved on climatic grounds;
- forests managed for the supply of useful timber;
- minor forests to supply woodfuel, fodder, grazing and small timber for agriculturists;
- pasture lands or grazing lands managed by the Forest Department.

The primacy accorded by the resolution to agriculture is evident from its stipulation that wherever an effective demand for cultivable land existed, and could only be supplied from forest areas, the land should be relinquished. Brandis did not like the idea of Forest Departments losing control of poor quality forests since he felt very strongly that forestry for rural welfare must be state directed to the same degree as commercial or environmental forestry.[89] What is not clear, and needs a little more exploration, is the extent to which the interdependence of forestry and agricultural production was clearly perceived and considered a priority item on Forest Department agendas. This becomes important when considering the silvicultural systems that were being developed. We should expect to see different systems being devised to meet the rather different output goals set in the policy of 1894, something that really does not happen. These policy currents, or turbulences we might say, constituted the dynamic political-economic context for the development of scientific forestry in India and Bengal at the turn of the century, which is the subject of the next two chapters.

Notes

1. OIOC P/9549 BORP (For), Jun–Dec 1914, A progs 39–41, Jun 1914, file IIF/43 of 1914, no. 77TR/IX3, 27 Feb 1914 Camp Gobindpur, Commr Chotanagpur, to Chief Secy GOBO, p. 193.

2. Ibid., no. 1222R, Chaibassa, 2 Nov 1913, B.C. Sen, DC Singhbhum, to Commr Chotanagpur, pp. 194–7.

3. de Certeau 1988: 34.

4. OIOC P/9549 BORP (For), Jun–Dec 1914, A progs 39–41, Jun 1914, file IIF/43 of 1914, no. 4318R, Ranchi, 9 May 1914, H. Coupland, Secy GOBO Rev, to Commr Chotanagpur, p. 198. In a similar decision, prescribing closed and open seasons for hunting in protected forests, the government exempted from these restrictions 'tribal hunts of the aboriginal races'. OIOC P/9549 BORP (For), Jun-Dec 1914, A progs 1-23, Jun 1914, file IIIF/47 of 1914, no. 4286R/IIIF-47, Secy GOBO Rev notification dated 6 May 1914, p. 47.

5. I have taken up this matter in some detail in Sivaramakrishnan 1996a: chs 8 and 9.

6. OIOC P/9548 BORP (For), Jan–May 1914, A progs 4–11, Mar 1914, file IIIF/1, no. 1818-254, Ranchi, 17 Sep 1913, H. Carter, CF B&O, to Secy GOBO Rev, p. 48.

7. OIOC P/9286 BORP (For), Jan–Jul 1913, A progs 1–5, Apr 1913, file IIIF/10, no. 1473-212, Ranchi, 8 Oct 1912, CF B&O, to Secy GOBO Rev, pp. 17–18.

8. OIOC P/432/69 BRP (For), Jun–Aug 1867, A Progs 1, Jul 1867, T.B. Lane, Secy BOR LP, to Secy GOB, no. 2232A, 8 Jun 1867, no. 33, 4 Jan 1867, Commr Chotanagpur to Secy GOB; no. 490, 9 Jan, Commr Burdwan, to Secy GOB, pp. 2–5.

9. OIOC P/432/72 BRP (For), Jan–Apr 1868, A progs 13–18, Feb 1868, Col E.T. Dalton, Commr Chotanagpur, to T.B. Lane, Secy BOR LP, no. 2927, 23 Dec 1867; L.R. Forbes, Asst Commr Palamau, to H.L. Oliphant, DC Loharduggah, no. 93, 30 Jul 1867; Capt. R.C. Money, DC Pooruleah, to Commr Chotanagpur, no. 1235, 31 Jul 1867; W.H. Hayes, Dy Commr Singhbhoom, to Commr Chotanagpur, no. 148, 22 Jul 1867.

10. Girdling was done by ringing the bark of the tree entirely round it, almost level with the ground. The tree died and in a couple of years was drained of resin. The value of timber from girdled trees remained a point of controversy.

11. OIOC P/432/71 BRP (For), Nov–Dec 1867, A progs 2, Nov 1867, W.H. Hayes DC Singhbhoom, to Commr Chotanagpur, no. 121, 20 Jun 1867, p. 1.

12. WBSA P/162 BRP (For), A progs 12–13, Feb 1872, no. 754, Fort William, 6 Feb 1872, H. Leeds, CF Bengal, to Secy GOB Genl Dept, pp. 7–9.

13. WBSRR Rev (For), A progs 5–19, Feb 1927, file 4-A/1 of 1926, no. F83-I-26A.C, Simla, 8 Sep 1926, W.T.M. Wright, Jt Secy GOI Legislative Dept, to Secy GOB Legislative; no. 5634-2M-131, Darjeeling, 13 Nov 1926, CF Bengal, to Secy GOB Rev; no. 87 For, Cal, 18 Jan 1927, F.A. Sachse, Secy GOB Rev, to Secy GOI, pp. 3–41.

14. Ruma Chatterjee 1990-1.

15. WBSRR Rev (For), A progs 1–4, Feb 1908, file 9-R/1, no. 141, Darjeeling, 27 Sep 1907, CF Bengal, to Chief Secy GOB, p. 94.

16. OIOC P/8136 BRP (For) Sep–Dec 1909, A progs 5–12, 13–17, Sep 1909, note dated 20 Apr 1908, W.H. Thomas, SDO Dumka, p. 139.

17. Ibid., A progs 1–4, Sep 1909, file 24/1, no. 1164R, Rajmahal, 12 Nov 1907, D.C. Patterson, SDO Rajmahal, to DC Santhal Parganas, p. 134.

18. Ibid., A progs 5–12, 13–17, Sep 1909, no. 1R, Dumka, 1 Apr 1909, H.W.P. Scroope, DC Santhal Parganas, to Commr Bhagalpur, p. 161.

19. Ibid., A progs 1–4, Sep 1909, file 24/1, no. 1164R, Rajmahal, 12 Nov 1907, D.C. Patterson, SDO Rajmahal, to DC Santhal Parganas, p. 134.

20. Ibid., no. 1806T-R, Darjeeling, 6 Oct 1908, A.H. Ley, US GOB Rev, to Commr Bhagalpur; A progs 5–12, 13–17, Sep 1909, no. 66R, Rajmahal, 14 Apr 1908, SDO Rajmahal, to DC Santhal Parganas, pp. 135–6.

21. Ibid., A progs 5–12, 13–17, Sep 1909, report, 12 Mar 1909, A.N. Stark, Dy Magistrate, p. 163.

22. Ibid., no. 1802T-R, Darjeeling, 6 Oct 1908, F.W. Duke, Offg Chief Secy GOB, to Commr Bhagalpur; no. 79MR, Bhagalpur, 17 Apr 1909, E.H.C. Walsh, Commr Bhagalpur, to Chief Secy GOB, pp. 146–50.

23. Ibid., report dated 12 Mar 1909, A.N. Stark, Dy Magistrate, pp. 166–7.

24. Ibid., p. 167.

25. OIOC P/8693 BRP (For), 1911, A progs 1–6, Jan 1911, file 2G/5 of 1910, and 2G/1, no. 466MR, Bhagalpur, 16 Jun 1910, Offg Commr Bhagalpur, to Chief Secy GOB Rev; no. 1569R-11-2, Dumka, 5 Jun 1910, H.L. Allanson, DC Santhal Parganas, to Commr Bhagalpur, p. 33.

26. OIOC P/11666 BORP (For), 1928, A progs 5–27, Jul 1928, file IIIF/19 of 1928, no. 3328R, Ranchi, 2 Sep 1927, H.T.S. Forrest, Commr Chotanagpur, to Secy GOBO Rev, p. 4; no. 1465R, 30 May 1927, J.R. Dain, DC Singhbhum, to Commr Chotanagpur, p. 7.

27. Ibid., no. 889-13E, P.O. Hinoo, Ranchi, 18 May 1927, A.J. Gibson, CF B&O, to Secy GOBO Rev, pp. 5–6.

28. Ibid., no. 1465R, 30 May 1927, J.R. Dain, DC Singhbhum, to Commr Chotanagpur, p. 8.

29. OIOC P/10498 BORP (For), 1919, A progs 6–29, Oct 1919, file IIIF/32 of 1919, no. 2562XXV-P-22, Ranchi, 16 Jul 1919, F. Trafford, CF B&O, to Secy GOBO Rev; no. 3871XXV-P-24, Ranchi, 11 Nov 1918, Trafford to Secy GOBO Rev, pp. 7–13; OIOC P/11666 BORP (For), 1928, A progs 5–27, Jul 1928, file IIIF/19 of 1928, no. 3750-13E, P.O. Hinoo, Ranchi, 13 Sep 1927, J.H. Lyall, CF B&O, to Secy GOBO Rev; no. 1383IIIF-212RR, Ranchi, 4 Oct 1927, R.E. Russell, Offg Secy GOBO Rev, to Commr Chotanagpur; no. 1532-13-E, Ranchi, 6 Jun 1928, A.J. Gibson, CF B&O, to Secy GOBO Rev, pp. 9–23; OIOC P/11752 BORP (For), 1929, A progs 1–19, Dec 1929, file IIIF/25 of 1929, no. 132RR, Ranchi, 3 Oct 1929, R.E. Russell, Secy GOBO Rev, to CF B&O, p. 97.

30. OIOC P/11904 BORP (For), 1931, A progs 24–40, Jan 1931, file IIIF/150, no. 9750-13E, Ranchi, 22 Mar 1930, E. Berrskin, Offg CF B&O, to Secy GOBO Rev; no. 1496R, Ranchi, 15 Apr 1930, E.H. Berthond, Commr Chotanagpur, to Secy GOBO Rev; no. 61IIIF/150RR, 22 Apr 1930, R.E. Russell, Secy GOBO Rev, to CH B&O, pp. 19–26.

31. Stebbing 1924: 376.

32. Brandis, *Suggestions for Darjeeling and Jalpaiguri*, pp. 24–5.

33. NAI Rev and Agri (For), A progs GOI 60–2, file 27 of 1912, Feb 1912, no. 125, Cal, 9 Jan 1912, J.H. Kerr, Secy GOB, to Secy GOI, p. 172.

34. Brandis, *Suggestions for Darjeeling and Jalpaiguri*, p. 26.

35. Ibid., p. 25.

36. Stebbing 1926: 209–10.

37. Stebbing 1924: 548–9.

38. NAI Home Rev and Agri (For), progs 1–2, Nov 1882, no. 506Rct, Darjeeling, 23 Aug 1882, Lord Ulicke Browne, Commr Rajshahi and Cooch Behar Div, to Secy GOB Rev, pp. 3–4.

39. OIOC P/2489 BRP (For), Jun 1885, A progs 10–17, no. 146TP-4, Darjeeling, 27 Sep 1884, A.L. Home, CF Bengal, to Secy GOB Rev; no. 10R,

Rampore, 16 Feb 1885, Lord Ulicke Brown, Commr Rajshahi, to Secy GOB Rev, pp. 97–103.

40. NAI Home Rev and Agri (For) A progs 1–2, Nov 1882, no. 511T-R, Darjeeling, 2 Oct 1882, A.P. Macdonell, Offg Secy GOB Rev, to Commr Rajshahi and Cooch Behar Div, pp. 3–4; OIOC P/2489 BRP (For), Jun 1885, A progs 10–17, GOB Rev notification, Cal, 9 Jun 1885, p. 104.

41. A similar problem was created when the development of a civil station or cantonment in a region increased the goala population dramatically to cater for the rapid increase in demand for fresh milk. The forests of the area then came under greater grazing pressures than when merely serving a neighbouring rural population.

42. Notes, queries and extracts relating to the meeting about the grazing question in Darjeeling, reproduced in *IF* 1884, 10(7): 332.

43. Thompson 1975: 60–2.

44. WBSRR GOB Rev (For), A progs 4–13, Jul 1898, file 10-R, notes in keep with papers of Wheeler and Finucane, pp. 1–2.

45. WBSRR GOB Rev (For), file 3W/4, A progs 4–16, Dec 1904, no. 347-119-2F, Cal, 11 Mar 1904, L. Robertson, US GOI Rev and Agri, to Secy GOB Rev; note dated 12 Jan 1904 by S. Eardley Wilmot, IGF; no. 130, Darjeeling, 10 Aug 1904, A.L. McIntyre, CF Bengal, to Secy GOB Rev, pp. 1–6; OIOC W 3022, W. Schlich, *Memorandum on the Management of the Forests in the Darjeeling Division of Bengal* (Simla, 1882), pp. 5–7.

46. WBSRR GOB Rev (For), Dec 1904, file 3W/4, no. 4956, 2 Dec 1904, Secy GOB Rev, to Commr Rajshahi.

47. WBSRR GOB Rev (For), file 3-W/1, A progs 6–12, Nov 1905, no. 681RCT, Darjeeling, 5 Jul 1905, C.R. Marindin, Commr Rajshahi, to Chief Secy GOB; no. 2713T-R, Darjeeling, 9 Sep 1905, R.W. Carlyle, Chief Secy GOB, to CF Bengal, pp. 1–5.

48. NAI GOI Rev and Agri, Selection of papers relating to measures taken in the various provinces for the utilization of fodder and forest grasses from reserved, protected and unclassed government forests, note dated 17 Sep 1914, M. Hill IGF, p. 8; no. 56C, 16 Jan 1913, C.E. Muriel, CF Bengal, to IGF, p. 38.

49. Grieve 1912: 20.

50. WBSRR GOB Rev (For), file 5M-1 of 1917 and file 5M-1, A progs 4–10, Sep 1918, no. 1563-1W-1, Darjeeling, 19 Jul 1917, Sir H.A. Farrington, Offg CF Bengal, to Secy GOB Rev; no. 1799, Cal, 22 Feb 1918, L. Birley, Secy GOB Rev, to CF Bengal, no. 2042-1G-4, Darjeeling, 20 Jul 1918, H.A. Farrington, CF Bengal, to Secy GOB Rev, pp. 1–2.

51. Ibid., pp. 5–11.

52. WBSRR GOB Rev (For), file 2-G/1, A progs 14–16, Dec 1913, no. 1853-844, Darjeeling, 1 Sep 1913, C.E. Muriel, CF Bengal, to Secy GOB Rev; file 2-G/3, A progs 14–19, Nov 1916, no. 1857-885, Darjeeling, 30 Aug 1916, H.A. Farrington, Offg CF Bengal, to Secy GOB Rev; no. 241CWE, Dow Hill, 14 Aug 1916, H.S. Gibson, DCF Kurseong, to CF Bengal, p. 2.

53. GOI Rev and Agri, Selection of papers relating to measures taken in

the various provinces for the utilization of fodder and forest grasses from reserved protected and unclassed government forests, Note of M. Hill, IGF, 17 Sep 1914, p. 1.

54. Ibid, p. 8; WBSRR GOB Rev (For), file 7F/3 of 1914, A progs 1–2, Jan 1915, no. 197C, Camp, 12 Dec 1914, C.E. Muriel, CF Bengal, to Secy GOB Rev, p. 1. The annual incidence of grazing was reduced from 75,551 animals to 48,387 animals in Bengal forests in the decade 1914–24. See BFAR, 1925.

55. NAI GOI Education, Health and Lands (For), file 22-4/41-F&L, 1941, no. 14803/40-IV-130, 14 Nov 1940, S.H. Howard, President FRI, to Secy GOI, pp. 9–10.

56. Stebbing 1926: 641–2; also see Prasad 1994: 150–5; Rangarajan 1992.

57. WBSRR BRP (For), A progs 26–43, Jun 1907, files 10-R/2&5 of 1905–07, no. 713-O, Cal, 22 Nov 1905, W. Parsons, Secy Indian Tea Association, to CS GOB, p. 1.

58. Ibid., no. 186G, Darjeeling, 17 Apr 1906, L.S.S. O'Malley, DC Darjeeling, to Commr Bhagalpur, p. 2.

59. Ibid.

60. Ibid., pp. 3–7; no. 2524T-R, Darjeeling, 24 Sep 1906, R.W. Carlyle, CS GOB, to Commr Bhagalpur, p. 9.

61. WBSRR BRP (For), A progs 19–35, Feb 1910, file 10-R/1 of 1910, no. 38, 23 May 1909, G.S. Hart, CF Bengal, to CS GOB, pp. 1–3.

62. Ibid., l/d 24 Sep 1909, G.W. Chistison, Lebong Tea Co., to DC Darjeeling; no. 144, Cal, 13 Jan 1910, F.W. Duke, CS GOB, to Commr Bhagalpur, pp. 6–9.

63. WBSRR BRP (For), A progs 11–14, Sep 1911, file 10-R/2, Report of the Committee on Deforestation in Darjeeling, pp. 25–30.

64. Ibid., p. 30.

65. Ibid., no. 40M Rct, Darjeeling, 20 Jun 1911, E.H.C. Walsh, Commr Bhagalpur, to Secy GOB Rev, pp. 1–3.

66. Ibid., pp. 4–5; no. 2622, Cal, 24 Aug 1911, J.G. Cumming, Secy GOB Rev, to Commr Bhagalpur, p. 47.

67. Ibid., no. 803G, Darjeeling, 14 May 1911, H.T.S. Forrest, DC Darjeeling, to Commr Bhagalpur, pp. 7–9.

68. Ibid., pp. 10–14. Tea covered 175 square miles in revenue-paying and freehold tenures, reserved forests extended over 445 square miles, cinchona plantations were on 59 square miles. This left 335 square miles of khas mahal lands and 77 square miles of revenue-paying and freehold lands under cultivation. Another 73 square miles of Darjeeling were under non-agricultural uses or political grants.

69. Ibid., p. 16.

70. This matter of the changing legal environment for land management in Bengal, and its increasing reliance on fortifying the rights of tenants and cultivators, is discussed in detail in Sivaramakrishnan 1996a: ch 8.

71. WBSRR BRP (For), A progs 11–14, Sep 1911, file 10-R/2, no. 803G, Darjeeling 14 May 1911, H.T.S. Forrest, DC Darjeeling, to Commr Bhagalpur, pp. 17–20.

72. WBSRR BRP (For), A progs 1–8, Jul 1917, file 10-R/1 of 1912, no. 250MPI, Cal, 19 Feb 1912, B.K. Finnimore, Secy PWD GOB, to Secy GOB Rev, p. 1; no. 1508 MR-A, Bhagalpur, 30 Mar 1912, E.H.C. Walsh, Commr Bhagalpur, to Secy GOB Rev; no. 6297G, Darjeeling, 19 Feb 1912, H.T.S. Forrest, DC Darjeeling, to Commr Bhagalpur, pp. 5–37. The quote is from the preamble, to the draft legislation enclosed with the last letter, p. 35.

73. OIOC BRP (For), A progs 1–4, Jan 1920, file 10R/1 of 1919, no. 813T-R, Darjeeling 18 Jul 1919, M.C. McAlpin, Secy GOB Rev, to Secy GOI Legislative Dept, pp. 2–7.

74. Ribbentrop 1900: 123–4.

75. Menzies 1994: 19. North Bengal presented a more moist, densely vegetated and erratically cultivated landscape where transhumant herding and occasional jhum combined with climate, elevation and soils to produce a different spatial arrangement of forest and field.

76. The historian of scientific forestry in Germany notes similar constraints when he says, 'traditional privileges and the continued use of the forest for such agricultural purposes as grazing or mast (windfall nuts) had long discouraged a conceptually precise demarcation of the forest.' See Lowood 1990: 320.

77. Stebbing (1926: 503–5) describes cases from north India where planning was abandoned due to the complexity of locally admitted forest usage.

78. In Japan too, forest regulation at the local level was effective during the fourteenth to seventeenth centuries and conflicted with higher levels of regulation because of agricultural dependence on woodlands. See Totman 1989: 33–8.

79. Rangarajan 1998.

80. See Parker 1923; Pearson and Brown 1981 (1932); Marriott 1928; Mobbs 1941; Warr 1926.

81. Ludden 1992b: 177.

82. See Cline-Cole (1994), for a comparable African case that makes a similar point about separation of agriculture and forestry–internal territorialization.

83. McEvoy 1986: 13.

84. Arguably, Gregory (1994), Haraway (1991), and Latour (1993) are notable among recent scholarship on nature, science and technologies of power for suggesting ways to move beyond nature/culture dualisms.

85. Ramachandra Guha (1989), Rangarajan (1992), Prasad (1994), Skaria (1992), Bhattacharya (1992) are all important and sophisticated studies from different parts of India that do not systematically examine the implications of what I have dealt with as regional variations. Other studies, operating at the national level, have been even more susceptible to a Fernowesque sweep of generalization. See Totman (1989) for Japan; Peluso (1992) for Indonesia; Vandergeest and Peluso (1995) for Thailand. Menzies (1994) does discuss regional variations, but only to report their erasure by late-nineteenth-century centralization.

86. D. Brandis, *Suggestions Regarding Forest Administration in Central Provinces* (Calcutta: GOI Press, 1876); idem, *Suggestions Regarding Forest Administration in Northwestern Provinces and Oudh* (Calcutta: GOI Press,

1881); idem, *Suggestions Regarding Forest Administration in Madras Prsidency* (Calcutta: GOI Press, 1883); Brandis 1879, 1897.

87. For a discussion of the Cleghorn/Gibson arguments and their diminishing influence on forest policy in India by the late nineteenth century, see Grove 1993a.

88. The architect of this view was J.A. Voelcker in his *Report on Indian Agriculture* (Calcutta, 1893); also see Taylor 1981.

89. Brandis 1897: 78.

7

Nature's Science: Fire
and Forest Regeneration

Introduction

Soon after the introduction of forest conservancy in Bengal during
1864, colonial foresters had their first notable success with fire
protection. Colonel Pearson, aided by two years of heavy monsoon
rains, had protected the Bori forests of the Central Provinces from
annual fires in 1865. Over the next decade the exclusion of fire
from reserved and other categories of forests became emblematic
of what modern management could accomplish in the forests of
India. By attempting to banish fire from the landscape, European
forestry distinguished modern forest management from the primitive techniques it claimed to supersede. The idea derived from
similar contrasts through which agriculture had been transformed
earlier in Europe.[1] By the 1870s, when forest policy was framed
and institutionalized in various parts of India, it certainly drew
on such a language of improvement, casting the project of
colonialism in terms of reclamation—of both colonized peoples
and lands. In 1875 the *Indian Forester* journal was established by
the Bengal Conservator, William Schlich, as the flagship journal
of scientific forestry. The very next year, Baden-Powell, one of
the architects of colonial forest policy wrote in its pages:

it is possible that the progress of knowledge has driven out this idea to
a great extent in India, but there are here and there some few who still
argue for forest fires, just as in England some eccentric individuals
occasionally question the fact that the earth is round.[2]

Certitude of this sort illustrates the confidence of imperial science. The assurance filling these words was inspired not only by what Whiggish notions of history predisposed British colonial administrators to think but also on their reading of nature in the colonies.[3] These representations moved swiftly across the diverse terrain of Bengal, cataloguing its non-agricultural resources and despairing at their mismanagement. The possible autochthonous remedies to these perceived problems, or the idea that any blanket view of nature's destruction by 'native' society was a misrepresent-ation, remained undiscovered. Political and ecological order then could only follow the application of external controls and foreign expertise. These predilections further ensured that fire, in the Indian forests, was firmly lodged in the colonial imagination as something rampant, random and reprehensible.[4] Yet, in the next fifty years, as British foresters in Bengal turned their attention to intensive forest management and regeneration of desired species, especially sal, they ruefully discovered that fire protection had created a situation in 'which in the moister tracts weighs heavily against the very species it was designed to assist.'[5] This remark summed up the lessons provided by growing evidence flowing from various sal bearing regions of eastern and northern India following a discovery made in Bengal two decades before these words were written. An itinerant expert had already drawn attention to the awkward conflict between prevailing scientific wisdom and the 'realities of nature' in sal areas across eastern India, saying:

anyone who considers the use of fire a breach of the laws of sylviculture need only see the excellent condition of a forest continually burnt . . . where you find all ages represented by different groups of sal . . . no creepers and remarkably rich patches of regeneration where you have an opening.[6]

This story about the case of fire and forest regeneration in sal regions of Bengal, provokes a re-examination of the processes through which colonial science was constituted. The term 'colonial science' is admittedly a lumped category. But like the term 'colonial knowledge', used in Chapter Three, it is useful to signal a network of ideas and practices that constituted the scientific corpus of colonial governance—in other words, technologies of power. My argument is that these ideas and practices remained under production in particular historical moments and in specific colonial locations. They bear rich testimony to the plurality and diversity of colonial situations that contributed to the heterogeneity and conflicted character of colonial science.

As a constellation of ideas, interests, and governmental procedures, colonial science aspired to coherence and achieved an integrated force. What was its relationship to imperial expertise? What does it show about the entanglements of knowledge and power? Did nature itself, or local politics, or regional history, or imperial design most definitively shape forest management in Bengal? How do we refine our understanding of scientific forestry in India? These are the organizing impulses for this chapter which concentrates on one of the most contentious issues, termed 'biotic interference' in forest management—fire. The chapter concludes with a brief discussion of the other biotic factor, grazing, and the indeterminacies that dogged scientific forestry into the last decades of colonial empire.[7]

Normalcy for Nature: Systematic Science and its Field of Vision

Clearly colonial science was about a style of imagining the world: a particular modern regime of representations. We have to focus, however, not merely on the content of these representations, but 'on the process of their composition and deployment.'[8] In what is one of the earliest systematic discussions of scientific forestry Ramachandra Guha shows the adaptation of colonial forest management in Uttarakhand to historical contexts of peasant resistance. But he does conclude, 'both legislation and silvicultural technique were designed to facilitate social control.'[9] This allows for no discordance between politics and science, nor does it explicate the nature of social control and the constant shifting of mechanisms. In this mode of analysis, colonial forest science is a received version of European models that is then subordinated to economic imperatives. Writing on the work of the Imperial Institute, Michael Worboys comes to a similar conclusion. In his account it is only in the 1920s that the decline of the institute signals emerging independent, scientific infrastructure in India and the Dominions.[10] Recently, other historians of colonial science have questioned such simplification. Challenging this perspective of Worboys, which also informs his work on malnutrition and diet, David Arnold asks, 'is it appropriate to look to metropolitan direction and design in the fashioning of colonial science or should we look instead to local processes of investigation and analysis?'[11] He goes on to demonstrate that between 1860 and 1914 famines and jails provided important contexts in India for widening knowledge and debate about nutritional sciences.

Similar controversy over centralist and peripheralist historiography of colonial science has broken out in other branches, notably the history of colonial geology in India.[12] For imperial historians empire was a captive laboratory for British geology, thus perpetuating a division of labour between metropolitan theorizers and provincial fact gatherers.[13] Stafford joins other historians of the Victorian era of exploration and bio-geography in proposing a congruent efflorescence in geographic sciences and maritime and colonizing traditions, where the goals of science were commensurable with the civilizing mission.[14] After all, Darwin, Forbes, T.H. Huxley and J.D. Hooker had all sailed on Royal Navy hydrographic expeditions, and 'obvious parallels exist between the belief of these scientists in the dominance of northern life forms and their faith in England's manifest destiny as an imperial power.'[15]

That said, while we may be closer to understanding parts of the imperial imagination, we remain far from explaining the processes in which it was formed. John Mackenzie draws our attention to this issue astutely when he points out that the nineteenth century was a period when amateurs dominated science.[16] So drawing sharp distinctions between doers and receivers may be problematic. We may misrecognize the process of scientific discovery, ignore failures and false moves in transfer of metropolitan science to the colonial province, and omit noting the conditions that empire often uniquely offered for scientific discoveries. In short, there is an interactive and contextualized production of knowledge that needs to be understood.

The history of natural science illuminates the interaction of imperial and indigenous ideas,[17] and the role of colonial experts, from forest guards to field foresters, in shaping this interaction. Evidently their expertise combined formal ideas and substantive local knowledge, and after a point trying to separate the two becomes an exercise in reification that provides no insight.[18] What do we lose sight of? First, as Sangwan perceptively points out, whether favouring local initiative or central fiat in explaining the conduct of colonial science, the debates referred to earlier and the 'two systems approach' share the common failing of treating the scientific establishment as a monolith that is a puppet either of ideologies of empire or colonial exploitative arrangements—science as a tool of revenue generation. He says, 'they ignore the internal culture of the *corps d'elite*, especially the making of a scientific discipline with its historical transition from amateurism to a professional form of activity.'[19]

Providing a welcome look at the absorption of indigenous knowledge in particular domains of science, and how local officials sought independence from metropolitan control, Satpal Sangwan notes further, 'though the colonial state pressed for a profit oriented scientific agenda, the scientific part was never completely set aside.'[20] In the same vein, Richard Grove has emphasized the role of colonial naturalists and surgeons in developing a discourse of environmental degradation in early-nineteenth-century India.[21] Mahesh Rangarajan has added by suggesting ways to comprehend the connections between conservationist ideas and specific policy changes in the early history of colonial forestry in India. Grove has also argued elsewhere that research done by botanists and medical specialists under the auspices of the East India Company facilities was important to establishing the relationship between global climate, drought, monsoons, hurricanes and El Nino and Southern Oscillation (ENSO).[22]

Tree planting and forest conservation in India began in the early nineteenth century based on such climatological knowledge collected by people like William Roxburgh, in early years of East India Company service, who then put it to use for promoting plantation forestry when he became the Director of the Calcutta Botanical Gardens. The famine of 1877–9 inspired a lot of work on the relationship between deforestation and rainfall under Edward Balfour, Surgeon-General, and the Inspector-General of Forests.[23] This resulted in prolonged controversy between different branches and levels of the colonial Indian government. When the research on relationships between climate and forest cover was given up in the early decades of the twentieth century, it had already contributed to a contentious and unsuccessful attempt to intervene in private forests in Bengal. A highly systematized and polarized view of western and local knowledge in the making of colonial science can underestimate the kind of productive forces unleashed by discord and conflict within colonial bureaucracies.

By separating the production of knowledge from the processes of statemaking in which it is embedded, the complex interlocking of governance issues with forms of knowledge and representation may be overlooked. We will do well to remember that discoveries made in colonies often 'consisted of a gesture of converting local knowledges into European national and continental knowledges associated with European forms and relations of power.'[24] What passes for local or modern scientific knowledge bears marks of such appropriation and counter-appropriation in specific colonial contexts. In an intriguing history of the management of air pollution in Calcutta through the nineteenth century, Michael Anderson has shown how the definition and regulation of smoke nuisances was peculiar to the colonial context of Calcutta as the seat of Indian empire. In addition to the valuable insight that 'many features that are now axiomatic in environmental regulation— systematic monitoring, reliance on technical experts, technological remedies, and close collusion between industry and bureaucracy— were consolidated under the aegis of the Smoke Nuisance Act,' Anderson illustrates how the conquest of smoke became a unique colonial enterprise.[25]

This chapter is also necessarily about the interplay of patterns in nature and schemes of representation. By 1905, making the first revision in the forest working plans for Jalpaiguri Division in the Bengal Duars, Trafford, the local forest officer, remarked upon the difficulty he faced in classifying the landscape into vegetation types. Vexed in particular by the mixed forest and savannah areas identified in the previous plan, he said, 'these types not only merge into one another but are constantly changing and a stock map is constantly getting out of date.'[26] In the next thirty years issues of classification gave way to the struggle over forest regeneration, reflecting a worldwide transition in the status and directions of ecological research and knowledge.[27] As Indian foresters grappled with sal regeneration, earlier forest classifications were revisited and altered in later years and forests were redefined to incorporate freshly gained awareness of ecological dynamics within them.[28] As a luminary of sal silviculture research from the period put it, 'the idea that a sal forest may be transient and in the time space continuum may fade away as a morning mist, will probably come as a shock to many foresters.'[29]

Such impermanence in the landscape bore no simple explanation. It was a result of both ecological processes and political inaction in north Bengal, something we shall examine in detail later in this chapter. At this point, I will merely anticipate and highlight the argument by pointing to the inadequacy of recent theorizing in cultural geography and environmental history for understanding sal regeneration in Bengal and its historical construction in the corpus of scientific forestry. The emphasis of cultural geography on representations, especially the inherent instability of meaning, is often taken to the point of making nature epiphenomenal or passive, a text, spectacle, icon or theatrical production.[30] An analysis that focuses only on a clash of representations and interests, howsoever the groups engaged in such conflict are defined, reproduces this tendency noted in cultural geography. Both in India and elsewhere, studies of forest management and resistance, though varying in their deployment of cultural analysis, have shared this neglect of ecological processes.[31] At the same time, we have to remain wary of the innate biocentrism and essentialism of environmental history, whether it takes the form of Worster's homoeostatic nature, Merchant's nested model of ecology, production, reproduction and consciousness, or Cronon's sophisticated analysis of 'first nature' and second nature.[32] All these fine exponents of environmental history assume at some point that nature simply is. My analysis, in contrast, wishes at no stage to disable 'consideration of the processes by which what passes for nature is actually determined.'[33]

At the time of forest demarcation and protection, grazing, fire, shifting cultivation and the collection of firewood rapidly became stock items on the standard list of ills to be curbed.[34] Forest fires, 'as the concomitants of both the nomad and the backwoods cultivator,' represented to senior foresters aggravating demands on forests from a primitive state of society, less legitimate than those made by 'rapidly spreading modern civilization.'[35] More immediately, during this phase, freedom from cutting and immunity from fires were central to the plans Brandis made for forest management in all provinces, including Bengal.[36] The rationalization of jungles (signifying wastes and wildlands) was part of a wider impulse to reform and manage the productive landscape. To politics British foresters added 'the symbolism of science'. To justify fire control they drew upon the history of transformations in European agronomy, where the divide between primitive and modern was instituted by the exclusion of fire.[37] As the noted historian of forests and fire observes, 'what emerged was a robust exemplar, an adaptation of European techniques to exotic woodlands and colonial politics.'[38] In his review of 1884, the successor to Brandis as Inspector-General of Forests, William Schlich, wrote, 'owing to the vast extent of the Indian forests and high cost of artificial operations, the natural system of reproduction must be chiefly relied on, that is to say forests must be properly protected, fires kept out . . . areas closed to grazing.'[39]

In Bengal, in the first flush of forest reservation almost everything was closed to firing. By 1900 the area under reservation had increased to 5,880 square miles, and the total managed area (including protected and unclassed forests) was up to 13,589 square miles in Bengal. This period also marked the beginnings of more scientific operations as the task of preparing working plans was taken up.[40] Each of these features of scientific forestry merits consideration in detail, especially since they are best understood through their interrelationships. Here only fire and grazing are discussed, tracing the attempts to accomplish fire protection in

Table 7.1. Fire Protection in Bengal Presidency (area in square miles)

Year	Reserved and Other Forests	Fire Protected Forests	Per Cent Excluded/Failure
1878–9	3,474	2,938	15
1879–80	4,072	3,709	09
1880–1	4,465	4,194	06
1881–2	4,897	4,283	13
1882–3	4,826	4,562	05

Source: Adapted from Schlich, 'Review of Forest Administration for British India', p. 376.

Bengal forests. Working plans and sal silviculture are taken up in the next chapter.

While southwest Bengal became the region where fire policies foundered on poor implementation, they were embarrassingly successful in the northern parts of Bengal, only to be confounded in their objects by this very fulfilment. The critique of fire exclusion from forest regeneration emerged from the confluence of these experiences, producing a range of strategies—both silvicultural and political—by which fire was incorporated in forest management. Distinguished both by regional political economy and agro-ecological characteristics, this north–south divide entered and shaped the classification of Bengal sal forests as dry (mostly southwestern) and moist (mostly northern).[41]

Fire in the Forest: Organizing its Exclusion

One of the major concerns of Indian silviculture, as it developed through the first decades of the twentieth century, was securing the natural regeneration of valuable timber species, which in eastern India were principally sal and teak.[42] Some niche partitioning between the two species does seem to occur given their different preferences for soil moisture regimes. Sal, not liking high residence times for soil water, avoids lowlands, hollows, plains and seeks highlands and hillslopes. These are typically areas that have nutrient-poor soils that are excessively drained and often high in sand content. Teak does better in mesic valley bottoms as it needs greater soil fertility and moisture which is available in the clayey floodplains. Thus sal forests seem to be in regions historically home to the socially marginal, like hill tribes, hunters, swiddeners and so on. None of these regions were free of anthropogenic disturbances; they were not virgin forests. As Brandis puts it, 'in the wildest forest regions of India we constantly come across evidence that the land at one time had been under cultivation—fruit trees, ruins of large buildings and terraces of old fields.'[43] Old-field sal forests less than two hundred years in age often contained old mango and tamarind trees, presenting evidence of former cultivation that had usually receded due to pestilence, famine or raids by hill tribes.[44]

Initially forest regeneration was seen as unproblematic,

something that would naturally follow protection. So all forest management in Bengal and other comparable regions in India began with keeping biotic influences like fire and grazing out of identified forest tracts. Fire protection thus became a marker of forest conservancy, serving not only to distinguish state forests from others but exemplifying care and farsightedness in management. Fire protection involved clearing fire lines and paths around the protected areas, and early dry-season burning of leaves and grass in broad belts surrounding the forest.[45] It also consisted of clearing internal and external fire lines and sweeping them regularly, for which village headmen and raiyats provided free labour in exchange for privileges in the forests. The earliest forest guards were primarily appointed for this purpose. Faced with both a daunting task and one of dubious wisdom, these guards were among the earliest sceptics. They often burned forests surreptitiously, if only to improve the prospect of limited protection.[46] Sometimes, almost unwittingly, policy and practice coincided. Brandis, in his first tour of north Bengal forests in 1879 had recommended that savannah around forest areas be burned early in the season when the grasses are drier than the forests around them. He also envisioned a future of more complete fire protection where both savannah and forest would be protected to facilitate extending the forest margins into grasslands.[47]

The importance of fire protection, especially in north Bengal, also grew with the noticeable failure to establish plantations of exotics like mahogany, chestnut and teak. By 1876, Schlich, the Conservator of Forests, Bengal, had started planting 100-foot sal belts in clearings around the Buxa reserve to protect it from fire. Curiously, while fresh sowings of sal invariably failed, transplants from the forest did much better.[48] Brandis acknowledged the role of fire in this outcome:

the vigour of these transplants is doubtless due to the large knobs of wood which form the underground stem, the result of forest fires, which kill the annual shoots year after year, while the underground portion increases steadily in bulk.[49]

But during the early years of forest conservancy such evidence was largely disregarded. Fire became an agent of deforestation, through the reckless acts of nomadic tribes or the extension of cultivation and pasturage, for all the nineteenth-century colonial foresters and scientific professionals.[50] Surveying the newly acquired north Bengal forests, the Bengal Conservator complained of frequent forest fires caused by shifting cultivation. A struggle ensued to establish Forest Department control in these areas by the restriction of firing. This was especially important as these forests in the Darjeeling foothills were mainly young sal regeneration, for anything above six feet in girth had already been taken out by timber contractors.[51]

While fire protection was introduced in Bengal, the administration soon encountered various problems, since it interfered with the agrarian economy and lifestyles in a variety of ways. For instance goalas used fire to secure fresh supplies of pasture grasses, while collective village fires in the early summer months were usually intended to facilitate the hunting of small game, foraging and gathering. In some upland tracts, wood from the forests was burned for manuring neighbouring fields.[52] In different parts of Bengal, fire would smoke out bees for honey, burn under mahua trees and other fruit trees to facilitate fruit gathering, and burn back tall grasses in the proximity of hamlets to keep away tigers and panthers.[53] The Deputy Commissioner of Darrang, Major Graham and Major Lamb of Kamrup argued that forest fires helped clear land for cultivation, extirpate wild beasts and regenerate pastures. Another district officer pointed to the rank and dense undergrowth that comes up in a fire-protected forest, which increased malarial parasites.[54] In one intriguing case, a forester reported that fire conservancy had increased the tiger menace since fires traced the routes taken by deer and tigers preying on them, thus indicating their place in the forest and making hunting easier.[55]

So objections to fire protection arose almost immediately after its introduction, but at this stage, in the 1870s, these contestations of policy pitted the field officer against the forestry expert, with the former doubting both the practicality of the exercise and its desirability in political and environmental terms. Arguments about fire protection revealed the territorial aspirations of the Forest Department, though they were conducted in the language of scientific certitude. The Deputy Commissioner of Darjeeling suggested that the boundaries of government forests be revised to include a smaller area. This in turn would interfere less with the spread of cultivation which would obviate the need for fire protection as more people took to settled farming in preference to mobile pasturing. To this officer, as to most other district officers in hilly and forested areas, material progress of the district was linked to the increasing sedentarization of peasantry in his jurisdiction.[56]

Ramachandra Guha and Madhav Gadgil describe several instances of struggle over forest use between peasantry and the colonial Indian state, and in many, fire played a role.[57] But one of the best discussions of forest fire and protest is the case of Punjab discussed by Neeladri Bhattacharya, which led to proposals for what could be called the colonial version of joint forest management. Bhattacharya points out, 'it was in the heart of the forest, where all use rights were expropriated that fires often originated. And frequently they broke out at three or four places simultaneously in the dead of the night.' Local forest officials feeling beleaguered, sometimes brought forth proposals that would return some rights.[58] Such are the instances of pressures from below that create a gap between plan and practice, or even lead to policy amendment, without necessarily being reflected in legislative reform.

These crimes of anonymity, as E.P. Thompson once called them, were not, however, the only way fire was used in negotiating fire conservancy and protection policies.[59] In the eastern Duars, zamindars had always levied a revenue on sal timber and villagers who practised shifting cultivation would not damage certain sal groves near settlements and river banks. In these forests the grass would be beaten down with the rains and burned soon after, causing no damage to regenerating sal poles. Cattle grazing in these woods and pigs rooting alongside (in the case of tribal hamlets) would leave sal seedlings alone. In other fields the grass was grown to three or four feet, called batha bun, and this would be harvested as thatch material. Jhumming was practised in more interior forests, where young second-growth sal was slashed and burnt for growing cotton. In the 1870s, with the decline in demand for Indian raw cotton and increase in the timber value of sal forests, this complex fire regime was simplified into an unalloyed evil for the forest. Notices on fire protection of sal forests were circulated, but by the admission of the Conservator of the region himself, they were a dead letter, grass was too extensive and the population too scanty to make supervision of such a policy feasible.[60] In this situation the conservator developed a system that combined intense local ecological knowledge and a series of concessions to the hardy Mechi people of the area (woodcutters by profession) to implement a selective fire protection programme.

In forest reserves, grazing was generally prohibited, but concessions to gather edible roots and fruits were usually granted. These collections took place in the very season that the forest was most vulnerable to fire—the dry, hot summer months. According to Wild, the Bengal Conservator, most fires were caused by these collectors and their carelessness with home-made cigarettes.[61] As the Deputy Conservator pointed out

in Singhbhum . . . the villages are small. The inhabitants are members of a community, perambulate the forests, reserved as well as protected, in small parties . . . in their search for fruits, honey, roots, sometime game and many parts to collect sabai grass.[62]

A communal responsibility for fire prevention was envisaged in these situations because such gathering activities were permitted only informally, as a local relaxation of a strict wider policy, thereby creating an obligation on the community as a whole. Another proposal was to give specific Mankis and Mundas responsibility to supervise fire protection in assigned forest blocks. This would draw on the Manki and Munda pattas where, as local police officers, they were charged, among other things, with reporting instances of incendiarism.[63]

In 1903–4, fire damage to forests in Singhbhum reached an all-time high of 57,330 acres as opposition to fire protection grew and gained a sharp edge. The increase in forest firing was a direct response to the Working Plan, which had made villagers habituated to gathering firewood from any part of the forest, restrict their collection activities to four specified coupés. This was exacerbated by a change in policy from 1900 when, instead of selective early burning, complete fire prevention was introduced.[64] Early burning was often confined to those mixed forests (as opposed to sal forests) where grazing was considered the more important revenue source than trees.[65] The failure to reduce the incidence of forest fires caused by gathering activities of tribal communities, however, remained a lasting concern.[66]

Fire control focused on the protection of 'valuable trees' for their timber during the hot months, which was the very season that villagers around the forest used fire to derive from the forest grass, fruit, flowers, small game and other non-timber products that supplemented their diet and income in the agricultural lean season. But in parts of north Bengal there were other problems that occurred during and outside the fire season. Seedlings of planted tun (*cedrela toona*) and walnut were cut and used to stiffen bundles of grass.[67] Streams were diverted in the cold season by small dams to catch fish, leading the Working Plan for Tista Division to observe that changes in beds of streams affecting valuable lands had been traced 'directly to this dangerous practice'.[68] At least in this division, elephants were more a menace to the valued forests than local human populations. So fire posed varying degrees of hazard to forests in different parts of Bengal, but in all locations fire control attacked local livelihood directly. It was particularly resented in the protected forests which were considered to be outside reserves, and hence not so readily subject to restrictions. Moreover, fire extinguishing work, in which proximate villagers were hired as coolies, was a source of wages in the lean season.[69] In Bengal as a whole, while the area covered by fire protection steadily increased, the annual figures of areas burned showed no steady decline, though they did exhibit distinct regional variations. Thus in north Bengal by 1900 fire protection was effective and quite complete. In southwest Bengal much more of the protected area continued to burn, and the extent to which fire protection would fail was quite unpredictable from year to year.

Despite its climatic range, and different provenances, sal in all ecological zones is critically dependent on fire for regeneration. Fire plays a role in both initiating and limiting sal. This means that the managed use of fire in the days before scientific forestry and fire protection was important not only to the production of the subsistence and other annual crops for local communities of the Bengal highlands, but also ensured the perpetuation of the mixed sal forests the British found and chose to exploit. Scientific forestry was a matter of changing the fire regime in the agro-silvical environment natural to sal, with a view to producing pure sal stands in multiple-use fields.[70] This produced unintended consequences and later started a process of structuring scientific silviculture with local knowledge about the use of fire. This kind of learning and reform based on field experience mitigated some of the more radical transformations of the landscape that would have followed prolonged fire protection.[71]

Discrepant results from fire protection produced regionally varied responses as scientific forestry was inflected by particularities of ecology and politics. In southwest Bengal, the region of dry sal, fire protection was confounded in various ways, so problems arose from its incompleteness. In north Bengal, the domain of moist sal, fire protection was too successful, and its completeness unleashed ecological transformations that were unsought.[72] What happened in the north is of great salience because, along with a parallel critique of fire protection in the teak forests of Burma, the experience of sal regeneration there permanently undermined the consolidation of scientific forestry and sent its planners scurrying back to the drawing board, arguing furiously among themselves. What emerged from those frantic decades was a critique from within, a privileging of local knowledge, a fractured wisdom tersely expressed by a participant in the All-India Sal Study Tour of 1953 thus, 'for the success of sal natural regeneration close individual attention and intimate knowledge of local conditions are essential.'[73] Bengal foresters learnt the political pertinence of local knowledge most often in southwest Bengal and the dry sal zone, but the moist sal and savannah complex of north Bengal was the landscape where ecological aspects of local knowledge became crucial to the construction of the discourse of sal silviculture.[74]

Scientific Forestry Fractured: Bringing Fire Back in

The introduction of fire protection as a forest management and regeneration strategy was contested almost from the very beginning. By the early 1870s district officers were voicing their discontent and objection to this policy. We have already encountered some of the cases from Bengal. In some ways these tensions arising from fire protection are easy to anticipate when we examine the most common causes of forest fires. These turn out to be, according to a report prepared on the subject in 1875, taungya or other forms of shifting cultivation (jhum, kurao, bewar), hunting small game, refuse burning, cleaning paths and tracks to avoid snakes, burning back forest fringes of villages to keep off tigers and bears, cleaning the air and killing of insect pests, and securing fresh pasture and thatch grass crops.[75] Brandis himself had not failed to note the beneficial fertilizing effects of wood ash from forest fires on teak taungyas. In a later report on the forests of north Bengal, Brandis expressed doubt whether fire protection was going to yield sal regeneration.[76]

Despite these early misgivings, fire protection proceeded apace, gathering momentum to such a degree that in the decade 1896–7 to 1906–7, the area covered in India increased from 1,856 square miles to 8,153 square miles.[77] But even the colonial forest historians were later compelled to admit that indiscriminate fire protection served more a political purpose of asserting unequivocal Forest Department control over newly reserved areas, than any scientific plan for regeneration of these areas. But since it was done in the name of both conserving and enhancing forest wealth, fire protection, initially attacked by local officers of civil administration, soon came under the critical scrutiny of those charged with the management of forests, namely, field foresters.[78]

Slade, a Burma forester, fired the first salvo when he argued that low ground fires removed slash, minimized fungoid and insect pests and had in fact produced the teak forests that were later reserved. A few years later, R.S. Troup published the results of experiments in Tharrawady where he had found that unprotected plots of teak had more sound stems, and ten times more seedlings, and that stems in these plots were in no danger of suppression as were half those in protected plots. He was led to remark, 'we are most certainly exterminating our teak by fire protection.'[79] Other foresters in eastern India corroborated. In 1907 the CCF of Burma, Beadon Bryant, offered the dire prognostication that the combination of selection felling and fire protection was killing out teak in most Burma forests.[80] The first two decades of the twentieth century saw the pages of the *Indian Forester* filled with debate on fire, grazing and forest regeneration. This did not presage the overturning of former attitudes to these 'biotic factors'. But it certainly prevented any consensus on these issues and forced a re-examination of categories through which both scientific forestry and its Other were defined.

By 1925 fire had entered the working and management of

forests as an element of taungya in north Bengal, parts of United Provinces and of course Burma.[81] In Assam fire protection had been given up by 1915, while the introduction of taungya provided a mediation of the needs of forest regeneration and the growing demands for land by Kukis, Lushais and Cacharis.[82] The intensification of regeneration work that was involved in taungya was resented not only by jhumiyas but also the subordinate staff accustomed to less strenuous revenue station work.[83] In other parts of eastern India, as elsewhere, it appeared in the form of 'early burning', where the forest floor was burned just before the advent of the dry season to prevent unmanageable conflagrations at a later time.[84] The growing scepticism about fire exclusion as a viable regeneration strategy for teak and sal forests in eastern India was supported by similar contestations of fire protection policy in United Provinces and Central Provinces.[85] Fire protection had become silviculturally unsound, financially unviable and impractical in labour-scarce forest areas. But more importantly the discovery of a positive role for fire in forest regeneration meant that one foundational feature of scientific forestry, its radical separation from anything unscientific by the definition of fire as bad for forests, came to be questioned. The destabilization of scientific forestry that ambiguity about fire entailed, ultimately led the Forest Department away from natural regeneration of endemic species to a regime of planted exotics in the schemes of post-colonial forest development.

In its optimal habitat (Duars, Assam), the continued fire protection of sal for a period of about forty years (1890–1930) altered the character of the forest, introducing a previously non-existent evergreen undergrowth, increasing the soil moisture and decreasing the soil aeration. As a consequence, even though the previously established crop continued to flourish and the tree seeded freely and seeds germinated, new regeneration could not establish itself.[86] At this stage regeneration of a forest too damp to burn posed a problem and the only way out seemed to be clear-felling, burning the dry refuse, soil working and artificial regeneration (Figure 11).

The case of the Duars is instructive about how sal profits from adverse conditions to establish itself gregariously. The corollary wisdom is that altering conditions for sal regeneration, apparently for the better, can cause sal to recede.[87] The sal forests of Assam

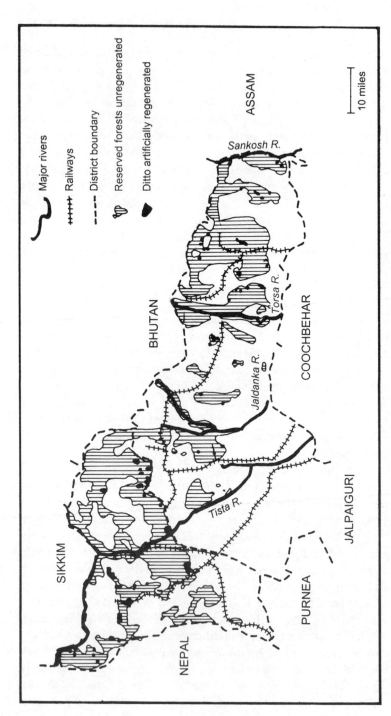

Fig. 11. North Bengal forests and areas under artificial regeneration, 1928–9

were originally thatch lands, mainly *imperata cylindrica* with sal seedlings nestling in the grass. Fire protection released the sal seedlings and as the canopy of the sal forest closed, the grasses were suppressed.[88] The sal forests of north Bengal had a similar origin. This suggests that successful regeneration of sal depended on a regulated use of fire and the knowledge of the ecology of the sal forest. In Nepal, without fire protection, sal regeneration was good. In the United Provinces, after fire protection sal regeneration failed due to seeds not reaching mineral soil through a thick mat of evergreen undergrowth. Where the seed did penetrate the forest floor excess moisture and poor light retarded germination.

The quest then was for reversal of conditions to re-obtain the grassy conditions favouring sal regeneration. In north Bengal, the path of taungya was taken, but in Assam some success in reversal was achieved by the Conservator Milroy. In Assam extensive burning was done during 1916–26 to release the sal seedlings and hasten their establishment as also to facilitate germination. As Milroy noted, the only alternate method was hoe cultivation: something the introduction of taungya acknowledges. A subsidiary local effect of burning was that it encouraged stool shoots to be thrown out by even quite large stumps, stimulating them in a way that fire protection had not done. This had to be combined with canopy opening to simulate 'Boko' conditions.[89] Reversal regeneration, as this came to be known, combined regulated canopy opening and controlled fire to induce in the forest floor a carpet of *microstegium ciliatum*, then *imperata cylindrica*.[90] The latter grass acts as a seed nurse for the sal seedlings which, as they establish, suppress the grass.[91]

The case of fire provides an excellent example of the revision of the categories of colonial discourse in the process of their historical articulation.[92] Indian field foresters were discovering that agro-pastoral peasantry used fire in preparing land for periodic low-intensity agriculture, drive game in the hunt and release fresh tender herbage for pasture-dependent cattle. Savannah is often 'a fine balance between the forces of society and the forces of nature.'[93] Scientific forestry, however, had developed a whole pejorative vocabulary for savannah. While the jungle aspect of savannah had been the conservatory in disorder, an obstacle to political goals and human progress in the early nineteenth century, the more closely observed bush savannah revealed overgrown grass lands with sal seedlings that appeared trapped in a sub-climax state to the scientific forester. Fire protection released these seedlings, and as they grew to climax they suppressed grass. This notion of climax, based on the most hygrophilous composition of the forest, was based on the exclusion of anthropogenic influences, and was expected to yield the most valuable woods, once the thorny issue of evicting tribals and other peasant users from the forest was resolved.

On both these counts scientific forestry faced a serious challenge from within and without.[94] Having considered some of the problems created by indiscriminate fire protection and the emergence of a critique from both within the forest bureaucracy and outside, it is possible to surmise that controlling fires proved to be one long lesson in forest ecology for scientific forestry. The representation of fire was closely related to the representations of grass. The perception of grass had to do with the agro-forestry system in vogue, and the desirability of annual or perennial grasses in that system as a fallow-period crop. Conflicts over this perception led to the incendiary use of fire, the critique of scientific forestry from without.[95]

Scientific forestry began, to restate our argument briefly, with Forest Departments locating and demarcating forest reserves, then regulating local demands for forest produce. Intervention in the ecological regimes of forests was primarily directed at managing natural regeneration to maximize certain timber values, though a complementary plantation programme was usually carried on. It is useful to distinguish the Bengal and Central Provinces cases from the Uttarakhand region because in zamindari areas reservation succeeded a period of private forest management where forests had been classed with other wastes in landed estates. In these situations reservation was superimposed on a more complex hierarchy of rights and localized arrangements than in the temperate evergreen or deciduous forests of the Himalaya. Early management was heavy-handed: foresters saw only the timber they needed, while villagers saw a variety of use values being destroyed. But a more fragmented discourse of scientific forestry emerged from the crude regulatory strategies initially employed. Neeladri Bhattacharya sums up well when he says:

the demands of different social classes and their attempts to protect and assert their rights had to be reconciled with what the colonial state perceived as its interest . . . conflict of opinion among different officials reflected the different ways in which reconciliation was sought.[96]

In the mixed deciduous plains forests, forest officers concentrated on one or two valuable species like sal, with the result that no silvicultural knowledge was accumulated on the large number of other species that might have value in the future.[97] This emphasis on the production of pure sal and teak forests, and pure strands of other valuable species (the trees varied regionally), was clearly the upshot of managerial compulsions, the necessary regimen of departmental functioning. Persisting with such a simplified species composition approach had adverse consequences for silvicultural knowledge. But it certainly promoted sustained research, in the period 1900–45, on sal regeneration. This research brought fire back into forestry in selected ways.[98]

A range of practices incorporated fire into sal regeneration. Here we need merely recapitulate the variety of silvicultural benefits that came to be identified with the use of fire. We should also note that these reports spanned a multiplicity of political-ecological settings in eastern India. Fire was found not only to suppress weed undergrowth but continued large-scale burning destroyed *Clerodendron infortunatum* and other 'noxious weeds' in their flowering season and thus prevented recurrence.[99] The introduction of early burning of the forest floor for soil aeration and exposure of mineral soil to falling seed has already been discussed.[100] Regular burning not only possessed the advantage of avoiding heavy occasional fires, but flocculated and sterilized damp ground.[101] But most importantly, reconsideration of the relationship between fire and sal forest landscapes recalls our discussion of nature and representation, the conundrum of 'first nature' and 'second nature.' Struggling with the natural regeneration of sal in Bengal and other parts of eastern and northern India, foresters redefined primary and secondary landscapes. In their recognition that fire protection established *an entirely artificial condition of affairs*, we can detect the interplay of knowledge, power and nature.

Scientific forestry thus retained a botanical focus on individual tree species which combined with a territorial approach to forest control. The forest-savannah transition zone in which all advance growth of sal was found in Bengal, at the inception of forest reservation and protection, was pictured as a regression from the climatic climax.[102] In southwest Bengal, what was considered a biotic and edaphic subclimax vegetation had been carefully created by years of complex agroforestry. Many valuable fruit trees of the region like mahua and kendu regenerated only in this zone. Shrubby vegetation not only encouraged the proliferation of micro-fauna, but when burnt on the forest floor, the ash was washed down by monsoon rains to fields at the bottom of slopes and fertilized the otherwise shallow and nutrient-poor soils.[103] An associate considered valueless, *Lagerstroemia parviflora*, and clero-dendron which became the noxious weed of later struggles for sal regeneration, were indicator species for favourable sal sites.[104]

Troup had demonstrated that the progression from savannah to sal to evergreen that he described for moist forests of north Bengal was the first indication of a wider phenomenon. Mobbs and Bourne wrote in detail about the succession of landscapes that produced sal forests, pointing out how fire and grazing produced forests through complex transformations.[105] When we imagine many years of such transformations we can picture forest islands in fields and savannah, but not 'first nature'. An interest in specific trees and not the forest, combined with a misconceived notion of 'first nature' in colonial science to produce a narrative of degra-dation.[106] One forester summed up aptly when he said, 'we talk glibly about following nature and forget that the nature we are visualizing may be a European nature inherited from our training and not an Indian nature.'[107]

Postlude: Grazing and Forest Regeneration

In the previous chapter grazing regulation in Darjeeling has already been discussed as an example of the ways in which forest management required the disarticulation of forests from agrarian economies. This chapter will conclude with some brief points on the disputes about grazing and sal forest regeneration that comple-ment the story told about fire. The regeneration debates around grazing were vigorous and divisive, between foresters working in different parts of eastern India. Conflicts with the other departments concerned with land administration also emerged around the issue of grazing in forests. The space for such contention had been created by the forest policy of 1894 and the different experiences of forest reservation and exploitation that were already building up in varied ecological, socio-political regions of India.[108] Like fire, grazing had been a major biotic influence on the forests of eastern India, and with fire protection had come the restriction of grazing. The initial view, buttressing reservation, was that next to bamboos the forest product that formed a most important factor in the management of Indian forests was grass as it appeared to hold considerable potential as a source of revenue.

The universal practice of grazing cattle at will in the forests, accompanied as it was by the firing of areas with the object of getting up a new crop of young grass as the early rains of the monsoon fell, however, had been ruled inimical to the reproduction of young tree crops in the early years of forest management that have been discussed in previous chapters as the period of conservative lumbering.[109] This decision proved quite unsupportable, at least in the context of sal regeneration, over the next three decades as regeneration replaced restriction as the most troublesome concern. The problem appears to have been similar to that relating to fire. Blanket restriction of fire and grazing compounded the regeneration problems that each created, especially in the moist sal forests.

In its second volume, the *Indian Forester* carried a piece presenting evidence of the beneficial role played by grazing in reducing fire hazards and assisting sal regeneration. Writing of the eastern Duars, Fisher described how the Mechi people settled every three-four years in a patch of grassland adjoining a sal forest. Their cattle would trample down the grass, minimizing fire and releasing sal regeneration, which in turn would suppress the grass, causing the Mechis to move on.[110] Disregarding such findings, open mixed sal forests in north Bengal, habitually grazed quite extensively, were closed to grazing. In the period 1900–20, this produced considerable ecological change. Densely stocked even-aged sal was recruited in large areas, unlike anything that had existed before. Under the heavy shade of these crops, continued protection from fire and grazing killed grasses and brought in shade-tolerant evergreen weeds like *Clerodendron infortunatum* and *Mallotus phillipinensis*. This inhibited sal seedling establishment.

The question then was not so much the fundamental opposition between grazing and forest regeneration but understanding the ecological role of grazing in the natural regeneration of sal. This required foresters to distinguish the phases of forest growth (establishment, building, canopy closure) and the part played by grazing and its regulation in each. But the kind of issues that control of grazing raised were not purely silvicultural. Building on the territorialization and zoning that characterized forest reservation, protection, and the pattern of forest working under conservative lumbering, forest regeneration was approached from a perspective of confining discrete activities to distinct blocks. The idea of the scientific forest as completely controlled and exclusively manipulated by the government expert could only be realized in a particular combination of managerial and productive strategies. This constraint of representation and practice created the ironic situation where biotic influences like grazing were excluded from forests in which they did not have degrading effects, thus leading to protracted struggles between graziers, foresters and the civil administration.

In addition to these problems of implementation, grazing restrictions were thrown into greater disrepute and scientific uncertainty when the research of Smythies in the United Provinces challenged the unqualified elimination of grazing in sal forests. Both he and Mobbs concluded that grazing could reduce weed competition in the second and third stages of a four-stage regeneration process outlined in their field research and that of R.S. Hole.[111] It appears that at specific stages in sal regeneration, in certain ecological zones, and in particular combinations with the use of fire, grazing played the part of cost-efficient weeding, at the very least.[112]

Mixed forests rarely facilitated a neat division between commercial and subsistence species, nor was such a distinction possible based on regeneration stages in the forest, which was not even-aged or segregated in discrete species-specific blocks. This meant that access to land for grazing or cultivation, the control of trees, and divergent interests in the timing of forest product extraction became issues for clashes between peasantry and foresters. With the increasing emphasis on working plans, which signalled the professionalization and institutionalization of scientific forestry, the resolution of these conflicts got caught up in these processes of introducing greater procedural formality and long-range planning.

The elaboration of planning contained its own dynamic of contest and negotiation between different branches and levels of government, which meant that issues of peasant-state relations became entangled in these internal arguments of colonial bureaucracies. The story of working plans and sal silviculture in Bengal is therefore taken up next.

Notes

1. Pyne 1994: 11.
2. Baden-Powell 1876–7: 2.
3. Anderson and Grove (1989: 4–6) make this point in the African context.
4. Sir Dietrich Brandis, *Suggestions Regarding the Management of Forests in Jalpaiguri and Darjeeling Districts* (Calcutta: Home, Rev and Agri Dept Press, 1881). Recently, this point and its implications for policy in parts of Africa have been documented by Fairhead and Leach (1995).
5. Ford-Robertson 1927: 511.
6. Moore 1909: 218–19.
7. The term biotic intereference, signifying human impacts on forests, gained currency in forestry project reports, evaluations of deforestation and reforestation proposals prepared in the 1970s–80s. But its definition in terms of firing and grazing practices draws upon on a long colonial history. In Bengal, examples from the earliest and terminal years of colonial forestry are Brandis, *Suggestions Regarding the Management of the Forests in the Jalpaiguri and Darjeeling Districts*, p. 19; Chakravarti 1948: 58.
8. Prakash 1992a: 164.
9. Ramachandra Guha 1989: 35–60, 59.
10. Worboys 1990: 164–86.
11. Worboys 1988; Arnold 1994: 2.
12. Kumar 1980, 1984; Sangwan 1992; Sheets-Pyenson 1986.
13. Stafford 1984, 1990: 70.
14. A small sample of this historiography would include Stafford 1989; Browne 1983; Desmond 1984.
15. Stafford 1990: 82.
16. John Mackenzie 1990: 5.
17. Ibid.: 3–4.
18. As Arun Agrawal explains,

a classification of knowledge into indigenous and western is bound to fail not just because of the heterogeneity between the elements . . . it also founders at another possibly more fundamental level. It seeks to separate and fix . . . in time and space systems that can never be thus separated or so fixed. Such an attempt at separation requires divorced historical sequences of change for the two forms of knowledge—a condition evidence simply does not bear out.

See Agrawal 1995: 421–2.
19. Sangwan 1994: 292–3.

20. Ibid.: 309. The critical, if unacknowledged role of local or indigenous knowledge in the formation of imperial science in the context of irrigation technologies in the Indus basin is discussed by Gilmartin (1994: 1130–2). Such scientific systematization and classification of colonial knowledge through the transformation of local information has been ably discussed for agriculture and rural life by Amin (1989).

21. Grove 1993a; Rangarajan 1994: 148.

22. Grove 1995b: 1–7.

23. Ibid.: 2, 5.

24. Pratt 1992: 34–5, 202.

25. Anderson 1995: 296.

26. Trafford 1905: 11.

27. For standard accounts of this transition see McIntosh 1988; Odum 1971; Ricklefs and Schluter 1993. Leopold (1986 [1933]: 3–22) and McEvoy (1986: 118) note this movement in connection with game and fisheries management in the US.

28. Though schematized and erased to some extent in his presentation the transition may be discerned through a careful reading of Troup 1921. Also see Champion (1936) for a reconsideration and elaboration of existing forest typology.

29. Smythies 1932: 199.

30. See for instance, various essays in Barnes and Duncan 1992; Daniels 1993; Daniels and Cosgrove 1988; and Duncan and Ley 1993. A critique of this trend is offered by Gregory 1994.

31. For India, the pioneering work on colonial forestry and resistance is Ramachandra Guha 1989. A short list of other studies where resistance has been used as an index of alternate views of the forest (in itself an uninterrogated and somewhat static category) would include Anderson and Huber 1988; Shiva 1991: 61–122; Gadgil and Guha 1992: ch. 5; Bhattacharya 1992; Skaria 1992: ch. 8; Prasad 1994: ch. 3; Sundar 1995; Baviskar 1992; Pathak 1994. For other parts of Asia, see Menzies 1992; Bryant 1994; Peluso 1992. A notable exception to this body of work, and where the processes of ecological change penetrating political and social interventions and imaginations in forest management have been examined to some degree, is Totman 1989.

32. Worster 1985, 1990; Merchant 1989; Cronon 1991: 56, 149–50.

33. Demeritt 1994: 175.

34. Notification in the Calcutta Gazette, 3 Nov 1879, under section 45 of the Indian Forest Act 1878, demarcated the entire province of Bengal, from Darjeeling to 24 Parganas and Jessore as the area within which 'all unmarked wood and timber shall be the property of the government.' Reproduced in *IF* 5: 406, 1879–80.

35. Ribbentrop 1900: 61.

36. Brandis, *Suggestions for Jalpaiguri and Darjeeling*, p. 19; Anon 1885: 349–50.

37. Pyne 1994, 1982.

38. Pyne 1994: 22.

39. Schlich 1884: 377.

40. Speaking of agricultural settlement reports which began to be prepared extensively on an individual plot basis in the 1870s, David Ludden, says, 'they succeeded in their most important goal, to channel copious information and thus power to define the terms of revenue transactions into urban official hands.' See Ludden 1992b: 176. The forest working plan accomplished similar things in the forest areas, though in the absence of annual cultivation and decennial renewal, as in the case of land settlements, there was no continuous arrangement to renew this information interactively with village communities. The other important aspect is that the tilling and production operations on forest land were a direct departmental effort with peasantry reduced to the role of skilled labour. This shifted expertise more completely away from local institutions to state science than in the case of agriculture.

41. The distribution and characteristics of sal are discussed in Sivarama-krishnan 1996c: 154–6.

42. BFAR, 1900–38, selected years. It is important to note, as Rangarajan (1992: 11-12) does for Central Provinces, and I find in Bengal—both cases of mixed, dry deciduous forests dominated by sal—that conservation in the early incarnation was directed at control of traders, rural tree and land users. With the 1878 Act, exclusion of fire and cattle came to be see as a prerequisite for forest management, and by the early twentieth century, both silvicultural wisdom and scarcity of labour for forest operations led to a series of accommodations that modified the law in practice.

43. Brandis 1897: 77.

44. Ribbentrop 1900: 39.

45. OIOC P/9548 BORP (For), Jan-May 1914, A progs 4–11, March 1914, file IIIF/1, no. 1818-254, Ranchi, 17 Sep 1913 H. Carter, CF B&O, to Secy GOBO Rev, pp. 42–3.

46. As at least one Working Plan was to confess, half the cases of illegal fire investigated, implicated protective staff. See Minchin 1921: 38–40.

47. Brandis, *Suggestions for Jalpaiguri and Darjeeling*, p. 2.

48. Stebbing 1926: 198.

49. Brandis, *Suggestions for Jalpaiguri and Darjeeling*, p. 6.

50. Birdwood 1910: 25; Ball 1880: 68; Schlich 1906.

51. OIOC P/432/65 BRP (For), Nov-Dec 1866, A progs 4, no. 62, 5 Dec 1866, Anderson to Secy GOB, pp. 16–18.

52. OIOC P/7033 BRP (For), 1905, A progs 14–16, Jun 1905, file 3W/9, no. 11WP, Cal 23 Jan 1905, S. Eardley Wilmot, IGF GOI, to CF Bengal; no. 1233T-R, Darjeeling, 15 Jun 1905, M.C. McAlpin, US GOB Rev, to CF Bengal, pp. 149–55.

53. Pyne 1994: 7.

54. OIOC P/243 BRC (For), 1873-5, A progs 18, Jan 1873, memo dated 4 Jul 1872 by H. Leeds, CF Bengal.

55. An Aged Junior, 'Some Remarks on Titles and Tigers', *IF* 16(1–3): 182-4, 1890.

56. OIOC P/234 BRP (For) Oct–Dec 1872, A progs 10, Oct 1872, no. 77C,

Darjeeling, 31 Jul 1872, Major B.W. Morton, DC Darjeeling, to Commr Cooch Behar, p. 7.

57. Gadgil and Guha 1992: 146–80.

58. Bhattacharya 1992: 130–2. In 1884, Sir William Wedderburn wrote that forest degradation had arisen from disregard of old village methods of conservancy in which forest preservation had been closely integrated into the agrarian economy. See his 'The Indian Raiyat and the Village Community,' *Indian Agriculturalist*, 12 Jul 1884, extracted in *IF* 10(11): 511–15, 1884. In 1903, Whittal, the Conservator of Forests in the Punjab recommended active involvement of villagers in the spirit of Wedderburn's suggestions. 'According to him, fire lines and fire watchmen were of little use: a vigorous enforcement of the principle of joint repsonsibility of villagers could provide a possible solution'. Quoted in Bhattacharya 1992: 132.

59. As Thompson says, this kind of occurrences 'gave publicity to a curious kind of dialogue between the authorities and the crowd.' See Thompson 1979: 270.

60. Fisher 1880. The extent of disruption that fire protection caused to the agrarian economy is astounding. Hunting of small mammals and micro-fauna became arduous; pastures became overgrown; villagers could not move about to gather small timber for fuel or construction; tigers would use high grass in the vicinity of settlement to menace cattle; elephants, buffalos and deer would ravage crops; and lastly a fire in a contiguous forest could devastate a large area uninterrupted by open spaces.

61. WBSA P/582 BRP (For), Jul 1900, A progs 1–4, file 4-A/3-1, no. 104, Darjeeling, 3 Jun 1900, A.C. Wild, CF Bengal, to Secy GOB Rev, p. 3.

62. Ibid., no. 532C, Singhbhum, 9 Feb 1900, E.G. Chester, DCF Singhbhum, to CF Bengal, pp. 5–6.

63. OIOC P/6794 BRP (For), Aug–Oct 1904, A progs 4–12, Sep 1904, file 3F/3, no. 473LR, Ranchi, 21 Jun 1904, W. Maude, Offg Commr Chotanagpur, to Secy GOB Rev; no. 313R, Chaibassa, 15 Jun 1904, H.F. Samman, DC Singhbhum, to Commr Chotanagpur, p. 131. The use of Mankis and Mundas was not easy. Quite often they gave evidence in cases of incendiarism in a way that subverted the prosecution of apprehended villagers from their jurisdiction. OIOC P/8693 BRP (For), 1911, progs 1–4, Dec 1911, file 7F/2, no. 701R, Chaibassa, 11 Aug 1911, A.W. Cook, DC Singhbhum, to Commr Chotanagpur, p. 4.

64. OIOC P/6794 BRP (For), Aug–Oct 1904, A progs 4–12, Sep 1904, file 3F/3, no. 3180, 8 Aug 1904, L.S.S. O'Malley, US GOB Rev, to Commr Chotanagpur; no. 115, Darjeeling, 28 Jul 1904, CF Bengal, to Secy GOB Rev; no. 77C, Chaibassa, 1 Jun 1904, J.W.A. Grieve, DCF Singhbhum, to CF Bengal, pp. 135–7.

65. Trafford 1905: 19.

66. NAI Rev and Agri (For), A progs GOI 41–3, file 11 of 1908, Feb 1909, GOB Res no 5110, Cal, 11 Dec 1908, p. 596.

67. Tinne 1907: 4.

68. Ibid.: 4.

69. For a similar discussion of forest firing as resistance in the Punjab, see Bhattacharya 1992: 131–2. Bhattacharya does not, however, make the distinction between reserved and protected forests that is important to my analysis.

70. The silvicultural literature on sal regeneration in India has a large debate on the use of fire. Champion (1933) provides the first comprehensive review of this literature. More recently, Joshi (1980) provides a revision and updated review of the issues. Some other important colonial writings are listed here. Sengupta 1910; Bailey 1925; Mobbs 1936; Osmaston 1935; Shebbeare 1930; Milroy 1930. One recent article on fire and sal regeneration is Maithani et al. 1986.

71. A brief consideration of the role of fire in sal regeneration is provided in Sivaramakrishnan 1996c: 163–4.

72. In Buxa, success in fire protection (measured by taking area burnt as a percentage of area protected) hovered between 50 and 70 per cent in the period 1875–80. It rapidly improved to 90–95 per cent in the next three years and by 1905 was always between 98 and 99 per cent. See, Hatt 1905: Appendix V, xvi–xviii. Similarly, in Jalpaiguri, fire protection was 99 per cent successful between 1896 and 1905. Trafford 1905: Appendix VII.

73. Soni 1953: 47.

74. The details of how fire was reintroduced into sal regeneration may be found in Sivaramakrishnan 1996c: 165–6.

75. Slym 1876: 1–2. A significant sidelight that indicates the extent to which this author in recommending regular forest firing was being counter-hegemonic within the forestry establishment is the fact that his paper, when first read out at a conference in Rangoon was denied a place in the conference proceedings, leading to its independent publication. It was later republished in the *Indian Forester*.

76. D. Brandis, Report on Teak Forests of Pegu, 1856, in NAI GOI Foreign Dept (Foreign Consultations), 108–11, 29 Aug 1858; idem, *Suggestions Regarding the Management of Forests in Jalpaiguri and Darjeeling Districts*, 1879, in NAI GOI Home, Rev and Agri (For) progs 98–100, 1881.

77. Stebbing 1926: 392–3. Ribbentrop, IGF 1884–1900, was the architect of this policy, who overruled a growing critique of fire protection.

78. Slade (1896) launched the attack on fire protection from Burma, where Brandis had formed his initial ideas.

79. Troup 1905a: 143; 1905b.

80. Braithwaithe 1905; Fisher 1905; Stebbing 1926: 393; Moore 1909; Sengupta 1910; Beadon-Bryant 1907: 538.

81. See the discussion of north Bengal taungya later in this chapter. Burma taungyas are discussed by H.R. Blanford 1925; memo dated 30 Mar 1925, from H.R. Blanford, CF Working Plans Circle, to EP Stebbing, quoted in Stebbing 1926: 395. United Provinces taungyas are discussed by Hussain 1925; and Wood 1922.

82. Tracey 1956; Rowbotham 1924.

83. Rowbotham 1924: 356.

84. For introduction of early burning in the Himalayan forests of Uttar Pradesh, see Ramachandra Guha 1989: ch. 3. For similar rethinking of fire protection and use of fire in forest regeneration in central India, see Rangarajan 1992.

85. Rangarajan 1992: 39–67, 78–89.

86. As early as 1876-7, Brandis had observed in a tour note from the Central Provinces, that prolonged fire protection in certain tracts had led to large quantities of wood of the less valuable kind crowding out teak and bijasal. See Brandis 1879.

87. The concept of gregariousness refers to survival skills of woody plants in conditions of water and nutrient scarcity. The related adaptive mechanism of sal to the adversities in its habitat, is the ability to dieback. The first fire often makes the sal seedling dieback and develop a strong root structure. This provides shoots the energy to grow rapidly in the second phase and acquire a tough bark on the young sapling that can then resist fire at the establishment stage, when fire removes the competition. This cycle can take four to five years. A similar behaviour is exhibited by oak and douglas fir in temperate zones. Eucalyptus in Australia does the same.

88. OIOC P/243 BRC (For), 1873-5, A progs, file 3-2/3, Jul 1873, no. 94C, Dibrugarh, 10 Jul 1873, William Schlich, to Secy GOB; no. 2819, Cal, 19 Sep 1873, Secy Rev GOB, to CF Bengal, pp. 103–16, p. 132; Stracey 1956.

89. Milroy 1936. The reference is to the region known as Boko in Kamrup district in lower Assam, where the first forest surveys discovered large fields of thatch grass with fleshy sal shoots, within reach of mother trees. Reservation followed by fire protection had produced a sal forest and led in good measure to the spread of the legend of sal as an obliging tree that grew abundantly if mere fire protection was afforded.

90. Similar recommendations came from Uttar Pradesh, where elimination of mallotus (an evergreen weed) was crucial to seedling growth. See Champion 1933; Mobbs 1936.

91. Smythies (1931: 198) says, 'It is a curious but well established fact that whereas vigorous grass in an artificial plantation is an indication of probable failure, the appearance of mild grass in a natural (shelterwood) regeneration area is the first indication of possible success.'

92. Prakash 1992a: 172.

93. Dove 1992: 241.

94. Dove (1992) has argued that the state tends to mystify and obfuscate the nature-culture relationship and this merits attention in the context of widespread forest degradation.

95. See Dove (1986: 182) for a fine discussion of these conflicts over trees and grasses in southeast Asia.

96. Bhattacharya 1992: 124–5.

97. Stebbing 1924: 582.

98. For more on this topic, see Sivaramakrishnan 1996a: chs 6 and 7.

99. Ford-Robertson 1927: 508–9.

100. Chaturvedi 1931: 158–60.

101. Ford-Robertson 1927: 562–4; Hole 1914–16.

102. Troup 1921; Chakravarti 1948.

103. Similar practices, fertilizing fields with burnt forest biomass, have been discussed by Skaria (1992: ch. 5).

104. Khan et al. 1961: 120–6; Seth and Bhatnagar 1960.

105. Smythies 1931: 201–2; Milroy 1930: 442.

106. A fascinating study of the forest-savannah transition zone in Guinea has made the same point. It details the various reasons why villagers actively produced forest islands, amidst swampy fields and savannah uplands. See Fairhead and Leach 1995: 57–63.

107. Greswell 1926.

108. Despite the best efforts to produce a single colonial discourse about the role of grazing in forest conservancy, as Prakash (1992a: 154) has pointed out in the context of agriculture, 'in its enunciation European knowledge and institutions emerged pursued by the shadow of colonial birth'; a birth, we should add, that created distinct progeny at different times and places.

109. Stebbing 1926; Ribbentrop 1900.

110. Fisher 1876–7: 204.

111. Smythies, the experienced forester of the United Provinces, had found it conceivable that cattle grazing, instead of being a confounded nuisance that foresters unwillingly tolerate and eliminate as far as they can, is in fact a useful factor in the production of gregarious sal crops. Also see Mobbs 1936.

112. Critics within the Forest Department in the Central Provinces echoed the developments in eastern India and began to argue that fire, cattle and *dhya* (swidden) cultivation could actually upgrade timber productivity and regeneration. For instance silvicultural research showed that grazing favoured teak renewal by suppressing other species when teak seedlings were being released. See Best 1909; Malcolm 1917.

8

Science and Conservation: Hybrid Genealogies of Development

As the animals went lumbering off, we could not help smiling at the comical appearance presented by the old guard, seated on the leading elephant. Picture to yourself a thin, withered, little old man, clad in what remained of a faded khaki suit, two sizes too small even for him. His legs bare from below the knee and innocent of calves, ending in ammunition boots of the largest size procurable. His close cropped grizzled head surmounted by a check deer-stalker's cap, erstwhile the property of his master, and now adorned with a green worsted fringe ...[1]

They (mahouts) make the forest seem so much more interesting, they know all about trees, the one from which they collect leaves to smoke ... then there is the tree that produces a little shoot that makes a good curry ....[2]

I'm very hard at work here laying out twenty experimental plots on sal regeneration. It takes a lot of time as they are a series of plots in our experiment so each plot must be exactly similar ... this entails a lot of tedious walking about searching the very extensive hilly forests ....[3]

Introduction

The last three chapters have demonstrated that scientific forestry invented regimes of restrictions that were intended to secure state control over forest Bengal, and introduce standard techniques derived from European forestry knowledge to harvest and regenerate the forests. More significantly, these chapters have shown the regional diversity of these regimes and their varying degrees of success or failure. The gap between outcomes and intent, I have argued, was a product of the constraints imposed by particular geographies, histories, and cultures on statemaking, which in turn transformed the spatial, material and ideal contexts in which it happened. This chapter will take that story further. By concentrating on the most formal and technical aspects of scientific forestry in Bengal—working plans and silviculture—I shall show that even the domain of scientific forestry most amenable to 'continental models' was a product of regionally varied processes outlined in earlier chapters.

I am forced, by recent scholarship in very different genres, to insist on this point. These scholars consistently conflate policy intent with practical outcome. They do so by reading forestry manuals, development project reports, and histories of scientific or development ideas as knowledge separated from the world of production. The best example of such writing on forestry in colonial India and its relationship to European models is the work of Ravi Rajan. He writes that in the late nineteenth century, 'the Indian Forest Department, staffed by personnel trained in Germany and France, systematically erected a framework of resource use modelled on European lines'. He asserts further that 'the emergence of the Continental forestry tradition . . . led to the creation of a . . . technocratic elite with a monopoly over decision making on forest use.'[4]

Ravi Rajan not only claims that Indian forestry was entirely imported, he also assumes that those in charge of the Indian professional forestry establishment successfully carried out whatever forest management they planned. So the argument offered by him not only homogenizes ideologies, policy intent, and representations of forest landscapes across widely divergent times and spaces, it allows no room for the transformation of these ideas, intentions, and representations by the force of experience. For instance, the case of sal regeneration through taungya and the reversal of ideas about fire in the Bengal forestry experience that was discussed in the last chapter would be inexplicable in Ravi Rajan's framework of analysis.

The continental tradition becomes, in this approach, a powerful system by 'which local knowledge systems were consequently discredited and replaced'.[5] Two problems arise from this narrative of colonial forestry as continental forestry cloned. First, the impact of planned forest management is overestimated. Second, a corollary slippage occurs when a technical lexicon of terms like deforestation, regeneration, and environmental risk, is accepted without adequately exploring their cultural construction. We may then fail to recognize that these terms become polyvalent resources for struggles over nature, often changing their meanings over time. Second, by wrongly diagnosing the impact of particular models of science, and exaggerating the influence of plans on outcomes more generally, we also foreclose the opportunity to ascertain the multiplicity of factors that always interact in any situation. Unpacking constructs like scientific forestry, state, and society reveals complex agency at work in ideological, social, ecological, and political structures. Suggesting that Bengal forests were not transformed into mirror images of stereotypical German or French forests,[6] in this chapter I shall explore the production of the contested domain of silviculture, and argue that it has been wrought in several histories— of ecology, politics, and scientific discourse.[7]

Scientific forestry was constructed in colonial Bengal by valorizing certain kinds of knowledge, and thereby, privileging attendant modes of forest management. Colonial rule in India was shaped by the flow of information and the categories through which it was absorbed and transformed into what might be lumped together as colonial knowledge.[8] In this chapter I examine the formation of colonial knowledge (information, ideologies, perceptions, and the lessons of practical experience) in the realm of forest regeneration by concentrating on the points of knowledge production, codification and transmission. By considering the conflicted and experimental ways in which sal silviculture was sought to be standardized in colonial working plans, this chapter advances two arguments. First, that the scientific and technical discourse surrounding forest management today is shaped by an historical legacy. Second, that such environmental development discourses are continuously under production through entanglement with debates about the locus of governance.

Through representations of the forest, definition of expertise, and manipulation of local structures of authority, colonial foresters created a body of knowledge about natural regeneration for sal forests in eastern India. They worked with, and sometimes against, the vanguard parties charged with establishing the colonial state on its frontiers. In Bengal these frontiers were often the forested fringes of settled cultivation. Managerial knowledge was defined and the landscape partitioned by the colonial state into discrete jurisdictions, which were then subject to different modes of management.[9]

My discussion thus locates the history of scientific discourse in the related histories of politics and landscapes. A lot of the politics involved was internal to the forestry and larger colonial bureaucracy but the unravelling of this politics is educative because it tells us about enduring mechanisms of state building. For the same reason I am less concerned with drawing distinctions between local and foreign knowledge and more interested in the production of knowledge in specific political-ecological-historical settings. The term local knowledge is used not to denote the distinct cultural categories of an indigenous people. I use it in a manner that implies that local knowledge is situated practice and, to that extent, is neither a system nor an alternative rationality. I refer more to the intimate acquaintance with a locality, its landscape, social relations of production, and environmental management, that people develop as they work to change its appearance. Such special knowledge of a place and its spatial history is produced by those who traverse it intensively to make something of the place for their purposes.[10]

I am concerned with such localized production of information, and the processes of translating it into standard terms, that made possible a project like scientific forestry. Historians and sociologists of science have frequently explored the ideological and intellectual processes in which particular branches of science developed.[11] I would suggest that such an approach pays insufficient attention to sites of application. That is, to places where these sciences become technological practices. The terrain of implementation leaves a strong impression on the production and transformation of scientific knowledge.[12] When these sites enter the processes of knowledge production in any specific domain like forestry, they bring with them much else that is going on there.

One of the important things that creeps into generating scientific knowledge is the issue of government. This happens in at least two ways. First, forestry as land management gets entangled in wider issues of land administration—agriculture, revenue, and stable local arrangements of production. Second, the pressure on Forest Departments to develop, standardize, and disseminate universal and replicable scientific management models which mesh with larger bureaucratic forms of government influences their selection and codification of procedures. There is then a tension between fitting forestry into a wider universe of managed landscapes of production and identifying it as a distinct, separate, professionalized activity. The work done by this tension suggests a constant production and transformation of science in its applications, often the context being development. We need to track these changes.[13]

In *Representing and Intervening*, Ian Hacking provided a landmark study that shifted the focus of science studies towards practice, by stressing the doing aspects of science as much as representing. He later emphasized the multiplicity, patchiness, and heterogeneity of the space in which scientists work.[14] More research in the history of science has moved in this direction of looking to the social to understand the way rules and practices shape each other.[15] The discussion of interests I have suggested allows us to understand how the stabilization of science takes certain routes and not others as the open-ended processes of experimentation unfold. But when scientists continually explore their way out of a problem, with experience as their guide, their interests intersect with ecological processes.

Foresters were interested in growing sal in large contiguous blocks along convenient conversion and transport networks for timber. Their silvicultural options were soon limited to natural regeneration. Ecology here is itself a product of human perception and intervention, not fully autonomous, but not entirely imagined either. So scientific forestry focused on devising silvicultural systems where concentrated natural regeneration of sal could easily be obtained. This condition of science as historical practice is what this chapter will explore and illustrate using the case of forestry in Bengal.[16]

The institutionalization and professionalization of forest management through its engagement with issues of governance, resource conservation, and enhanced productivity also brought scientific forestry into the realm of emerging development discourse in the late nineteenth century. It may well be true that 'we do not yet know enough about the global, regional and especially local historical geographies of development—as an idea, discipline, strategy or site of resistance—to say much with any certainty about its complex past.'[17] We then need histories of development to trace the recurrence of ideas, imagery and tropes of development across a range of nineteenth- and twentieth-century contexts.

Forestry as Science and Practice: A Development Regime

In 1884 Sir Dietrich Brandis, the first Inspector-General of Indian Forests, drew a distinction between a modular science that could be transposed from the European laboratory, so to speak, into the Indian field site; and a location-specific forest protection programme where local knowledge was the best asset of the forest guard.[18] But this neat division could not always be sustained in practice. Cultural operations were often crucially dependent on local ecological knowledge and understanding the social mechanisms by which scarce labour could be secured for silvicultural tasks. Performing the routines of forestry and through their sporting pursuits, forest officers assumed the mantle of authority and scientific expertise that became central to their functioning by the end of the nineteenth century.[19]

But at the moment of translating silvicultural prescription into action in any coupé, compartment or block, the mantle sometimes slipped, tugged away by a shift in the locus of necessary knowledge to reveal the precarious relationship between assumptions and practice. For in the words of a novel written in 1909 and set in woodland Bengal, 'when it is time to work the forest, *burra sahib* will need Dulall to mark the trees, as no one can tell which can be cut.'[20] These were the travails of moving from lumbering to forest regeneration, which meant the Bengal Forest Department had to change from building a regime of restrictions around the forests to detailing a set of interventions in them.

Forestry, in this respect, was part of a wider process noticeable in the late nineteenth century when increasing intervention in agricultural production and its justification by appeal to a rhetoric of conservation went hand in hand.[21] Following the Famine

Commission reports and the creation of an Agriculture Department in 1885, a new managerial assertiveness was discernible in the whole rural landscape of production.[22] The possibility of transforming the floral and crop composition of this landscape was perceived through the powerful lens of institutionalized science.

Working plans came to symbolize this confidence of the scientific forester. But visualizing a terrain where science could plan unimpeded the manipulation of the forest, compelled a more complete enumeration and disposal of local rights that might obscure the vision. Forest settlements therefore became a necessary prerequisite and where they could not be concluded, working plans remained an ideal little realized.[23] The lasting irony of this situation was that these plans aspired to define a universal code but their implementation was always caught up in securing and controlling local ecological and political knowledge.[24] The more working plans were reworked in an attempt to render them into modular prescriptions, their successive revisions became disconcertingly aware of specific regional ecological histories which slowed the drive to modularity. These paradoxes of working plans also reflect the ways in which forest management was caught up in a wider tension in colonial governance between central direction and local autonomy.

Forest administration emerged at a time when the colonial state (by 1880) had become a huge investor in India, and was manned by a large and disciplined bureaucracy that conceived itself the custodian of public welfare.[25] This increased governance created a massive documentation project that has now been widely discussed in the literature.[26] As a result, to use David Ludden's words, 'India's development regime evolved on coherent, consistent lines after 1870, the trend being towards more ramified and centralized state power.'[27] To a limited extent this evolutionary view of the state centralization process is salient. To this we must add the redefinition of expertise in terms of high technology, complicated science and environmental uncertainty that privileged metropolitan agency in the identification of environmental problems, outlining solutions and mooting federal legislation even while bringing into sharp relief local conflicts over resources. Yet, as the previous chapters have shown in the context of Bengal forestry, and as Sugata Bose argues for the diverse origins of the ideas of colonial and nationalist development, trends of state expansion were not inevitable or inexorable. Where development regimes emerged they were marked by patterns of opposition and resistance that cannot be ignored.[28]

Crafted in an expanding world of empire and information, the development regime for sal forests in Bengal began to take shape during the period 1893 to 1937 when political consolidation converged with technocratic assertion. This chapter, confined as it is to the latter process, necessarily dwells on the way expertise in forest management was defined by different elements in the colonial state, in whom it was seen to repose and how it was deployed. The natural regeneration of sal, the most valued tree in Bengal forests, became an elusive goal for scientific forestry by the end of the colonial period, despite tremendous research effort. During those frustrating years the production of detailed reports, compendia, manuals, calendars, agricultural censuses and of course working plans, created a textual basis for fixing and transferring expertise.[29] Such textualization in agriculture was a managerial act that established an instrumental attitude toward farming. From a situation where farm practice was expertise, peasant wisdom was moved to the category of folklore even as it was appropriated and transmogrified into the scientific format of statistical tables or maps and reports.[30] Similar things happened in forestry.

Offering a general explanation of such processes and an agenda for their analysis, Arturo Escobar says:

the demarcation of fields and their assignment to experts . . . is a significant feature of the rise and consolidation of the modern state. What should be emphasized however is how institutions utilize a set of practices in the construction of their problems through which they control policy themes, enforce exclusions and affect social relations.[31]

But the 'modern state' was not as much of a steamroller as projected here. If we look closely within the broad contours provided by Escobar's framework, we find that the constitution of expertise was always conditioned by the exigencies of particular contexts. Realized at an uneven pace, and in diverse forms, the processes identified by Escobar were constituted through significant regional variations in Bengal.

The emergence of scientific forestry and rational management in late-nineteenth-century Bengal is then best analysed as a product not only of intellectual revolutions and transformations in the organization of knowledge in Europe or the US, but also the practical circumstances of controlling land, labour, and manipulating tree species that became valuable at different points in the history of forest management in India.[32] However, assigning to knowledge or material conditions the determining role in policy is likely to prove unsatisfactory. Representations are intimately connected to the production of knowledge and, thereby, its codification in government into legal and policy instruments. At the same time law or policy as instruments of power are not only shaped by this knowledge, but also by practice and experience.[33] The British administrators engaged in framing forest policy clearly brought with them ideas about what was appropriate from European experience, but they also encountered and described a pattern of forest use that they displaced only partially, after disparaging it. Grounded thus in regional histories of conquest and ecology, forest management still emerged as a development regime. We are then required to see how 'a corpus of knowledge, techniques and scientific discourses is formed and becomes entangled with the practice of power.'[34]

An allied feature was the urban bias in determining forest value and use. Forest science flourished where central states began to rationalize administration, as in eighteenth-century Germany and later in France. This process narrowed the focus of management on wood. As one historian of German forestry notes, 'identifying wood mass as the crucial variable of forestry set the stage for quantitative forest management.'[35] As the measures of wood mass and volume were perfected, the urge to grow the accurately measurable forest increased. But the measurable forest needed more accurate estimation also because its principal products, greatly reduced in diversity, participated in world markets. Plantations of valuable trees were created through an impressive network that supplied seeds and planting material across continents.

But initially forest conservancy in Bengal, as in the rest of India, commenced with the urge to directly control, systematize and regulate the extraction of timber from what was perceived as the rapidly dwindling hardwood forest wealth of the subcontinent. The rhetoric of conservancy espoused both the 'environmental' tones of watershed management, species conservation and wildlife protection, and the strident political-economic realities of territorial expansion, the establishment of British rule in strategic regions, and laying down infrastructure for administering empire.[36] While both strains of conservancy ultimately facilitated disempowering local communities in the forests, and expedited capital accumulation through forest exploitation, they created in their discordances, interstitial spaces for the modulation of forest policy.

Protection, reservation, extraction, and marketing of timber were gradually followed by a growing interest in regeneration, particularly by natural methods, of the principal timber species, which were teak and sal. The period broadly coinciding with the first three decades of the twentieth century witnessed burgeoning research on structural qualities of many hardwoods and a few softwoods like *pinus longifolia*.[37] Such research and the dissemination of intensive silvicultural systems through working plans were complementary aspects of scientific forestry that placed new demands on forest management. There was a conflicted expansion of knowledge and a contested growth of managerial arrangement through which scientific forestry was professionalized and institutionalized in the sal forests of Bengal.

The rest of this chapter will, therefore, consider certain programmatic aspects of scientific forestry that illuminate its constitution through demarcation, inventory, protection, regeneration, working plans, and silviculture. In particular I shall take up the vicissitudes of preparing and implementing working plans, and the concomitant reduction of forestry to silvicultural models, to illustrate the two main features through which scientific forestry developed. The first, which may be called *management by demarcation and exclusion*, aimed at simplifying land use in state forests by regulating local access. The second, which may be called *management by inventory and controlled regeneration*, complicated the earlier regulation of people with the added regulation and transformation of tree growth.

In combination these types of managerial aspirations sought to move the locus of expertise and direction up and out into the higher echelons of the forest service. At their most ambitious, working plans were to serve as powerful instruments that would permit the Inspector-General of Forests to dictate what trees were grown where, how, and when they would be harvested. Through the cases of working plan preparation and the tentative, often baffled, silviculture they document, the following sections evaluate the making and unmaking of scientific forestry against the standard of this grand ideal.

Working Plans as Instruments of Remote Control

The first, albeit simple, working plan in India was introduced by Munro, Superintendent of Travancore forests, in 1837. This was largely an exercise in enumeration of trees by size class and thus an estimate of the harvestible timber in any year. Linear surveys introduced by Brandis in the 1870s to estimate the growing stock were the basis of early working plans. In 1874 Schlich drew up a preliminary working plan for Buxa reserve, where the problem of water in the dry season had necessitated reconsideration of timber conversion and removal operations.[38] But progress was slow and by 1884-5, only 109 square miles of reserves were under regularly sanctioned working plans. In 1884, Schlich centralized in the office of Inspector-General of Forests the control and preparation of working plans. This hastened things somewhat and by 1899, 20,000 square miles of government forests were covered by working plans, of which 1,900 square miles were in Bengal.[39]

Local governments were to use the working plan as an instrument that would balance the 'reasonable requirements of neighbouring populations' with the 'exigencies of sound forest conservancy'. Forest settlements that would ascertain and finally record all admissible rights of village communities and private persons in forest lands and their produce were introduced as a necessary corollary of working plans.[40] The plans themselves were declared the most important job of the forest officer as the Forest Department quickly recognized them as the key to professional continuity in forest management.[41] Struggles to perfect these plans reflect the transformation of forest management into predominantly a scientific question. But such recasting of the terms of argument did not necessarily alter the elements that were the subject of contention. A stable silvicultural system and the locus of its governance remained the most pressing issues.

There are many definitions of working plans, but their

important feature was the effort to reduce control into a matter of holding local agencies to plan prescriptions. As one of the last colonial Inspectors-General of Forests wrote, 'a working plan is a forecasted framework for the management of a forest over a considerable period, often 120 years, with a detailed plan of what to do in the next 10–15 years to achieve the ultimate result.'[42] These working plans were seen as restoring normalcy to the state forests, which would consist of normal age classes, normal increase and normal growing stock in compartments, blocks and coupés.[43] The object of the plan was often limited to showing the quantum of timber and firewood that could be removed without detriment to a continuous output and what works of improvement were desirable.[44]

These plans were not, however, introduced as soon as a forest was taken over or reserved. For any forest area the initial decade after its reservation was one of limited forest management. Operations were confined to forest protection, selection and improvement felling, creeper cutting to release advance growth and conversion of sapling sal into timber. In Bengal the general pattern was one where reservation mostly took place in the 1880s, initiation of planning in the 1890s, and approval of short-term (ten to twenty year) plans in the period 1900 to 1910. During this period, non-timber products like bamboo, lac, and mahua flowers, and grazing fees in southwest Bengal were seen as a significant component of forest revenues.[45] Quite often the working plan for a particular division was taken up when competing demands grew, like the spread of tea gardens in Darjeeling district which caused a sharp increase in the demand for firewood.[46]

Within twenty years of preparing the first working plans in north Bengal, however, the systems of management prescribed under them came under a cloud as they were not yielding the desired regeneration of sal. The Inspector-General of Forests, after inspecting Jalpaiguri and Buxa forest divisions, observed that the 'forest conditions in Bengal are more diverse than any other province and with more intensive methods of management plans must be revised more frequently.'[47] Yet the revision to the Jalpaiguri plan simplified the landscape classification of its predecessor from eight to four—sal, mixed, evergreen, and savannah.[48] To ensure central control over these revisions the Inspector-

General of Forests had deputed the Imperial Superintendent of Working Plans to collect 'necessary local knowledge' from Bengal. The state government refused to organize the tour, suggesting that it would impair service discipline and arguing further that providing local knowledge was the task of the regional government, since such knowledge pertained to demand and supply of timber and extraction facilities available.

The real argument soon narrowed to the silvicultural prescriptions. While the Inspector-General of Forests was perturbed by Bengal adopting annual plans of operation that smacked of a lack of professionalism, the Bengal Conservator was adamant that these were appropriate as 'through silvicultural control centralized control of forest management can go too far.'[49] This official went on to forcefully present the case for granting local officers full freedom to experiment with various silvicultural techniques as working plans were not to be cluttered with prescriptions not justified by the poor levels of local knowledge.[50] The outcome was a curtailed tour by the Imperial Superintendent of Working Plans. The controversy admirably illustrates the way planning and professionalization, valorized by scientific forestry, repeatedly became the issues around which more important disputes about regional autonomy and central control in forest management got re-enacted.

Let us take another example. The first Singhbhum working plan was prepared in 1903. By then there were several plans already under operation, or in advanced stages of preparation, for different north Bengal areas.[51] These plans came up for revision, often prior to the end of their initial duration. The changes made in the plan usually incorporated the practical deviations that had been carried out, and sometimes noticed in the annual reports. In Singhbhum, for instance, the plan had been to continue selection felling of sal timber from a thirty-two square mile area of good valley type forests, half of which had been taken out in 1895–8 to supply sleepers to the East India Railway. For the rest of 693 square miles of reserves in Singhhum the plan was fire protection for sal regeneration, something that had already covered 64 per cent of the reserves, increasing from a mere 12 per cent in 1895.[52] At the time the projected revenue from timber operations (major produce) was expected to exceed that from sabai grass (minor produce) sale.[53] But the Inspector-General of Forests ruled in 1915 that the poorer quality hill forests, being of little value in the production of timber, should be given up to sabai production. This would entail burning of the forest floor and creating large blocks for sabai production.[54]

At the same time, with the introduction of concentrated regeneration blocks, the working plan was also to be revised to include a regular engineering scheme covering extraction, conversion and transport of timber. The elaboration of such a scheme hinged on the prior completion of procedures allotting specific areas to blocks and determining a sequence of coupés in the first block. Mechanized timber extraction hinged on a prior reorganization of silvicultural practice and thus a reordering of the landscape into more systematic and compact blocks of even-aged tree crops.[55] As management became more intensive, Singhbhum was divided into Saranda, Kolhan, Porahat and Chaibassa, with separate plans for each division by 1925. The Uniform System, under which partial clear-felling was prescribed, was introduced under all these plans. But the monopoly contractor for timber, the Bengal Timber Trading Company, was mostly in arrears with fellings, thereby largely keeping prescriptions at bay.[56]

During the same period, the Inspector-General of Forests and silviculturists visiting Jalpaiguri and Buxa also recommended taking up artificial regeneration since the evergreen undergrowth in fire-protected areas was suppressing sal.[57] Three hundred acres of experimental sowings of sal and fifty acres of other species annually was taken up as approved deviations from the working plans in 1917. The revision of working plans paid great attention to problems of regeneration. Removal and exploitation of both major produce and minor products remained under a regime of permits and concessions to contractors. Shaw Wallace and Company, for instance, had the monopoly for collecting nettle fibre from the reserved forests of Darjeeling; while Burn and Company had the timber and bamboo concessions in Darjeeling and Jalpaiguri forests.[58]

The significant patterns that emerge in working plan revisions are revealed by the case of the Darjeeling plan, which was altered several times in the space of thirty years. Darjeeling forests ranged in elevation from 12,000 feet in Singalila to 600 feet above sea level in the Tista valley. The ground was steep with deep gorges and rapid mountain torrents. Hill and valley forests covered an area of 115 square miles.[59] The upper hill forests (above 9,000 feet) were mainly silver fir and rhododendron. The middle hills (5,000–9,000 feet), chiefly oak, walnut, tun, laurel, maple, champ and alnus. Sal was found in the valley forests (600–3,000 feet).[60] The middle and upper hill forests had been worked departmentally for three years from 1865 and then under the permit system till they came under Manson's plan of 1892.[61]

The Manson plan prescribed a 160-year rotation in five thirty-two-year periods, with the first and last block closed to grazing. The first block was to be regenerated by concentrated felling over one-sixteenth of the area annually under the shelterwood system. Soon after the introduction of the plan, a summary forest settlement extinguished all private rights in all the forests.[62] Osmaston revised this plan in 1902 and shelterwood was removed in all the ten original coupés, but no new regeneration felling was undertaken. Due to inadequate removal of overwood in the Manson plan, plantings had failed to establish.[63]

In 1912 Grieve's plan was introduced to avoid the second felling that destroyed much of the regeneration following the first felling. Grieve excluded all areas open to grazing, put all regenerated areas under a plantation working circle and divided the rest of the forest into High and Coppice working circles. We can see the emerging separation of the managed forest into discrete compartments subject to distinct treatments. More significantly, a clear separation came to mark forests open to grazing and those managed for the production of fuelwood and timber. The plan proposed selection felling in groups, relying on natural regeneration after mature stems had been removed. In areas open to grazing, regeneration was assumed to be impossible and so green felling was confined to the closed areas, from which all stems over two feet in diameter would be removed in fifty years. Overruling the local officers, the Inspector-General of Forests changed this plan to one of regeneration felling in groups which would remove all first and second class trees from closed areas in fifty years and rely on open forests and plantations to yield produce for the second and third fifty-year periods in the 150-year rotation.[64]

By 1920 the plan was revised again. The failure of natural regeneration and crop improvement methods had led to the adoption of taungya cultivation of sal. Clear-felling and taungya sowings remained the prescriptions in later working plan revisions in north Bengal sal forests of the Duars, though in some cases as in Buxa, the division of the regeneration areas into sal conversion and softwood working circles brought various blocks under rotations of different durations.[65]

Recognizing the local peculiarities of conditions for sal regeneration and a few other desired species certainly undermined the centralizing expertise claims of the working plan, hindering the use of these plans as an instrument of control by the Inspector-General of Forests. But there were other advantages and the gains from working plans were sometimes of a different nature. As the Inspector-General of Forests noted in 1942, working plans had become more exact and scientific with increased intensity of management, but chiefly in terms of superior surveillance and estimation of growing stock or merchantable timber. He said, 'those parts of India which have the best arrangements for working plans and the best control are just those parts which are providing the most supplies ... because they know what they have and where it is.'[66]

The overall pattern that emerged in working plans placed better quality forests under conversion to the Uniform System, with a rotation or conversion period of sixty to a hundred years. Three or four blocks were formed with roughly equal areas of teak and sal. The area allotted to the first block being regenerated under shelterwoods, while selection felling and thinning was continued in the third and fourth blocks. Forests deemed to be of poorer quality were placed under a coppice working circle to produce small timber and firewood on a thirty- to forty-year rotation.[67] As will become more apparent in the following section, by making conversion to the Uniform System the ultimate aspiration of any working plan, the entire forest management regime in Bengal was placed on a scale of approximation to scientific forestry with very little being close to the top of the scale. This then served both as a powerful urge to achieve the ideal and as a way of accommodating deviations or distortions.

Silvicultural Models and Elusive Particularities

From the beginning natural regeneration was the favoured method of crop development, and for this purpose Brandis favoured the training of recruits to the Indian forest service in Europe where this method was successfully used for high timber production.[68] This was not by any means a recent innovation since natural regeneration through coppice for fuelwood and selection felling for timber had emerged by the fourteenth century as standard forestry practices.[69] The main crop improvement activities were creeper cutting and improvement felling. Cleanliness of crops (absence of creepers) became a measure of whether foresters and guards were working hard in their beats. The main idea underlying improvement felling was to 'favour the valuable species and eliminate the less valuable and those interfering with the growth of the former.'[70]

Subsidiary work would encourage the younger classes by removing weeds and low growth that choked them, and lighten the cover overhead by girdling or removing inferior species of trees. Thinning aimed at achieving similar ends of improving light for valuable-species regeneration and producing even-aged crops.[71] Experienced foresters did often confess to their diaries that it was far from being an exact science,[72] yet for each species a model forest was envisioned through thinning alone.

The ideal was light and frequent thinning but lack of men and money made this unattainable in Bengal, where rapid forest growth made under-thinning the main problem.[73] The chief purpose of thinning was to attain the largest possible timber trees per acre with straight well-formed boles, an ideal as aesthetic as it was commercial. Noting the reluctance of marking officers to take out trees of large girth that were forked or crooked, Homfray wrote, 'it cannot be too carefully impressed on thinning officers that the chief point about the tree is its shape and not its size.'[74] Work progressed better where valued species appeared in unmixed stands, a condition that crop improvement aspired to create, imitate or stimulate. But by 1915, prescriptions for intensification in sal management included moving into the regeneration block system and here crop improvement was introduced into every felling series on a continuous basis as improvement felling became a management practice in forests of all age classes instead of a forest harvesting technique allied to selection felling.[75]

By 1925, selection and improvement fellings were being replaced in Bengal with concentrated regeneration under taungya or various coppice systems, aspiring to the shelterwood compartments or uniform system (Figures 12–17). But except for north Bengal taungyas, where clear-felling and artificial regeneration were essential to securing sal growth, in other parts the transition remained mostly an ideal. Movement toward the ideal was interrupted by wars which increased the demand for timber in sharp spikes, recurrent labour problems, and financial stringency. But the most significant thing about incomplete, unattained and constantly revised working plans and silvicultural arrangements was the repudiation by local governments of central direction and control after a point. Typically the local officials would emphasize the need for short-term and revisable schemes that were transformed gradually into long-term plans. Such rejection or resistance that challenged notions of absolute expertise residing at the centre emerged as part of the development regime in the forests of Bengal.

Changes in silvicultural practices necessitated the modification of all access rules to production and protection forests. With the shift to regeneration by blocks, which was considered permanent, provisional arrangements like selection and improvement felling were displaced and this change required closing protected forests for longer periods (twenty to thirty years).[76] As closer attention was paid to tree crops that maximized valuable species in even-aged stands, concomitant changes in silvicultural practice altered labour requirements. Subordinate departmental staff skilled in the organization of tasks like thinning and cleaning became necessary and were sought through the pool of trained personnel recruited in the Kurseong Forest School.[77] Silvicultural tasks like thinning always posed a challenge where the benefits of scientific prescription were realized only through local and intimate knowledge of the managed forest. This displaced the balance of expertise down the forester hierarchy, making the guards and rangers key personnel in identifying trees to be marked or determining the intensity of thinning in any season.[78] Periodic thinning thus became not only a management operation but the performative moment when expertise, like ritual knowledge, was passed on to younger generations of specialists in the making.

Till 1910 all the sal forests in Bengal were worked under the selection method with some improvement felling, a system

Fig. 12. Original high forests in the hills of Kurseong Forest Division

Fig. 13. A clear felling coupé in hill forest, Kurseong Division

Fig. 14. A taungya plantation area in the hills

Fig. 15. A plantation area at Sukna, Kurseong Division

Fig. 16. One and a half year old sal at Sukna, Kurseong Division

Fig. 17. Eight year old sal plantation at Sukna, Kurseong Division

necessitated by the supply of large sal trees being scant and dispersed through the forests.[79] In the decade following World War I, radical changes were introduced in the silviculture of sal. By the late twenties, Shebbeare, the Conservator of Forests in Bengal who fifteen years earlier had pioneered taungya experiments for Sal in north Bengal, could report that taungya had become the single most important means to regenerate sal forests. Taungya was a method of raising pure plantations that was labour-intensive and required land preparation by clear-felling and firing of scrub. But the advantage was that the outcome was easily managed for timber. Taungya working plans were simple. They prescribed clear-felling and restocking 1/r of the total area every year, where r was the length of rotation, usually eighty years.[80] One consequence of taungya and the switch to concentrated regeneration of any other sort was that trees like mahua, valued mainly for their non-timber forest product yields to villagers, were also removed under silvicultural prescriptions. Under earlier selection systems they were left alone in recognition of their utility to local people.

During 1926–39 most sal forests were brought under the Uniform System, but within the next ten years this was being given up as it was found that natural regeneration of sal could not be obtained through canopy manipulation.[81] As the sal study tour had noted, the conversion to Uniform System was predicated on the belief that regeneration *de novo* could be established in sal forests through the management of edaphic and light factors. But the highly mixed results had revealed that success was limited to regions with existing adequate advance growth, which was due to the past history of the forest, not the purposeful action of the forest officer.[82] Ironically, this was a condition that could obtain only under conditions not subject to the criteria of scientific management, namely, financial prudence and standardized silviculture. Field foresters in the late 1920s were finding that sal seedlings established in situations of varying light, possible only under an uneven-aged canopy. Felling to simulate these conditions would be both uneconomical and violate the ideal of concentrated regeneration blocks. Conversion to the Uniform System had failed on several counts. First, preparatory felling did not induce regeneration; second, final removal of overwood damaged established crops;

third, the debris of concentrated felling posed a pest and fire hazard; fourth, in damp areas weed infestation became rampant.

Thus the Uniform System came under critical scrutiny. On the eve of World War II, the national silvicultural conference was still calling for research on sal regeneration and the problems of pure teak plantations, pointing to the enduring problems in the silviculture of the two most valued species of Indian forests.[83] The same issues were revived as Indian foresters convened for sal and teak regeneration planning after India's first post-independence National Forest Policy was announced in 1952.[84] In short, scientific forestry remained a development discourse under production.

Development Regimes and Historical Processes

Forest guards—often assigned to lead off tricky silvicultural operations in their beat—mahouts, and the toils of the field forester, all generated information that was schematized to assemble scientific forestry in Bengal as a complex of knowledge and technologies of power.[85] In this way they contributed to processes redefining forest conservancy and providing it with several distinct elements. These were: a clear delimitation of its domain—*demarcation*; identification of indigenous vegetation and its valuation—*inventory*; enumerating, simplifying and circumscribing the bundle of usufructuary and other rights of local people—*reservation/protection*; devising a scheme of management that would produce the most desirable woods in the quickest time—*regeneration/plantation*; formalizing the arrangements through codes, manuals and division of responsibility among forestry officials, to ensure the establishment of routines—*working plan*. These elements were and remain crucial to the definition of scientific expertise and privileging its role in forest management, but they also contain traces of what has now widely come to be called development discourse.[86]

I am not speaking of development here as a received doctrine or cultural schema.[87] Dogged by uncertainty, conflict within its agencies, and resistance from local political configurations that it sought to coopt, scientific forestry remained a discourse continually under production. I have argued, therefore, that we should examine the historical processes in which such production occurs. One benefit of such an approach is that 'by tracking development historically, one can appreciate the complex origins of what came to be the unitary meaning of development that seemed to surface in the late colonial period in and around the second world war.'[88] I have further suggested that identifying the locus of production for any particular variant of a discourse should be integral to any such historical inquiry. The discourse of development has a long history then, precisely because it is not a received doctrine.[89]

The argument advanced here recognizes that recent scholarship on development, like that of Arturo Escobar or James Ferguson, acknowledges a comparability between aspects of colonial discourse and development. Yet both these powerful analysts of development firmly locate the production of a distinct development discourse in the decolonization period after 1945 and in the corridors of western development agencies like the World Bank.[90] They also focus on the effects of development discourse working as a system of knowledge. They conform thus to the currently dominant mode of analysis where development is treated as a schematic represent-ation of the third world and thereby the agent of certain political consequences including those whereby local 'targeted populations' absorb and manipulate the discourses of development.[91] In contrast, I propose a deeper history for development in general and suggest that the particular case of conservation and development reveals a conflicted and contested production of development discourse.

Scientific forestry, with its focus on the efficient and systematic production of timber for the 'public interest', clearly enunciated the productionist agendas that we take as characterizing develop-mentalist state policies in the twentieth century. On the other hand, the authors and propagators of scientific forestry launched a sustained critique of shifting cultivation, deplored private forest management, and inquired into the relations between deforestation and desiccation. This aspect of their work helped formulate the discourse of conservation so carefully identified and traced back to the seventeenth century in *Green Imperialism*. I am suggesting that by conflating conservation with the civilizing mission of colonial-ism by the end of the nineteenth century, scientific forestry took on another important feature of development discourse—the notion of progress. Emerging thus as a development regime, colonial scien-tific forestry became a complex of changing institutions and ideas informing post-colonial forestry.

Notes

1. Gouldsbury 1909: 7.

2. CSACL, Meikejohn Papers, undated journal entry by Florence Meiklejohn, accompanying her husband William Meiklejohn on tour in the forests of Bengal, pp. 78–9.

3. CSACL, Osmaston Papers, l/d 12 Feb 1929, F.C. Osmaston to his parents, from Saranda, Singhbhum, box 11.

4. Rajan 1998: 1–2 (page numbers here and in subsequent notes refer to the manuscript version).

5. Ibid.: 22.

6. Rajan (1998: 38) insists that this was indeed the case in India as a whole.

7. Silviculture as used here refers to the art of producing and tending a forest. It comprises, at the very least, the theory and practice of controlling forest establishment, composition, structure, and growth. See Smith 1986: 1–28.

8. Bayly 1993.

9. A fine discussion of this historical process as functional territorialization of state resource control may be found in Vandergeest and Peluso 1995.

10. Local knowledge is most usefully recognized by its 'inseparability from a particular place in the sense of embeddedness in a particular labour process.' Kloppenburg 1991: 522. This formulation is also central to much feminist analysis of science which emphasizes the importance of producing knowledge through sensuous activity, experience that is specifically local. See for instance, Haraway 1988; Harding 1986; Dorothy Smith 1987.

11. Scholarly discussion of science as representation has flowered into the sociology of scientific knowledge. Notable exemplars being Barnes 1977; Bloor 1976; Collins 1992; Gooding 1990.

12. The relationship between science and practice, and the practitioner debates in which institutionalized 'basic science' is shaped, are well discussed in the context of late-nineteenth-century American medicine by John Harley Warner, 'Ideals of Science and their Discontents in Late Nineteenth-Century American Medicine', *Isis* 82(313): 454–78, 1991. I am grateful to Warwick Anderson for alerting me to this work and its endorsement of my approach.

13. For recent work that stresses that scientific ideas were not imported into colonies and were more often in a process of continuous construction, reconstruction and transformation there, see Chambers (1987: 297–322), and several other essays in Reingold and Rothenberg 1987.

14. Hacking 1983; Hacking 1992: 29–64.

15. Pickering 1992: 1–28.

16. I am thus arguing that 'scientific knowledge has to be seen as intrinsically historical, in that its specific contents are a function of the temporally emergent contingencies of its production.' The phrase is from Pickering 1995: 209.

17. Crush 1995: 8.

18. Brandis 1884.

19. The memoirs of many foresters reveal this. See Stebbing 1930; Best 1935; Forsyth 1889.

20. Gouldsbury 1909: 243.

21. Here the distinction made between conservation and preservation is useful to bear in mind. The former combined utilitarian and developmentalist ideas in environmental management, underpinning soil conservation, water management, sustained yield forestry and so on; while the latter inspired more directly the creation of wilderness areas, parks, and sanctuaries. See Hays (1959) for a discussion of these ideas in the context of American environmentalism. Grove 1989; Anderson 1984; Beinart 1989; Beinart 1984; and Peters 1994: 78–80, are among a growing body of scholarship on Africa that suggests the same trends there in the late nineteenth and early twentieth centuries.

22. OIOC P/2934 BRP (Agri), Jan 1887, Misc/1/29-30, A progs, no. 510T, Cal, 19 Oct 1886, M. Finucane, Director Agri, to Secy GOB Rev.

23. OIOC P/2800 BRP (For), Jan–Feb 1886, A progs 32, Jan 1886, head I, col 1, no. 21F, Simla, 31 Aug 1885, res. by GOI, Home Dept, p. 19.

24. Both Adas (1990: 95–108) and Ludden (1989: 101–30) have pointed out the translation of local knowledge that was basic to the creation of colonial science and its codification as discourses of rule through classification, standardization and textualization.

25. Ludden (1992a) makes the argument for the process of colonial state formation in general; Brandis (1884) recognizes similar trends in forest administration in particular.

26. Ludden (1989) has described the process in respect of agricultural surveys and settlements; Dirks 1992 for colonial anthropology; Sivaramakrishnan 1995a for forestry; while the wider theoretical implications of the process have been discussed in Cohn and Dirks 1988 and Prakash 1990b.

27. Ludden 1992a: 264.

28. Bose 1997: 480.

29. Commenting upon the discursive strategies whereby expertise is constituted as the exclusive preserve of development agencies, Mitchell (1991a: 19) says, 'the discourse of international development constitutes itself . . . as an expertise and intelligence that stands completely apart from the country and people it describes.' This theme of objectification and depoliticization recurs in all critiques of development discourse.

30. Ludden 1992a: 270. Similarly scientific forestry in Europe had already declared the restoration of forests a task beyond mere preventive laws and something that called for scientific expertise. See R.P. Harrison 1992: 117; Lowood 1990.

31. Escobar 1988: 435. Harvey (1989: 244-46) calls this perspectivism, and points out that it led to rules of rational practice and the idea that the expert as a creative individual was always capable of a view from the outside—a totalizing vision.

32. Peluso (1992) has done an admirable job of analysing forest policy in terms of these three aspects of control in Java. Rangarajan (1996a) has looked at forest management in nineteenth-century central India from the same perspective. He attends carefully to changes in the silvicultural agenda—cast in terms of scientific advance—as the political-economic conditions changed.

33. Thompson (1975: 28–9) gives us an excellent discussion of this complex

fusion where forest law takes shape both in agrarian practice and conflicting representations when he says, 'the forest in fact was so by virtue of legal and administrative designation rather than by any unitary organization.'

34. Foucault 1979: 23.

35. Lowood 1990: 326.

36. 'Forest Conservancy in Bombay', minutes by Sir Richard Temple, Governor of Bombay on the forests of different districts and states of the province; reproduced in *IF* 5(3): 335–67, 1880, is a good reflection of the shared thinking in different provinces. Temple had been Lieut Governor of Bengal before this assignment. Cleghorn et al. 1852; Clutterbuck 1927.

37. Champion 1975.

38. Stebbing 1926: 199; BFAR 1875-6; Hatt 1905: 5. Schlich's plan had proposed annual removal of 5,785 trees over five feet in girth in the next eight years. This turned out to be a rankly optimistic estimate.

39. Stebbing 1924: 592–8.

40. OIOC P/2800 BRP Jan-Feb 1886 A progs 32, Jan 1886, head I (RR), col 1, no. 21F, Simla, 31 Aug 1885, GOI Home Res.

41. NAI GOI Rev and Agri (For), A progs 12–16, file 45 of 1901, Jul 1901, no. 128For, Cal, 5 Jan 1901, GOB Rev Res, p. 1026.

42. NAI GOI Education, Health and Lands (For), file 13-3/42-F&L 1942, S.H. Howard, IGF inspection note for the forests of Bengal, Dec 1941 and Jan 1942, p. 1.

43. W. Schlich, 'Notes on Preliminary Working Plans', in Brandis and Smythies 1876: 104–7.

44. OIOC P/7034 BRP (For), 1905, A progs 71–7, Dec 1905, file 9R/1, no. 271, Darjeeling, 11 Jan 1905, A.L. McIntire, CF Bengal, to Secy GOB Rev, p. 122.

45. OIOC P/7033 BRP (For), 1915, A progs 14–16, Jun 1905, file 3W/9, no. 11WP, Cal, 23 Jan 1905, S. Eardley Wilmot, IGF, to CF Bengal; no. 1233T-R, Darjeeling, 15 Jun 1905, M.C. McAlpin, US GOB Rev, to CF Bengal, pp. 150–3; OIOC P/7034 BRP (For) Oct–Dec 1905, A progs 60–4, Dec 1905, file 3W/4, no. 297, Darjeeling, 6 Feb 1905, CF Bengal, to Secy GOB Rev; no. 527, Cal, 7 Jan 1905, A.L. McIntire, CF Bengal, to IGF, pp. 107–9.

46. OIOC P/6561 BRP (For), A progs 19–25, Jul 1903, file 3W/3, no. 14, Cal, 29 Jan 1903, R.C. Wroughton, IGF, to Secy GOB Rev; no. 584T-R, Darjeeling, 22 May 1903, A. Earle, Offg Rev Secy GOB, to CF Bengal, p. 17.

47. NAI GOI Rev and Agri (For), A progs 28–32, file 162 of 1915, Jul 1915, Inspection note dated 28 Mar 1915 of Buxa and Jalpaiguri by G.S. Hart, IGF, p. 4.

48. Trafford 1905: 4.

49. WBSRR GOB Rev (For), file 6D/4, B progs 20–4, Nov 1907, no. 1857/320-7, Simla 6 Sep 1907, S. Eardley Wilmot, IGF, to Chief Secy GOB; no. 2219JR, Cal, 2 Oct 1907, Chief Secy GOB, to IGF; no. 11/C, Simla, 11 Oct 1907, IGF to Chief Secy GOB, pp. 1–5.

50. Ibid., note dated 24 Sep 1907 by A.L. McIntire, CF Bengal, in keep with papers, pp. 9–10.

51. OIOC P/7034 BRP (For), A progs 71–7, Dec 1905, file 9R/1, no. 271, Darjeeling, 11 Jan 1905, A.L. McIntire, CF Bengal, to Secy GOB Rev, pp. 122–3; Haines 1905; Trafford 1905; Hatt 1905; Tinne 1907; and Grieve 1912.

52. OIOC P/7034 BRP (For), Oct–Dec 1905, A progs 60–4, Dec 1905, file 3W/4, no. 297, Darjeeling, 6 Feb 1905, CF Bengal, to Secy GOB Rev; no. 527, Cal, 7 Jan 1905, McIntire to IGF, pp. 108–9.

53. The distinction made between major and minor produce indicated the priorities of forest management, especially in the matter of transforming the character of the forest under working plans. The idea was to increase the yield of major produce. This was further classified into valuable and less valuable or jungle trees. I have discussed this schematization emerging as an all-India feature of forest management elsewhere. See Sivaramakrishnan 1995a. Prasad (1994: 78–90) has discussed these classifications and their implications for forest management in Central Provinces.

54. NAI GOI Rev and Agri (For), A progs 35, file 136 of 1916, May 1916, no. 376/24-WP, Simla, 17 May 1916, G.S. Hart, IGF, to Secy Rev GOBO, p. 1.

55. Ibid., p. 2.

56. Stebbing 1926: 524.

57. NAI GOI Rev and Agri (For), A progs 24–6, file 311 of 1915, Feb 1916, no. 12319 GOB Rev Res, Cal, 17 Dec 1915, p. 2.

58. Ibid., A progs 23–5, file 286 of 1916, Jan 1917, no. 10091-For, Cal, 18 Dec 1916, GOB Rev Res, p. 1.

59. Grieve 1912: 1.

60. Ibid.: 5.

61. Under the permit system of forest working, the sale of individual stems at a fixed price per tree led to the removal over time of the best trees, leaving behind the defective trees.

62. No. 1449, 26 Mar 1896, DC Darjeeling to Commr Rajshahi; no. 909G, 12/14 Sep 1896, DC Darjeeling to CF Bengal, cited in Grieve 1912.

63. GOB 1935: 29.

64. NAI GOI Rev and Agri (For), A progs 5–7, file 54 of 1917, Feb 1917, inspection note on the Darjeeling and Kurseong Hill Forests by G.S. Hart, IGF, 28 Dec 1916, pp. 3–4. According to one official history, this alteration of the plan was based on a misunderstanding. The author (unknown, but probably the CF Bengal of the time, E.O. Shebbeare) notes that the idea of selection felling in groups was unworkable, but Grieve's proposal being confused with the group method of Europe the IGF approved the modified plan. See GOB 1935: 31.

65. NAI GOI Education, Health and Lands (For), file 13-3/42-F&L 1942, inspection note by IGF on the forests of Bengal, Dec 1941 and Jan 1942, p. 5; GOB 1935: 31.

66. NAI GOI, ibid., p. 2.

67. See, for example, J.N. Sinha 1962; Phillips 1924. This does not apply to sal forests of north Bengal where taungya had been introduced and other forests of Bengal like those in Chittagong and Sundarbans where sal was not the valuable species.

68. Stebbing 1924: 47.

69. Fernow 1907: 38.

70. Stebbing 1924: 576–8.

71. OIOC P/10122 BRP (For), 1917, A progs 4–6, Mar 1917, file 3I/1, inspection note by IGF, G.S. Hart, on Darjeeling forests, 28 Dec 1916; no. 85F/54-1, Simla, 14 Feb 1917, A.E. Gilliat, US GOI Rev and Agri, to Secy GOB Rev, pp. 16–17.

72. One forester was candid enough to assert that 'no two officers would mark exactly the same tree in a given area.' CSACL, Wimbush Papers, undated typescript entitled 'Life in the Indian Forest Service, 1907–1935,' pp. 88–9.

73. Homfray 1936: 4–6.

74. Ibid.: 9. Since the marking of trees was often left to village mandals, forest guards and even intelligent coolies, omission of crooked trees may have been deliberate as these were needed for making ploughs.

75. NAI GOI Rev and Agri (For), A progs 28–32, file 162 of 1915, Jul 1915, inspection note of Buxa and Jalpaiguri, 28 Mar 1915, by G.S. Hart, IGF, p. 17.

76. Ibid., A progs 43–6, file 124 of 1916, May 1916, no. 174For, 3 May 1916, Offg Secy Rev Punjab to GOI.

77. Ibid., A progs 5–7, file 54 of 1917, Feb 1917, inspection note on the Darjeeling and Kurseong Hill Forests by G.S. Hart, IGF, 28 Dec 1916, pp. 2–3.

78. The respect that senior foresters had for the local knowledge of the venerable guard of long standing is well conveyed in Gouldsbury 1909. Homfray (1936: 11) writes, in his authoritative manual on thinnings, 'village mandals, guards and especially intelligent coolies . . . know a good deal about silviculture.' Stebbing (1926: 460) notes that in France too the inspector would carry out thinnings after assembling all guards and rangers and the oldest forest guard would be the acknowledged expert.

79. McIntire 1909: 6. For selection felling the minimum prescribed diameter was 2 feet, that is, a girth of 6 feet.

80. OIOC P/11712 BRP(For), 1928, A progs 20–4, Apr 1928, file 9R/19 of 1927, no. 5191/R-53, 26 Sep 1927, E.O. Shebbeare, CF Bengal, to Secy GOB Rev, p. 39.

81. Stebbing 1926: 84–6; the work of R.S. Hole and E.A. Smythies initially, and the supplementary research of a host of other field foresters had been the basis for moving into the shelterwood compartment system, notably, Hole 1919; Smythies 1931; Makins 1920; Bailey 1924; Sen and Ghose 1925; Ford-Robertson 1927.

82. Anon 1934; Smythies 1940; Warren 1940; Raynor 1940; Warren 1941; De 1941; Griffith and Gupta 1948.

83. NAI GOI Education, Health and Lands (For), A progs, file 22-4/41-F&L, 1941, no. 14803/40-IV-130, 14 Nov 1940, S.H. Howard, President FRI, to Secy GOI, pp. 6–7.

84. *Proceedings of the All-India Sal Study Tour and Symposium* (Dehradoon: Forest Research Institute, 1953; *Proceedings of the All-India Teak Study Tour and Symposium* (Dehradoon: Forest Research Institute, 1953).

85. For instance, in thinning operations forest guards were required to work ahead of the officer to present a clear view. See Homfray 1936: 29.

86. See, for definitions of discourse of development, several essays in Sachs 1992, Ferguson 1994, Escobar1995, Mitchell 1991a. The characteristic of importance to us is the ability of development discourse to comprehend any situation requiring 'improvement' or 'development' through a non-local technical expertise that can then offer modular and generalized solutions.

87. My usage of the term follows Ortner 1989.

88. Watts 1995: 49.

89. I am extending here Ludden's (1992a) argument that post-colonial Indian development discourse was shaped by the colonial governmental forms that emerged in the later part of the nineteenth century. More recently a longer European history of development has been presented by Cowen and Shenton 1995: 29–33. In this context also see Hettne 1990.

90. Ferguson 1994: 67–8, 86, 264; Escobar 1995: 9, 23. The invention of development in the 1940s and 1950s is also argued with historical evidence from British colonial policy debates by Cooper (1997).

91. Appfel-Marglin and Marglin 1990, Dubois 1991, Parajuli 1991, Esteva 1992: 6–25 are some of the general writings promoting this approach. Pigg 1992 and Woost 1993 are instances where development as a received and systematic discourse engages other discourses in a local setting in south Asia.

9

Conclusion

Rethinking Governmentality

I set out to study the emergence of modern forest management in Bengal as a form of governmentality or governmental rationality. This was to be an exploration of statemaking—the formation, modification, and maintenance of multiple regimes of government. The dominant patterns of statemaking that emerged in Europe in the nineteenth century and influenced colonial statemaking from the top were those that Foucault has discussed as emerging from liberal ideologies, which recognized that 'the finitude of the state's power to act is an immediate consequence of the limitation of its power to know.'[1] This resulted in a reliance on structures of micro-power which were formed through contextually specific devolution of power to localities and formations often outside the formal apparatus of state power, but always serving an overarching principle of inspectability.

Use of the term devolution may suggest a degree of control from the top that did not exist in practice. Even before Foucault used Bentham's ideas about the Panopticon to develop his theory about the micro-physics of power, we were alerted by Polanyi that the panoptic gaze was turned on everything, including ministries and civil servants.[2] This was one reason why 'the state' as a monolithic entity could not always be in charge of how and what devolved. Fractures in the colonial Indian state, particularly in the constitution of forest management regimes, emerged precisely when dealing with questions of inspectability at different levels and across functional jurisdictions. This led to patterns of governmental devolution that did not emerge from one all-controlling centre.

Devolution was also context-determined and this meant that space and time factors shaped the precise arrangement of interlocking power structures through which governmentalities emerged. We then have to attend to issues of historicity and spatiality, temporal and regional variation, that influenced the everyday forms of power. The study of these variations brings into view the role of everyday resistance in delineating the contours of power.[3] Integrating notions of resistance, space, and history into an examination of cultures and structures of power required a focus on the complex interplay of representations and practices, their modification and maintenance over long periods, and thus analysis of the relationship between past and present in the realms of ideas and institutions. Consequently, in my larger research project, I sought to trace the specific regional histories of agrarian relations, forestry and statemaking in Bengal and thus document the changes in peasant-state relations, while recounting the ways in which state perspectives of forest peoples were constructed over time and how these have influenced contemporary developments in Bengal.

Drawing on part of the outcome of the larger research project, this book has considered governance, politics, and forest management as practice.[4] If the interplay of politics and governance, discussed as statemaking, has been the focus of its inquiries, forest management in Bengal has been the site of those inquiries. Apart from my own interest in forestry, environmental politics of south Asia, the history of state building in India and other post-colonial nations, and the anthropological study of development, all of which certainly directed my research, I was also moved by two convictions when choosing particular sites of inquiry. The first was that our present problems of forest conservation, human rights and the equitable distribution of development opportunities in what have been called 'low income democracies'[5] cannot be diagnosed, and hence cannot be prescribed for, without a sophisticated and situated understanding of the historical processes moulding them. The second was that forest management offered a unique and instructive window on processes of statemaking.

The need for historical study of contemporary issues has been recognized in several social science disciplines. There are now, as a consequence, numerous subdisciplines like historical sociology, historical demography, historical anthropology, economic history, historical ecology, and so forth. I have described in the introduction how I conceive of my version of historical anthropology. It has also been developed and demonstrated in this book. I will, therefore, refrain from dilating again on that topic. But there is need to say a little more about why forestry in India offers a special opportunity to study statemaking.

More than agriculture, industry, and other key sectors of the economy, forestry in India, as was the case in many other European colonies in the nineteenth century, witnessed the most sustained efforts by both colonial and subsequent nationalist governments to establish direct state control over an important natural resource. These efforts resulted in vast areas of land, large human populations, and the entire regime of production through which the resource was transformed into goods and services for national interest and international commerce, being subjected to disciplines and procedures of government. If we think of the central state as a master magician, forests were the place where this prestidigitator *par excellence* put on the greatest show.[6]

Given their location, most often on the fringes of settled agriculture, which has been variously discussed as the civilized arable or the place of sedentary development-friendly plains-people, some forests also proved a difficult and somewhat unpredictable place for modern statemaking to ramify. In that sense, to persist with our metaphor of the magician, they frequently compelled the central state to use every trick in the bag. Obviously, a magician giving an extended display, and running through the full repertoire of her craft, is also most likely to edify the audience, especially students of the subject.

A fully stretched state machinery, or the diverse play of resistance, accommodation, and expropriation, are not the only reasons why I suggest that forests in India provide a rewarding field or archive for the historical anthropologist of statemaking. I was, and continue to be, intrigued also by the growing global concern for tropical environmental resources, first world commitments to the spread of democratic governance in the developing countries where these resources are concentrated, and the possible contradictions between the pursuit of international conservation and local self-determination. I felt, further, that a study of forestry in India that examined in one sweep and with equal care, the past, the present, and their mutual interpenetration, could begin to illuminate these contradictions. Conservation, development and the political empowerment of citizens at all levels, are goals that nations pursue with different degrees of commitment. When, as in India, each of these lofty ideals is equally valued, knotty problems of governance presented themselves, and they were expressed most often in terms of the contrary pulls of central direction and local autonomy in government.

Recent studies of democracy in India have commented upon the paradoxes of proliferating political institutions and widespread powerlessness among them; or the accentuation of urban–rural inequities in a framework of enhanced devolution.[7] Curiously enough, the increase in central environmental regulations and their implications for economic liberalization and democratic devolution, proceeding apace in India since 1991, have not generated the same scholarly interest. Defining a sector of governmental practice that may be out of step with others, environmental management has also become a field of heated controversy, where nationalists, localists, and advocates of global free enterprise contest key issues of governance, environmental impacts, and human rights. In India this debate is particularly rich and long-standing around the question of forest management and modern agricultural development, though it is beginning to encompass issues of pollution and public health.[8]

Nationalist ideas are also being fuelled by the rapidly growing international governmental system for the environment.[9] In some cases, like the opposition to Intellectual Property Rights clauses in GATT (General Agreement on Tariffs and Trade) and the resultant agitation generated in parts of India, such elite coalitions around nationalist anxieties promote a favourable climate for generating a grid of national environmental regulations. Periodic pressure from events like the Earth Summit of 1992 are a powerful impetus for a stringent national forestry code which would serve as an effective platform for staking out an 'Indian' position that diverges from the international dispensation. So the authors of this position are actively crafting and claiming a national consensus. In contrast, India also has a vigorous environmental movement, long divided over the complicated issue of where the locus of environmental management should be. Many influential environmentalists have recommended the empowerment of local communities for effective conservation and sustainable development of scarce natural resources.[10]

Such contradictions reveal several fascinating aspects of the relationship of environmental management to democracy in India. At the ethical level, the debate is certainly about contested definitions of equity. Should inter-generational fairness be stressed over fairness across different segments of society? How do people living in different parts of the country share the costs of environmental management? Seeking answers to these questions threatens to institute new divides, or accentuate existing ones, between town and country, locality and nation. Much of the tension arises from the definition of expertise, and identifying where it reposes in society.

In engaging substantially with the other two issues raised above I have contended with the debate on the roots of colonial forest management. Ramachandra Guha and his political-economic argument have stressed the post-1878 developments in state forestry as the formative influence on regimes of forest management in India. Richard Grove and his ideological approach have stressed the earlier work of surgeon-naturalists in formulating and disseminating a desiccation discourse that shaped colonial forest policy and global environmentalism. The seductive appeal of this 'origins' controversy continues to be very powerful.[11]

My effort has been less to pinpoint the problematic, sharp distinctions between pre-colonial and colonial forestry (and the related issue of when they become discernible). I have endeavoured more to delineate the processes by which forest management emerges as an arena of statemaking, where the past, both as historically shaped institutions and contradictorily imagined conditions, influences the present. Such an approach takes a more modest and contingent view of state capability in any situation because it attends to the very micro-physics of power, the routines that implement policy and standardize governance. By studying the making of disciplines and the contested performance of procedures, we find there was always a gap between intent and practice, between representation and experience.

I have been at pains to show that scientific forestry, colonial discourse, and other such lumped categories need to be unpacked to find the distortions, transformations, and hybridizations that they undergo in the very processes of their construction and implementation in specific historical locations. My most important means to making this case has been the discussion of regional variations, the spatial differentiation of power—its modes and experience—and the emergence of differently configured but persistent zones of anomaly in the terrain of statemaking. We need to understand the significance of uneven impacts in diverse places, of procedures like forest demarcation, working plan preparation, resource inventory, reservation, and regeneration.[12] But I have not been satisfied with promoting such understanding only through well-worn analyses of resistance, property rights, and state territorialization of forested landscapes. I have suggested that forest management in Bengal has uniquely constructed the key categories of village or local community, forest control or managerial jurisdictions, and the matter of knowledge or expertise.

These are categories redolent with history, because they were formed, altered, challenged, and revised in the colonial processes of statemaking, becoming polyvalent and conflict-ridden legacies for post-independence nation building. For example, we find the questions of participation, local self-government, and regional autonomy coming to centre-stage with the Bengal Tenancy Bill in the 1880s, a period often discussed in the political history of modern India as a time of liberal experiments.[13] Similarly, in many areas of Madras, the government set up 'between 1890s and the 1920s local committees to enforce forest conservation, . . . select policemen, to settle communal disputes, and to control distribution of water from irrigation schemes'.[14] As we have seen in Chapter 5, which discusses the entire area of southwest Bengal where protected forests were established, in many places forest management was perforce dependent on local structures of authority even while they were imagined and reshaped by colonial power and practice to some degree.

Environment, Development, and Statemaking

In discussing this reshaping of local landscapes, and the social identities of people resident there, I have combined genealogical and chronological histories of forest management in Bengal, to document a veritable reorganization of nature that invalidated prior and ongoing reorganizations which might impede the dominant mode.[15] The early British state worked to create a woodland Bengal landscape that did not easily harbour guerrillas and other mobile populations. Whether it was police powers for zamindars or vermin eradication, the three-pronged approach of community construction, establishing regimes of control, and compiling bodies of standard expertise provided material that would enable transitions from forests perceived largely as a wild place, into managed forests that were duly incorporated into regimes of production. Initially the chief colonial strategies were the sedentarization of tribes, and policies to create conditions for the spread of peaceful and profitable agriculture. Where the expansion of settled agriculture did not easily combine with the legitimation of British rule, there was a tendency to go slow on curbing tribal food production strategies. These deviations from overarching principles established early in the nineteenth century made space for administrative exceptionalism and the creation of tribal places in woodland Bengal.

By the 1860s, India was well integrated into empire, and forests had emerged as an important revenue source, ensuring that conflicts around them were mainly played out in terms of the definition of property rights. The overriding concern soon became what form of property would best encourage government profits from forestry. Reservation was the dominant mode of securing state control over forest lands and revenues. Zamindari rights in the forests of the Permanent Settlement areas forced themselves on the attention of colonial administrators and revived the lingering belief in the salutary effects of protecting superordinate property rights. This meant that substantial forest areas were left in the charge of landlords. Forests covering around 6,000 square miles were reserved by 1905 but an equal area left within private estates.

By the early decades of the twentieth century forest regeneration had become a prime concern in reserved forests. This facilitated the centralization of authority and the distancing of management issues from local politics. A tougher attitude towards private forests, which came to be officially represented as ill-managed, also became possible. Demarcation, inventory, and planning in the reserved forests complemented cadastral survey and settlement in agricultural lands. These procedures collectively objectified land, classified it by prescribed land uses, and constructed property regimes that were considered best suited to promote the productive use of land.

An ironic and significant contrast that emerged out of these processes was that in the realm of agriculture, rights moved downwards within social hierarchies as policies were designed to provide more effective land control to the tiller. But in forestry, companion policies took land control in the opposite direction. These contrary trends have to be linked to outcomes of cadastral survey and settlement operations especially the definition of rights in village lands that classified and recorded landlord ownership on large tracts of village wastes and forests. They also point to the fast emerging formalization of forest management—professionalization and institutionalization—to use Arturo Escobar's terminology—that defined expertise as something exclusively controlled by the foresters, even though the constitutive knowledge was gained through active interaction with local practices on the ground.

We are then presented with an apparent two-part puzzle. First we learn (in contradiction to the prevailing wisdom of Indian historiography) that state forest control resulted not from the sweeping away of peasant rights across the board but from the clarification and elaboration of property rights, which resulted in their reorganization, often to the detriment of poor farmers and other lower classes in forest Bengal. Second, we are forced to recognize that, notwithstanding varied regimes of property rights, forest control was uniformly wrested from the locality and assigned to central and other superordinate agencies deemed to possess requisite technical expertise for forest management.

By arguing that the key issue in this welter of issues was the locus of effective governance, I have suggested that the post-colonial version of scientific forestry is a product both of its times and its past. It has a genealogy and a history. As development discourse it has instrumental effects, but my interest is not so much in these effects as it is in the historically specific sites and modes of the production of scientific forestry. Unpacking scientific forestry into specific concepts and then considering the production of these concepts in the history of forest use and management in Bengal allows us to understand the constitution and mobilization of scientific forestry as a development discourse. While scientific forestry is concerned with productive aspects like regeneration and revenue, it is also obliged to deal *in extenso* with questions of rights and property. To add to its woes, by the third quarter of the nineteenth century, scientific forestry was also deeply challenged by responsibility for environmental health, the preservation of nature, and thus with conservation. How these facets were dealt with cannot be separated from the representational schemes through which the forested landscape of India came to be known to its administrators and managers. I have, therefore, insisted throughout this book that 'we cannot take science out of its culture, out of the realm of meaning, value, and ethics.'[16]

This interaction of vision and action produced technologies of management that were, in their fully conceived form, also technologies of rule. Once visualized and translated into a programme of action these ideas were structures of governance or the channels through which power flowed. But knowledge and power are not always in a predefined and predictable relationship. The grooved routines of management, where knowledge and power flow into each other, are produced through a process of experimentation, of reading existing arrangements, of culling knowledge and shaping the basis of power. This turns out to be a tortuous and uncertain process where the perfect coincidence of knowledge and power cannot be presupposed.

The points of tension in the composite entity of scientific forestry are located in the conflicting representations of forests as refuge, resource, and regeneration that were generated in Bengal, and the uneasy fashion in which these representations co-existed at all times in the definition of forests and thereby shaped the programme of scientific forestry. This is where I take issue with much of the extant scholarship on forest history in India, because the system of forest management that was laid down did not flow from the dominance of any one scheme of landscape representation. So no simple relation can be established between the effective degree of central control that was established in any place, and the dominant notions of forest management that were aired in policy documents and technical forestry publications. I have argued that scientific forestry is a complex, multi-layered discourse formation that was historically and contingently produced. Such a processual approach to the heterogeneous and changing character of the corpus of expertise known as scientific forestry allows me to suggest theoretical refinements to several important areas of current historical-anthropological interest.

Theories of state formation that focus on collisions between society and state present a very unified account of discourses of rule, discourses of colonialism, and so on. I have shown that a close study of statemaking reveals the struggles and tensions through which these discourses are constituted, bringing into repeated contact the world of ideas and that of experience. This implies that the colonial project was not over-determined by a few powerful tropes of western culture, but was more an outcome of its location-specific histories. We may then explain the strong linkages of post-colonial projects to the colonial and place in perspective the changes and continuities we find in any sector of governance.

Approaches made widely influential by James Ferguson, Arturo Escobar, and Timothy Mitchell have provided valuable insights into discourses of development. The Bengal forestry case, however, presents evidence to question their rigid handling of the subject that causes them to lose sight of process, agency, and history. Colonial forestry, agricultural modernization, soil erosion control schemes, and large multi-purpose projects like the Tennessee Valley Authority in the US were not only precursors of a post–World War II notion of international development. They were divergent sites where the complex, polyvalent discourses of development were produced in historically specific ways. Multiple genealogies of development ideas are present in society. Uncovering these contested histories gives us clues to the actions of historically grounded agents as the politics of development enters new phases.[17]

Much environmentalist writing starts from a premise that the state was an unmitigated predator on natural resources, unremittingly centralizing, and totally at the service of wealthy, powerful, and wider commercial interests. They also contrast such a strong and simply motivated state with a romanticized village community, a natural local entity imbued with a conservation ethic. These communities are presented as basically free of conflict, or possessing the ability to resolve all issues in a stateless condition through community rules.[18] I have suggested, to the contrary, that ecological warfare and environmental conservation, as the extremes in a range of possible management outcomes in forest Bengal, were both more likely the product of statemaking than the preferred, or immanently logical, practice of communities and states.

Unpacking constructs like state and society to understand their mutual constitution in the politics of forest management has been one of the objectives of this study. To comprehend the varied cultures of governance in colonial forest management, and their change over time and space, this book has also moved beyond existing environmental histories of India in its attention to ecological factors. I have discussed the environment as a hybrid entity that is produced through the interaction of bio-physical processes that have a life of their own and human disturbance of the bio-physical. Human agency in the environment, mediated by social institutions, may flow from cultural representations of processes in 'nature' but we cannot forget the ways in which representations are formed in lived experience of social relations and environmental change. This dialectical formulation has been used here to explain the construction of scientific forestry. I have concurrently theorized statemaking, and the notion of political society as the realm of such statemaking, by integrating the study of conflict and cooperation, and by bringing resistance and power into a single field of analysis. Through the concept of zones of anomaly, that I defined in Chapter Two and used in subsequent chapters, we can begin to understand the spatiality of regional variations, the spatial history of uneven development, and thus the geographies of empire, development, and conservation.

The diversity of environments, and strategies to manage them in India, are a legacy that has both survived colonialism and been enriched by it. India is certainly faced with many difficult problems of environmental management. It also has unique opportunities for tackling these problems in ways that are likely to yield fair and responsible results. Selecting the right opportunities is no easy task. True opportunities have to possess two qualities. They must deliver benefits of environmental management even-handedly across society by overcoming disparities caused by all kinds of social inequality. They must also do such contemporary justice without mortgaging the future. We can only begin to assess opportunities, and seize the most appropriate ones, when their historical provenance is properly described and understood.

After fifty years of independence India's environment, forests in particular, is neither bearing a crippling burden of colonial mismanagement nor is entirely free from the quirks and predispositions of that history of British empire in the subcontinent. The relationship to colonial pasts is contradictory and gives cause for both discouragement and hope. Finding some of those threads of hope and weaving them together can be a useful enterprise. Scholars, activists, development agencies, and government departments have combined in unprecedented ways to embark upon this enterprise in forest management in India. It was the desire to study that exciting enterprise that led me to carry out the work that has resulted in this monograph and other related research. At this juncture my work only wishes to remind all those engaged in fashioning new forest management, and similar enterprises, that we have to create the fabric of future policy from the threads of past experience. But we must weave those threads from the past with care.

Notes

1. Gordon 1991: 16.
2. Polanyi 1957: 140.
3. My ideas on the place of social agency in statemaking are formed in engagement with Scott 1990, Joseph and Nugent 1994, and Migdal et al. 1994.
4. 'Practice is the site of a dialectic between opus operandum and modus operandi, of the objectified products and incorporated products of historical practice', Bourdieu 1990: 52.
5. Kohli 1994: 96.
6. Indeed the effort is in some ways comparable to the case where the Austrian Hapsburgs built a 1,000 mile long *cordon sanitaire*, to keep out the plague, with little effect other than a huge administrative effort that contributed to building. See Jones, 1981.
7. Jalal 1995; Varshney 1995; Kohli 1994.
8. Agarwal et al. 1982, 1985; Agarwal and Narain 1992; Shiva 1989; Guha 1989; Gadgil and Guha 1992; Gadgil and Guha 1995; Ahmed 1991; Arnold 1991; Harrison 1994.
9. I am referring to laws, treaties, conventions and organizations like the Montreal Protocol, Biodiversity Convention, laws for ocean pollution, Antarctic exploration, deep-sea fishing, and institutions like the Global Environment Fund, United Nations Conference on Environment and Development, and General Agreement on Tariffs and Trade.
10. In a series of essays, Ramachandra Guha has explored the various intellectual provenances and political affiliations of Indian environmentalists. See Martinez Alier and Guha 1997.

11. It continues to preoccupy scholars of modern Indian environmental history. See for instance, Rangarajan 1996b.

12. Recent writing on south and southeast Asia in the context of colonial and post-colonial forestry continues to discuss the question of regional variations at a continental level. Thus, Arnold and Guha (1995: 18–21) call for a distinctive environmental history of south Asia, suggesting again the very monolith that this book problematizes. Similarly, all varieties of modern state intervention in southeast Asian forests are uniformly presumed to strengthen state control of them. See Peluso et al. 1995: 207–12.

13. Gopal 1965: 144–51.

14. Washbrook 1976: 62.

15. For a fine discussion of the need for environmental history to turn to the study of the reorganization of nature, and recognize that such reorganization by human and other forces has gone on longer than conservationists readily grant, see Worster 1996: 7-9.

16. Worster 1996: 11.

17. See for instance, Adams (1995), who provides a succinct analysis of sustainable development in terms of the longer history of the development idea.

18. I have discussed this issue in detail elsewhere. See Sivaramakrishnan 1995a. See also, Rangarajan 1996b.

Note on Primary Sources

Archival research for this project was carried out at several locations. I began in various collections in the US, notably the Forestry Collections at Yale. Here I have used the serials *Indian Forester* (1875–1991), *Empire Forestry Journal* (1921–50), and Proceedings of the Empire Forestry Conferences. Technical publications generated soon after the founding of the Forest Research Institute, Dehradun, in 1906 are mostly available through various series known as *Indian Forest Records, Indian Forest Bulletin, Indian Forest Pamphlet,* and the *Bengal Forest Bulletin*.[1] Though interrupted, annual reports of the Bengal Forest Department are available in this collection after 1898. Earlier annual reports were read at various locations in England, like the Indian Institute, Oxford, and the India Office Library. The Forest Collection, supplemented by material garnered through inter-library loans from the Library of Congress, University of California, Duke University, University of Minnesota, New York Public Library and University of Pennsylvania proved a rich source of reports, legal compendia, gazetteers, Bengal district records, Bengal forest working plans, and many nineteenth-century publications. Not least among these treasured sources was an impressive collection of early-nineteenth-century books on hunting and fauna in Bengal, written by travellers, Company officials, and other itinerant Europeans.[2]

Then I worked in the Oriental and India Office Collections of the India Office Library; the Western Manuscript Collections of the British Library and the Bodleian Library, Oxford; the Private Papers Collection in the Cambridge South Asia Centre Library; and read the papers of missionaries in the School of Oriental and African Studies Library, all in England.[3] This research divided my time between four classes of records. First, the proceedings and consultations of the Bengal government. For the period 1780 to 1860 I have consulted Judicial (Criminal) Consultations, Judicial (Civil) Consultations, Revenue Consultations, Board of Revenue Proceedings (Lower Provinces), Committee of Revenue and Home Miscellaneous Records in a sporadic fashion. These are mostly manuscript records and very voluminous, so I have read them around dates identified as significant on the basis of preliminary research—like the 1793-9 period—which saw the introduction of the Permanent Settlement and the chuar disturbances in Midnapore.

For this East India Company raj period, I read many volumes of the Board's Collections, a compilation that draws upon the original sources from departments concerned and brings them together around something that was important enough to require detailed information to the Board of Control of the East India Company.[4] These I regard as my second class of records. The volumes I read pertained to the management of police in Midnapore, teak plantations in jungle mahals of Bengal, chuar disturbances in Midnapore, the Bhumij and Kol unrest, and related events covering the period 1783 to 1837. A companion series were the relevant volumes of Bengal Despatches.

The India Office Library has an extensive manuscript collection of private papers where I found my third class of records. The papers included those of several people who had served as magistrates and district officers of Midnapore, like Henry Strachey, and others who had served in other neighbouring jungle mahal districts.

My fourth class of records are a fascinating collection of journals that appeared erratically through the nineteenth century. For example the *Bengal Sporting Magazine* (1838-46), the *Journal of the Agri-Horticultural Society* (off and on for twenty-five years), the *Journal of the Scottish Arboricultural Society*, and so on. While some of these, like *Indian Antiquary,. Asiatic Researches* and *Journal of the Asiatic Society of Bengal* were reputed fora for the debates and scholarship of eminent Indologists and Orientalists, many others like *Calcutta Review*, or *Hickey's Gazette*, were useful for news of European life in the subcontinent. These magazines and pseudo-academic periodicals also provided a venue for reproducing important reports and documents (both government and non-official), as well as articles in serious and frivolous vein by various classes of Europeans living and working in India. One can read in these pages the work of boxwallahs, merchants, mercenaries, surveyors, scientists, Company officials, missionaries and brown sahibs who wrote frankly on the people, the shikar, each other and their purposes in empire. I hope that the fragmented but at times intriguing material encountered in this class of records has helped me to show the diverse strands in the viewpoint of 'colonial power' and how it was being created in a situation where expectation constantly met history and both were modified.

In Delhi I divided my time between the National Archives, Central Secretariat Library, and the Nehru Memorial Museum and Library. I was able to identify and read selectively the settlement reports for all the Jungle Mahal districts. There were six reports for Midnapore alone, over the period 1874 to 1919. At the National Archives, I read the Government of India proceedings in the Public Works Department (1861–71); Revenue, Agriculture and Commerce Department (1871–80); Home Department (1858–81); Revenue and Agriculture Department (1881–1923); Education, Health and Lands Department (1923–45). Although all these proceedings were principally consulted with a view to finding material on Bengal forestry, some allied material on forestry elsewhere (for comparative purposes) and land revenue matters was also collected.

Another series that I have gone through in the National Archives is the Foreign Department Consultations for 1820–60. These were expected to yield information on forestry and on the political developments of the period, which for my region were greatly overlapping subjects. The search was disappointing. The material was good for Burma and Central Provinces when pertaining to teak, and it showed me that 'forest as resource' was an idea that developed in the colonial governmentality as a species-specific thing, in addition to being a region-specific affair and, of course, temporally varied. Selections from the Records of Bengal and India Governments, annual reports of the Department of Court of Wards, Bengal and the Reports of the Political Agent for the South West Frontier Agency were also consulted chiefly in the National Archives.[5]

The principal locations for archival research in Calcutta are the West Bengal State Archives which house everything up to the beginning of the twentieth century, and the West Bengal Secretariat

Record Room, which has everything pertaining to 1908–47. This location also has government publications, statistics, census documents, and so on. In the nineteenth-century records I have gone through the proceedings relating to forestry, agriculture, and land revenue in Finance Department (1873–84), Agriculture Department (1873–4), Land and Land Revenue Department (1858–99), Judicial Department and General Department (1858–1908). The main difference with the India Office Library is that the B Proceedings were available here, and departmental consolidation of indexes and records made the tracking of individual themes a little more easy. Similarly, for the twentieth century, the A and B Proceedings were consulted for Revenue Department (1900–45), Home (Political) Department (1900–25), Agriculture and Industries Department (1921–37), Forests and Excise Department (1937–44), Agriculture, Cooperative Credit and Rural Indebtedness Department (1937–44), Agriculture, Forests and Fisheries Department (1944–47).

The library of the Anthropological Survey of India in Calcutta yielded the Bulletin of the Department of Anthropology, while the Centre for Studies in Social Sciences had some useful compilations of Bengal Census statistics. National Library, Calcutta was a wonderful, if sometimes frustratingly difficult source of published government documents. Many of the technical forestry reports that were missing in the Yale collection were located there, as well as most of my agricultural and land ownership statistics for Midnapore. Despite the fact that the National Library was closed for extensive periods due to a strike during my time in Calcutta, I was able to find there most of the relevant volumes of the serial *Journal of Bengal and Orissa Research Society*, which was otherwise difficult to locate.

Local archive research was carried out in the District Record Room, Midnapore; Subdivisional Record Room, Jhargram; Records of the Subdivisional Land Reforms Officer, Jhargram; Block Development Officer, Jhargram; Subdivisional Agricultural Officer, Jhargram; Divisional Forest Officer, West Midnapore; Range Forest Office, Bandharbola (Jhargram); and the Jhargram Raja's Zamindari Records at the Rajbari in Puratan (Old) Jhargram. The District Records were disappointingly scanty, but I was compensated by the excellent condition of cadastral survey and settlement records in Jhargram.

Notes

1. These are deposited in the Seeley Mudd Library, Yale University, where may also be found the 'blue books' or Parliamentary Papers reporting on the Proceedings of the British Parliament.

2. Mainly in the Beinecke Rare Books Library, Yale University.

3. The standard guide to western manuscript collections in Britain is M.D. Wainwright and Mathews, *A Guide to Western Manuscripts and Documents in the British Isles Relating to South and Southeast Asia* (London: Oxford University Press, 1965); see also J.D. Pearson, *A Guide to Manuscripts and Documents in the British Isles Relating to South and Southeast Asia*, Vol. 1 (London: Mansell Publishing, 1989).

4. This series continues to be fully indexed at the Oriental and India Office Collections, where the curator was kind enough to let me consult lists of unindexed volumes and call those that were needed.

5. Much of the printed sources available in India have been catalogued by V.D. Divekar (ed.), *Annotated Bibliography for an Economic History of India, 1500–1147* (Pune: Gokhale Institute of Economics and Politics, 1977).

Select Bibliography

Abbott, Andrew, 1991. 'History and Sociology: the Lost Synthesis'. *Social Science History* 15: 201–38.

Abrams, Philip, 1988. 'Notes on the Difficulty of Studying the State.' *Journal of Historical Sociology* 1(1): 58–89.

Adams, W.M., 1995. 'Green Development Theory? Environmentalism and Sustainable Development'. In Jonathan Crush (ed.), *Power of Development*. London, Routledge.

Adas, Michael, 1981. 'From Avoidance to Confrontation: Peasant Protest in Pre-colonial and Colonial South Asia'. *Comparative Studies in Society and History* 23(2): 217–47.

——, 1990. *Machines as the Measure of Men: Science, Technology and Ideologies of Western Dominance.* New Delhi, Oxford University Press.

Agarwal A. et al. (eds), 1982. *The First Citizen's Report on the Environment in India.* New Delhi, Centre for Science and Environment.

——, 1985. *The Second Citizen's Report on the Environment in India.* New Delhi, Centre for Science and Environment.

Agarwal, Anil and Sunita Narain, 1992. *Towards a Green World: Should Global Environmental Management be Built on Legal Conventions or Human Rights.* New Delhi, Centre for Science and Environment.

Agnew, John, 1987. *Place and Politics: The Geographical Mediation of State and Society.* London, Allen and Unwin.

——, 1989. 'The Devaluation of Place in Social Science'. In John Angew and James Duncan, *The Power of Place: Bringing Together Geographical and Sociological Imaginations.* Boston, Unwin Hyman.

Agrawal, Arun, 1995. 'Dismantling the Divide Between Indigenous and Scientific Knowledge.' *Development and Change* 26(3): 413–39.

—— and K. Sivaramakrishnan (forthcoming). 'Agrarian Environments'. In Arun Agrawal and K. Sivaramakrishnan (eds), *Agrarian Environments: Resources, Representations, and Rule in India.* Durham, Duke University Press.

Ahmad, Aijaz, 1991. 'Between Orientalism and Historicism: Anthropological Knowledge of India'. *Studies in History* new series 7(1): 135–63.

—— (ed.), 1992. *Theory: Classes, Nations and Literatures.* London, Verso.

Ahmed, S., 1991. 'Questioning Participation: Culture and Power in Water Pollution Control: The Implementation of Ganga Action Plan at Varanasi'. PhD thesis, Cambridge University.

Alavi, Seema, 1991. 'North Indian Military Culture in Transition'. PhD thesis, Cambridge University.

————, 1993. 'The Makings of Company Power: James Skinner in the Ceded and Conquered Provinces, 1802–1840'. *Indian Economic and Social History Review* 30(4): 437–66.

Alonso, Ana Maria, 1994. 'The Politics of Space, Time and Substance: State Formation, Nationalism and Ethnicity'. *Annual Review of Anthropology* 23: 379–405.

Amery, C.F., 1876. 'On Forest Rights in India'. In D. Brandis and A. Smythies (eds), *Report on the Proceedings of the Forest Conference Held at Simla, October 1875*. Calcutta, Superintendent of Government Printing.

Amin, Shahid, 1989. Introduction. *William Crooke's Glossary of North Indian Peasant Life*. New Delhi, Oxford University Press.

————, 1995. *Event, Metaphor, Memory: Chauri Chaura, 1922–1992*. Berkeley, University of California Press.

Anderson, Benedict, 1991a. *Imagined Communities: Reflections on the Origins and Spread of Nationalism*. London, Verso.

————, 1991b. 'Census, Map, Museum'. In *Imagined Communities*, London, Verso.

Anderson, David, 1984. 'Depression, Dust Bowl, Demography and Drought: The Colonial State and Soil Conservation in East Africa During the 1930s'. *African Affairs* 83: 321–43.

Anderson, David and Richard Grove (eds), 1989. *Conservation in Africa: People, Policies, and Practice*. Cambridge, Cambridge University Press.

Anderson, David, 1989. 'Managing the Forest: The Conservation History of the Lembus, Kenya, 1904–63'. In *Conservation in Africa: People, Policies and Practice*. Cambridge, Cambridge University Press.

Anderson, Michael, 1995. 'The Conquest of Smoke: Legislation and Pollution in Colonial Calcutta'. In David Arnold and Ramachandra Guha (eds), *Nature, Culture, Imperialism: Essays on the Environmental History of South Asia*. New Delhi, Oxford University Press.

Anderson, Perry, 1974. *Lineages of the Absolutist State*. London, New Left Books.

————, 1976–7. 'The Antinomies of Antonio Gramsci'. *New Left Review* 100: 5–78.

Anderson, Robert and Walter Huber, 1988. *The Hour of the Fox: Tropical Forests, the World Bank and Indigenous People in Central India*. New Delhi, Vistaar.

Anon, 1852. 'Results of Inquiries in Regard to the Different Kinds of Timber Procurable in Bengal for Railway Purposes'. *Journal of the Agricultural and Horticultural Society of India* 8: 116–32.

Anon, 1859. 'Geology in India'. *Calcutta Review* 32: 122–61.

Anon, 1885. 'Mr. Brandis's Work in Bengal'. *Indian Forester* 11(8): 349–55.

Anon, 1892. *Illustrations of Indian Field Sports, Selected and Reproduced From the Coloured Engravings First Published in 1807 After Designs By Captain Thomas Williamson of the Bengal Army*. London, Archibald Constable.

Anon, 1934. 'The Report of the Sal Study Tour'. *Indian Forest Records* XIX(3).

Appadurai, Arjun, 1993. 'Numbers in the Colonial Imagination'. In Carol Breckenridge and Peter van der Veer (eds), *Orientalism and the Post-colonial Predicament*, New Delhi, Oxford University Press.

Appfel-Marglin, Frederique and Stephen Marglin (eds), 1990. *Dominating Knowledge: Development, Culture and Resistance*. Oxford, Clarendon Press.

Arnold, David, 1991. 'The Indian Ocean as a Disease Zone, 1500–1950'. *South Asia* 14(2): 1–21.

————, 1994. 'The Discovery of Malnutrition and Diet in Colonial India'. *Indian Economic and Social History Review* 31(1): 1–26.

————, 1996. *The Problem of Nature: Environment, Culture and European Expansion*. Oxford, Basil Blackwell.

———— and Ramachandra Guha, 1995. 'Introduction'. In David Arnold and Ramachandra Guha (eds), *Nature, Culture, Imperialism: Essays on the Environmental History of South Asia*. New Delhi, Oxford University Press.

Atran, Scott, 1990. *Cognitive Foundations of Natural History: Towards an Anthropology of Science*. Cambridge, Cambridge University Press.

Avineri, Shlomo, 1972. *Hegel's Theory of the Modern State*. Cambridge, Cambridge University Press.

Baden-Powell, B.H., 1876-7. 'Forest Conservancy in its Popular Aspect'. *Indian Forester* 2(1): 1–17.

Baigent, Elizabeth and Roger Kain, 1984. *The Cadastral Map in the Service of the State: A History of Property Mapping*. Chicago, University of Chicago Press.

Bailey, W.A., 1924. 'Moribund Forests in United Provinces'. *Indian Forester* 50: 188–91.

————, 1925. 'Sal Coppicing and Burning'. *Indian Forester* 51(8): 404–6.

Baker, Keith, 1985. 'Memory and Practice: Politics and the Representation of the Past in Eighteenth Century France'. *Representations* 11: 134–63.

Baldwin, J.H., 1883. *The Large and Small Game of Bengal and the Northwest Provinces of India*. London, Kegan, Paul and Trench.

Balfour, E.G., 1849. 'Notes on the Influence Exercised by Trees in Inducing Rain and Preserving Moisture'. *Madras Journal of Literature and Science* 25: 402–8.

Balfour, Edward, 1862. *The Timber Trees, Timber and Fancy Woods as also the Forests of India and of Eastern and South Asia*. Madras, Union Press.

Ball, Valentine, 1880. *Jungle Life in India: Or the Journeys and Journals of an Indian Geologist*. London, Thomas De La Rue & Co.

Banuri, Tariq and Frederique Appfel-Marglin (eds), 1993. *Who Will Save the Forests: Power, Knowledge and Environmental Destruction*. London, Zed.

Barnes, B., 1977. *Interests and the Growth of Knowledge*. London, Routledge and Kegan Paul.

Barnes, T.J. and J.S. Duncan (eds), 1992. *Writing Worlds: Discourse Text and Metaphor in the Representation of Landscape*. New York, Routledge.

Barron, T.J., 1987. 'Science and the Nineteenth Century Ceylon Coffee Planters'. *Journal of Imperial and Commonwealth History* XVI: 5–23.

Bartolovich, Crystal, 1994. 'Spatial Stories: The Surveyor and the Politics of Transition' (manuscript).

Basu, K.K., 1956a. 'Early Administration of the Kol Peers in Singhbhum and Bamanghati'. *Journal of Bihar and Orissa Research Society* 42(3–4): 357–76.

———, 1956b. 'The History of Singhbhum'. *Journal of Bihar and Orissa Research Society* 42(2): 98.

Bates, Crispin and Marina Carter, 1992. 'Tribal Migration in India and Beyond'. In Gyan Prakash (ed.), *The World of the Rural Labourer in Colonial India, Themes in Indian History*. New Delhi, Oxford University Press.

———, 1993. 'Tribal Indentured Migrants in Colonial India: Modes of Recruitment and Forms of Incorporation'. In Peter Robb (ed.), *Dalit Movements and the Meanings of Labour in India*. New Delhi, Oxford University Press.

Baviskar, Amita, 1992. 'Development, Nature and Resistance: The Case of the Bhilala Tribals in the Narmada Valley'. PhD thesis, Cornell University.

Bayart, Jean-Francois, 1993. *The State in Africa: The Politics of the Belly*. London, Longman.

Bayley, H.V., 1902. *History of Midnapore*. Calcutta, Bengal Secretariat Press.

Bayly, Christopher A., 1990. *Indian Society and the Making of the British Empire*. Cambridge, Cambridge University Press.

———, 1993. 'Knowing the Country: Empire and Information in India'. *Modern Asian Studies* 27(1): 3–43.

———, 1996. *An Empire of Information: Political Intelligence and Social Communication in North India, c 1780–1880*. Cambridge, Cambridge University Press.

Beadon-Bryant, F., 1907. 'Fire Conservancy in Burma'. *Indian Forester* 33(12): 537–49.

Behal, Rana and Prabhu Mohapatra, 1992. ' "Tea and Money Versus Human Life": The Rise and Fall of the Indentured System in the Assam Tea Plantations 1840–1908'. *Journal of Peasant Studies* 19(3–4): 142–72.

Behar, Ruth, 1986. *Santa Maria Del Monte: The Presence of the Past in a Spanish Village*. Princeton, Princeton University Press.

Beinart, William, 1984. 'Soil Erosion, Conservationism and Ideas about Development: A Southern African Exploration, 1900–1960'. *Journal of Southern African Studies* 11(1): 52–83.

———, 1989. 'The Politics of Colonial Conservation'. *Journal of Southern African Studies* 15(2): 143–62.

———, 1991. 'Empire, Hunting and Ecological Change in Southern and Central Africa'. *Past And Present* 128: 162–86.

———— and Peter Coates, 1996. *Environment and History: The Taming of Nature in the USA and South Africa*. London, Routledge.

Bendix, Reinhard, 1977a. *Max Weber: An Intellectual Portrait*. Berkeley, University of California Press.

————, 1977b. *Nationbuilding and Citizenship*. Berkeley, University of California Press.

Berman, Bruce, 1990. *Control and Crises in Colonial Kenya: The Dialectic of Domination*. London, James Currey.

Bernier, Francois, 1891(reprint 1968, Delhi, S. Chand). *Travels in the Mughal Empire*. London, Archibald Constable.

Berry, Sara, 1993. *No Condition is Permanent: The Social Dynamics of Agrarian Change in Sub-Saharan Africa*. Madison, University of Wisconsin Press.

Best, J.W., 1909. 'The Effects of Cattle Grazing in the Bhandar Division, Central Provinces'. *Indian Forester* 35(11): 610–17.

————, 1935. *Forest Life in India*. London, John Murray.

Béteille, André, 1992. 'The Concept of Tribe with Special Reference to India'. In Andre Beteille (ed.), *Society and Politics in India: Essays in Comparative Perspective*. New Delhi, Oxford University Press.

Beverly, H., 1872. *Report on the Census of Bengal*. Calcutta, Bengal Secretariat Press.

Bhabha, Homi, 1984. 'Of Mimicry and Man: The Ambivalence of Colonial Discourse'. *October* 28: 125–33.

————, 1986. 'Difference, Discrimination and Discourses of Colonialism'. In Francis Barker, Peter Hulme, Margaret Iverson and Diana Loxley (eds), *Literature, Politics, Theory*. London, Methuen.

———— (ed.), 1990. *Nation and Narration*. New York, Routledge.

Bhatnagar, S.K. and H.P. Seth, 1960. 'Indicator Species for Sal Shorea robusta Natural Regeneration'. *Indian Forester* 86(9): 520–30.

Bhattacharya, Neeladri, 1992. 'Colonial State and Agrarian Society'. In Burton Stein (ed.), *The Making of Agrarian Policy in British India, 1770–1900*. New Delhi, Oxford University Press.

Bhowmick, Sharit, 1981. *Class Formation in the Plantation System*. Delhi, People's Publishing House.

Birdwood, H.M., 1910. *Indian Timbers: The Hill Forests of Western India*. London, The Journal of Indian Arts and Industry.

Blanford, H.R., 1925. 'Regeneration with the Assistance of Taungya in Burma'. *Indian Forest Records* XI(3), Calcutta, Government of India Press.

Blanford, W.T., 1888. *The Fauna of British India*. London, Taylor and Francis.

Blochmann, H., 1873. 'Contributions to the Geography and History of Bengal.' *Journal of the Asiatic Society of Bengal* 42(Part I): 209–310.

Bloor, D., 1976. *Knowledge and Social Imagery*. Chicago, University of Chicago Press.

Boag, Peter, 1992. *Settlement Cultures in Nineteenth Century Oregon*. Berkeley, University of California Press.

Bogue, Ronald, 1989. *Deleuze and Guattari*. London, Routledge.

Bose, Sugata, 1993. *Peasant Labour and Colonial Capital: Rural Bengal since 1770*. Cambridge, Cambridge University Press.

——————, 1997. 'Instruments and Idioms of Colonial and National Development: India's Historical Experience in Comparative Perspective'. In Frederick Cooper and Randall Packard (eds), *International Development and the Social Sciences: Essays on the History and Politics of Knowledge*. Berkeley, University of California Press.

Botkin, Daniel, 1990. *Discordant Harmonies: A New Ecology for the Twentieth Century*. New York, Oxford University Press.

Bourdieu, Pierre, 1990. *The Logic of Practice*. Stanford, Stanford University Press.

Boyarin, Jonathan (ed.), 1995. *Remapping Memory: The Politics of TimeSpace*. Minneapolis, University of Minnesota Press.

Bradley-Birt, F.B., 1905. *The Story of an Indian Upland*. London, Smith, Elder.

Braithwaithe, F.J., 1905. 'Fire Protection in the Teak Forests of Burma.' *Indian Forester* 31(7): 383–5.

Brandis, Dietrich, 1879. 'The Utilization of the Less Valuable Woods in the Fire Protected Forests of the Central Provinces, by Iron Making'. *Indian Forester* 5(2): 222–5.

——————, 1884. 'The Progress of Forestry in India'. *Indian Forester* 10(10): 452–62.

——————, 1897. *Indian Forestry*. Woking, Oriental Institute.

Breckenridge, Carol and Peter Van der Veer (eds), 1993. *Orientalism and the Postcolonial Predicament*. New Delhi, Oxford University Press.

Bright, Charles and Susan Harding, 1984. 'Processes of Statemaking and Popular Protest: An Introduction'. In Charles Bright and Susan Harding (eds), *Statemaking and Social Movements: Essays in History and Theory*. Ann Arbor, University of Michigan Press.

Brown, R.S. and H.P. Pearson, 1981 (1932). *The Commercial Timbers of India*. Delhi, Researchco Reprints.

Browne, Janet, 1983. *The Secular Ark: Studies in the History of Biogeography*. New Haven, Yale University Press.

Bryant, Raymond L., 1994a. 'From Laissez Faire to Scientific Forestry: Forest Management in Early Colonial Burma, 1826–1885'. *Forest and Conservation History* 38(4): 160–70.

——————, 1994b. 'Shifting the Cultivator: The Politics of Teak Regeneration in Colonial Burma'. *Modern Asian Studies* 28: 225–50.

Bunting, Robert, 1996. *The Pacific Raincoast: The Environment and Culture of an American Eden*. Lawrence, University Press of Kansas.

Burkhill, I.H., 1965. *Chapters on the History of Botany in India*. Calcutta, Botanical Survey of India.

Burton, R.W., 1952. 'A History of Shikar in India'. *Journal of the Bombay Natural History Society* 50(4): 845–69.

Calhoun, Craig, 1978. 'History, Anthropology and the Study of Communities: Some Problems in MacFarlane's Proposal'. *Social History* 3(3): 363–73.

——————, 1980. 'Community: Toward a Variable Conceptualization for Comparative Research'. *Social History* 5(1): 105–29.

Campbell, A., 1894. 'The Traditional Migration of the Santhal'. *Indian Antiquary* 23: 103–4.

Campbell, Justice, 1866. 'The Ethnology of India'. *Journal of the Asiatic Society of Bengal* 35(Part II): 1–153.

Carmack, Robert, 1972. 'Ethnohistory: a Review of its Development, Definitions, Methods and Aims'. *Annual Review of Anthropology* 1: 227–46.

Carr, David, 1986. *Time, Narrative and History*. Bloomington, Indiana University Press.

Carstairs, Robert, 1912. *The Little World of an Indian District Officer*. London, Macmillan.

Carter, Marina and Crispin Bates, 1993. 'Tribal and Indentured Migrants in Colonial India: Modes of Recruitment and Forms of Incorporation'. In Peter Robb (ed.), *Dalit Movements and the Meanings of Labour in India*. New Delhi, Oxford University Press.

Carter, Paul, 1989. *The Road to Botany Bay: An Exploration of Landscape and History*. Chicago, University of Chicago Press.

Cartmill, Matt, 1993. *A View to a Death in the Morning: Hunting and Nature through History*. Cambridge, Harvard University Press.

Cerny, Philip, 1990. *The Changing Architecture of Politics: Structure, Agency and the Future of the State*. London, Sage.

Chakravarti, R., 1948. 'The Natural and Artificial Regeneration of Dry Peninsular Sal'. *Indian Forester* 74(2): 57–67.

Chakravarty, Dipesh, 1992. 'Postcoloniality and the Artifice of History: Who Speaks for "Indian" Pasts?' *Representations* 37: 1–26.

Chakravarty, Lalitha, 1978. 'Emergence of an Industrial Labour Force in a Dual Economy: British India, 1880–1920'. *Indian Economic and Social History Review* 15: 249–328.

Chakravarty, M.M., 1908. 'Notes on the Geography of Old Bengal'. *Journal of the Asiatic Society of Bengal* new series 4(5): 267–91.

Chambers, David Wade, 1987. 'Period and Process in Colonial and National Science'. In N. Reingold and M. Rothenberg (eds), *Scientific Colonialism: A Cross-Cultural Comparison*. Washington D.C., Smithsonian Institution Press.

Champion, H.G., 1936. 'A Preliminary Survey of the Forest Types of India and Burma'. *Indian Forest Records* new series I(1).

Champion, Herbert, 1933. 'Regeneration and Management of Sal Shorea robusta'. *Indian Forest Records* XIX (3).

————, 1975. 'Indian Silviculture and Research over the Century.' *Indian Forester* 101(1): 3–8.

Chatterjee, Basudeb, 1980. 'The Darogah and the Countryside: The Imposition of Police Control in Bengal and its Impact, 1793–1837'. *Indian Economic and Social History Review* 18(1): 19–42.

Chatterjee, Nandalal, 1954. 'Jungles Mahals under Lord Wellesley'. *Bengal Past and Present* 73: 73–6.

Chatterjee, Partha, 1993. *The Nation and its Fragments: Colonial and Postcolonial Histories*. New Delhi, Oxford University Press.

Chatterjee, Ruma, 1990–1. 'The Tussar Silk Industry and its Decline in Bengal and South Bihar, 1872–1921'. *Indian Historical Review* 17(1–2): 174–92.

Chaturvedi, M.D., 1931. 'The Regeneration of Sal in United Provinces'. *Indian Forester* 57(4): 157–66.

Chaudhuri, Benoy Bhushan, 1976. 'Agricultural Growth in Bengal and Bihar, 1770–1870: Growth of Cultivation since the Famine of 1770'. *Bengal Past and Present* 95: 290–340.

——————, 1993. 'Tribal Society in Transition: Eastern India, 1757–1920'. In Mushirul Hasan and Narayani Gupta (eds), *India's Colonial Encounter: Essays in Memory of Eric Stokes*. Delhi, Manohar.

Chaudhuri, K.N., 1918. *Sport in Jheel and Jungle*. Calcutta, Thacker, Spinck & Co.

Cleghorn, Hugh, 1861. *The Forests and Gardens of South India*. London, W.H. Allen.

——————, R. Baird Smith, J. Forbes Royle, R. Strachey, 1852. 'Report of the Committee Appointed by the British Association to Consider the Probable Effects in an Economical and Physical Point of View of the Destruction of Tropical Forests'. *Journal of the Agricultural and Horticultural Society of India* 8: 118–49.

Cline-Cole, Reginald A., 1994. 'Political Economy, Fuelwood Relations and Vegetation Conservation—Kasar Kano, Northern Nigeria, 1850–1915'. *Forest and Conservation History* 38(2): 67–79.

Clutterbuck, Peter, 1927. 'Forestry and the Empire'. *Empire Forestry Journal* 6: 184–92.

Cobden-Ramsay, L.E.B., 1910. *Bengal District Gazetteers: Feudatory States of Orissa*. Calcutta, Bengal Secretariat Book Depot.

Cohen, Jean and Andrew Arato, 1992. *Civil Society and Political Theory*. Cambridge, MA, MIT Press.

Cohn, Bernard (ed.), 1987a. *An Anthropologist among the Historians and Other Essays*. New Delhi, Oxford University Press.

——————, 1987b. 'Notes on the History of the Study of Indian Society and Culture'. In Bernard Cohn (ed.), *An Anthropologist among the Historians and Other Essays*. New Delhi, Oxford University Press.

——————, 1987c. 'The Census, Social Structure and Objectification in South Asia'. In Bernard Cohn (ed.), *An Anthropologist among the Historians and Other Essays*. New Delhi, Oxford University Press.

——————, 1996. *Colonialism and its Forms of Knowledge: The British in India*. Princeton, Princeton University Press.

Collins, H.M., 1992. *Changing Order: Replication and Induction in Scientific Practice*. Chicago, University of Chicago Press.

Comaroff, John and Jean Comaroff, 1991. *Of Revelation and Revolution: Christianity, Colonialism and Consciousness in South Africa, Volume 1*. Chicago, University of Chicago Press.

——————, 1992. *Ethnography and the Historical Imagination*. Boulder, Westview Press.

Cooper, Fred and Ann Stoler, 1989. 'Introduction: Tensions of Empire: Colonial Control and Visions of Rule'. *American Ethnologist* 16: 609–21.

Cooper, Frederick, 1997. 'Modernizing Bureaucrats, Backward Africans, and the Development Concept'. In Frederick Cooper and Randall Packard (eds), *International Development and the Social Sciences: Essay in the History and Politics of Knowledge*. Berkeley, University of California Press.

Corbridge, Stuart, 1986. 'State Tribe and Region: Policy and Politics in India's Jharkhand, 1900–1980'. PhD thesis, Cambridge University.

—————, 1993. 'Ousting Singabonga: The Struggle for India's Jharkhand'. In Peter Robb (ed.), *Dalit Movements and the Meanings of Labour in India*. New Delhi, Oxford University Press.

Cosgrove, Denis, 1984. *Social Formation and Symbolic Landscape*. London, Croom Helm.

—————, 1993. *The Palladian Landscape: Geographical Change and its Cultural Representation in Sixteenth Century Italy*. Leicester, Leicester University Press.

————— and Stephen Daniels, 1988. *The Iconography of Landscape*. Cambridge, Cambridge University Press.

Coupland, H., 1911. *Bengal District Gazetteers: Manbhum*. Calcutta, Bengal Secretariat Book Depot.

Cowen, Michael and Robert Shenton, 1995. 'The Invention of Development'. In Jonathan Crush (ed.), *Power of Development*. London, Routledge.

Croll, Elizabeth and David Parkin (eds), 1992. *Bush Base, Forest Farm: Culture, Environment and Development*. London, Routledge.

Cronon, William, 1983. *Changes in the Land: Indians, Colonists and the Ecology of New England*. New York, Hill and Wang.

—————, 1990. 'Modes of Prophecy and Production: Placing Nature in History'. *Journal of American History* 76(4): 1122–31.

—————, 1991. *Nature's Metropolis: Chicago and the Great West*. New York, W.W. Norton.

—————, 1992. 'A Place for Stories: Nature, History and Narrative'. *Journal of American History* 78: 1347–76.

————— (ed.), 1995. *Uncommon Ground: Towards Reinventing Nature*. New York, W.W. Norton.

Crosby, Alfred, 1986. *Ecological Imperialism: The Biological Expansion of Europe, 900–1900*. Cambridge, Cambridge University Press.

Crush, Jonathan, 1995. 'Introduction: Imagining Development'. In Jonathan Crush (ed.), *Power of Development*. London, Routledge.

Dalton, E.T., 1865. 'Notes of a Tour Made in 1863–64 in the Tributary Mahals under the Commissioner of Chotanagpur, Bonai, Gangpore, Odeypore and Sirgooja'. *Journal of the Asiatic Society of Bengal* 34: 1–32.

—————, 1866a. 'Notes on a Tour Through Manbhum.' *Journal of the Asiatic Society of Bengal* 35: 186–94.

—————, 1866b. 'The Kols of Chotanagpur'. *Journal of the Asiatic Society of Bengal* 35(Part II): 153–200.

—————, 1974 (1872). *Descriptive Ethnology of Bengal*. Delhi, Today and Tomorrow.

Daniels, Stephen, 1993. *Fields of Vision and National Identity in England and the United States*. Cambridge, Polity Press.

Das, Binod S., 1973. *Civil Rebellion in Frontier Bengal*. Delhi, Mittal Publications.

Das, Binod, 1984. *Changing Profile of Frontier Bengal*. Calcutta, Punthi Pushtak.

Das, Bishnupada, 1971. 'Anglo Maratha Relations under Wellesley.' *Bengal Past and Present* 90: 224–30.

Dasgupta, Ranajit, 1981. ' Structure of the Labour Market in Colonial India'. *Economic and Political Weekly* special issue (Nov): 1781–1806.

————, 1986. 'From Peasants and Tribesmen to Plantation Workers: Colonial Capitalism, Reproduction of Labour Power and Proletarianisation in North East India, 1850s–1947'. *Economic and Political Weekly* 21(4): 2–10.

De, R.N., 1941. 'Sal Regeneration de novo'. *Indian Forester* 67(6): 283–91.

de Certeau, Michel, 1988. *The Practice of Everyday Life*. Berkeley, University of California Press.

Demeritt, David, 1994. 'The Nature of Metaphors in Cultural Geography and Environmental History'. *Progress in Human Geography* 18(2): 163–85.

Desmond, Adrian, 1984. *Archetypes and Ancestors: Paleontology in Victorian London, 1850–1875*. Chicago, Chicago University Press.

Dirks, Nicholas, 1987. *The Hollow Crown: The Ethnohistory of an Indian Little Kingdom*. Cambridge, Cambridge University Press.

————, 1989. 'The Invention of Caste: Civil Society in Colonial India'. *Social Analysis* Special Issue: 42–52.

————, 1990. 'The Original Caste: Power, History and Hierarchy in South Asia'. In McKim Marriott (ed.), *India Through Hindu Categories*. New Delhi, Sage.

————, 1992. 'Castes of the Mind'. *Representations* 37(56–78).

————, 1993. 'Colonial Histories and Native Informants: Biography of an Archive'. In Carol Breckenridge and Peter Van der Veer (eds), *Orientalism and the Post-colonial Predicament*. New Delhi, Oxford University Press.

———— and Bernard Cohn, 1988. 'Beyond the Fringe: The Nation-State, Colonialism and the Technologies of Power'. *Journal of Historical Sociology* 1(2): 224–9.

Donham, Donald, 1990. *History, Power and Ideology*. Stanford, Stanford University Press.

Douglas, Mary, 1986. *How Institutions Think*. Syracuse, Syracuse University Press.

Dove, Michael, 1986. 'The Practical Reason of Weeds in Indonesia: Peasant versus State Views of Imperata and Chromolaena'. *Human Ecology* 14(2): 163–89.

————, 1992. 'The Dialectical History of "Jungle" in Pakistan: an Examination of the Relationship Between Nature and Culture'. *Journal of Anthropological Research* 48(3): 231–53.

Driver, Felix, 1992. 'Geography's Empire: Histories of Geographical Knowledge'. *Environment and Planning D: Society and Space* 1: 23–40.

Duara, Prasenjit, 1995. *Rescuing History from the Nation: Questioning Narratives of Modern China*. Chicago, University of Chicago Press.

Dubois, Marc, 1991. 'The Governance of the Third World: A Foucauldian Perspective on Power Relations in Development'. *Alternatives* 16(1): 1–30.

Dunbar-Brander, A.A., 1931. 'Game Preservation in India'. *Journal of the Society for Protection of the Fauna of Empire* 14: 23–35.

Duncan, J.S., 1990. *The City as Text: The Politics of Landscape Interpretation in the Kandyan Kingdom*. Cambridge, Cambridge University Press.

———— and D. Ley, 1993. *Place/Culture/Representation*. London, Routledge.

Edney, Mathew, 1990. 'Mapping and Empire: British Trigonometrical Surveys in India and the European Concept of Systematic Survey, 1799–1843'. PhD thesis, University of Wisconsin, Madison.

Edwardes, S.M., 1925. 'Forest Tribes of India: Casual Notes'. *Empire Forestry Journal* 4: 86–91.

Eley, Geof, 1994. 'Nations, Publics and Political Culture: Placing Habermas in the Nineteenth Century'. In Nicholas Dirks, Sherry Ortner and Geof Eley (eds), *Culture, Power, History*. Princeton, Princeton University Press.

Emery, Frank, 1984. 'Geography and Imperialism: the role of Sir Bartle Frere, 1815–1884'. *The Geographical Journal* 150: 342–50.

Engels, Dagmar, 1993. 'The Myth of the Family Unit: Adivasi Women in Coal Mines and Tea Plantations in Early Twentieth Century Bengal'. In Peter Robb (ed.), *Dalit Movements and the Meanings of Labour*. New Delhi, Oxford University Press.

Escobar, Arturo, 1988. 'Power and Visibility: Development and the Invention and Management of the Third World'. *Cultural Anthropology* 3(4): 428–43.

————, 1995. *Encountering Development: The Making and Unmaking of the Third World*. Princeton, Princeton University Press.

Esteva, Gustavo, 1992. 'Development'. In Wolfgang Sacks (ed.), *The Development Dictionary*. London, Zed.

Fairhead, James and Melissa Leach, 1995. 'Reading Forest History Backwards: The Interaction of Policy and Local Landuse in Guinea's Forest-Savannah Mosaic'. *Environment and History* 1(1): 55–90.

————, 1996. *Misreading the African Landscape: Society and Ecology in a Forest-Savanna Mosaic*. Cambridge, Cambridge University Press.

Faragher, John Mack, 1986. *Sugar Creek: Life on the Illinois Prairie*. New Haven, Yale University Press.

Feierman, Steven, 1990. *Peasant Intellectuals: Anthropology and History in Tanzania*. Madison, University of Wisconsin Press.

Ferguson, James, 1994. *The Anti-Politics Machine: Development: Depoliticization and Bureaucratic Power in Lesotho*. Minneapolis, University of Minnesota Press.

Fernow, Bernhard E., 1907. *A Brief History of Forestry in Europe, the United States and Other Countries*. Toronto, University of Toronto Press.

Firminger, W.K., 1917 (1812). *Fifth Report from the Select Committee of the House of Commons on the Affairs of the East India Company—1812, 3 volumes*. Calcutta, R. Cambray & Co.

Fisher, W.R., 1876–7. 'On the Effects of Grazing in Sal Forests'. *Indian Forester* 2(2): 203–5.

——————, 1880. 'Fire Conservancy in the Sal Forests of the Eastern Duars'. *Indian Forester* 5(4): 428–44.

——————, 1905. 'Fire Protection in the Teak Forests of Burma'. *Indian Forester* 31(7): 385–7.

Ford-Robertson, F.C., 1927. 'The Problem of Sal Regeneration with Special Reference to the "Moist" Forests of the United Provinces'. *Indian Forester* 53(9, 10): 500–11, 60–76.

Forsyth, J., 1889. *The Highlands of Central India: Notes on their Forests and Wild Tribes, Natural History and Sports.* London, Chapman and Hall.

Foucault, Michel, 1973. *The Order of Things: An Archaeology of the Human Sciences.* New York, Vintage.

——————, 1979. *Discipline and Punish: The Birth of the Prison.* New York, Vintage.

——————, 1990. *The History of Sexuality: An Introduction, volume 1.* New York, Vintage.

——————, 1991. 'Governmentality'. In Graham Burchell, Colin Gordon and Peter Miller (eds), *The Foucault Effect: Studies in Governmentality.* Chicago, University of Chicago Press.

Fraser, Nancy, 1989. *Unruly Practices: Power, Discourse and Gender in Contemporary Social Theory.* Minneapolis, University of Minnesota Press.

Freeman, John F., 1994. 'Forest Conservancy in the Alps of Dauphine, 1287–1870'. *Forest and Conservation History* 38(4): 171–80.

Gadgil, Madhav and Ramachandra Guha, 1992. *This Fissured Land: An Ecological History of India.* New Delhi, Oxford University Press.

——————, 1995. *Ecology and Equity: The Use and Abuse of Nature in Contemporary India.* London, Routledge.

Ghani, Ashraf, 1995. 'Writing a History of Power: An Examination of Eric Wolf's Anthropological Quest'. In Jane Schneider and Rayna Rapp (eds), *Articulating Hidden Histories: Exploring the Influence of Eric Wolf.* Berkeley, University of California Press.

Ghosh, Kaushik, 1994. 'A Market for Aboriginality: Primitivism and Race Classification in the Indentured Labour Market of Colonial India' (manuscript).

Giddens, Anthony, 1981. *A Contemporary Critique of Historical Materialism: Power, Property and the State.* Berkeley, University of California Press.

Gilmartin, David, 1994. 'Scientific Empire and Imperial Science: Colonialism and Irrigation Technology in the Indus Basin'. *Journal of Asian Studies* 53(4): 1127–49.

Godlewska, Ann and Neil Smith (eds), 1994. *Geography and Empire.* Oxford, Basil Blackwell.

Gooding, D., 1990. *Experiment and the Making of Meaning.* Dordrecht, Kluwer Academic.

Goody, Jack, 1996. 'Man and the Natural World: Reflections on History and Anthropology'. *Environment and History* 2(3): 255–70.

Gopal, Sarvepalli, 1965. *British Policy in India, 1858–1905*. Cambridge, Cambridge University Press.

Gordon, Colin, 1991. 'Governmental Rationality: An Introduction'. In Graham Burchell, Colin Gordon and Peter Miller (eds), *The Foucault Effect: Studies in Governmentality*. Chicago, University of Chicago Press.

Gordon, Stewart, 1994. 'Bhils and the Idea of a Criminal Tribe in Nineteenth Century India'. In Stewart Gordon (ed.), *Maratha, Marauders and State Formation in Eighteenth Century India*. New Delhi, Oxford University Press.

Gouldsbury, Charles E., 1909. *Dulall the Forest Guard: A Tale of Sport and Adventure in the Forests of Bengal*. London, Gibbings & Co.

Government of Bengal, 1868–9 to 1904–5. *Annual Progress Reports on Forest Administration in the Presidency of Bengal*. Calcutta, Superintendent of Government Printing.

———, 1911–12 to 1940–1. *Annual Progress Reports on Forest Administration in the Presidency of Bengal*. Calcutta, Superintendent of Government Printing.

———, 1935. *The Forests of Bengal*. Calcutta, Superintendent of Government Printing.

Government of Bihar and Orissa, 1911–12 to 1940–1. *Annual Progress Reports on Forest Administration in the Presidency of Bihar and Orissa*. Patna, Bihar and Orissa Government Press.

Gregory, Derek, 1994. *Geographical Imaginations*. Oxford, Basil Blackwell.

Greswell, E.A., 1926. 'The Constructive Properties of Fire in Chir Pinus longifolia Forests'. *Indian Forester* 52: 502–5.

Grieve, J.W.A., 1912. *Working Plan for the Darjeeling Forests*. Darjeeling, Darjeeling Branch Press.

Griffith, A.L. and R.S. Gupta, 1948. 'The Determination of the Characteristics of Soil Suitable for Sal'. *Indian Forest Bulletin* Silviculture new series no. 138.

Griffith, W., 1842. 'Remarks on a Few Plants from Central India'. *Calcutta Journal of Natural History* 3(11): 361–8.

Grove, Richard, 1989a. 'Scottish Missionaries, Evangelical Discourses and the Origins of Conservation Thinking in Southern Africa, 1820–1900'. *Journal of Southern African Studies* 15(2): 163–87.

———, 1989b. 'Early Themes in African Conservation: The Cape in the Nineteenth Century'. In David Anderson and Richard Grove (eds), *Conservation in Africa: People, Policies and Practice*. Cambridge, Cambridge University Press.

———, 1990. 'Colonial Conservation, Ecological Hegemony and Popular Resistance: Towards a Global Synthesis'. In J.M. Mackenzie (ed.), *Imperialism and the Natural World*. Manchester, Manchester University Press.

———, 1993a. 'The European East India Companies and their Environmental Policies on St. Helena, Mauritius and in Western India, 1660–1854'. *Comparative Studies in Society and History* 36: 318–51.

——————, 1993b. 'Imperialism and the Discourse of Desiccation: The Institutionalisation of Global Environmental Concerns and the Role of the Royal Geographic Society, 1860–80'. In R. Butlin (ed.), *Geography and Imperialism*. Manchester, Manchester University Press.

——————, 1995a. *Green Imperialism: Colonial Expansion, Tropical Island Edens and the Origins of Environmentalism, 1600–1860*. Cambridge, Cambridge University Press.

——————, 1995b. 'The East India Company and the El Nino—The Critical Role Played by Colonial Scientists in Establishing the Mechanisms of Global Climatic Fluctuations and Teleconnections between 1770 and 1930' (manuscript).

——————, 1998. 'Introduction'. In Richard Grove, Vinita Damodaran and Satpal Sangwan (eds), *Nature and the Orient: Essays on the Environmental History of South and Southeast Asia*. New Delhi, Oxford University Press.

—————— and David Anderson, 1989. 'Introduction: The Scramble For Eden: Past, Present and Future in African Conservation'. In David Anderson and Richard Grove (eds), *Conservation in Africa: People, Policies and Practice*. Cambridge, Cambridge University Press.

Grove, Richard and JoAnn McGregor, 1995. 'Zimbabwe'. *Environment and History* special issue 1(3): 253–376.

Guattari, Felix and Gilles Deleuze, 1977. *Anti-Oedipus: Capitalism and Schizophrenia, Vol. I*. New York, Viking.

——————, 1988. *A Thousand Plateaus: Capitalism and Schizophrenia, Vol. II*. London, Athlone Press.

Guha, Ramachandra, 1983. 'Forestry in British and Post-British India: an Historical Analysis'. *Economic and Political Weekly* xvii: 1882–96.

——————, 1989. *The Unquiet Woods: The Ecological Bases of Peasant Resistance in the Himalaya*. New Delhi, Oxford University Press.

——————, 1990. 'An Early Environmental Debate: The Making of the Indian Forest Act of 1878'. *Indian Economic and Social History Review* 27(1): 65–84.

——————, 1993. 'Writing EnvironmentalHistory in India'. *Studies in History* 9(1): 119–29.

——————, and Madhav Gadgil, 1989. 'State Forestry and Social Conflict in British India: A Study in the Ecological Bases of Agrarian Protest'. *Past and Present* 123: 141–77.

Guha, Ranajit, 1963. *A Rule of Property for Bengal: An Essay on the Idea of the Permanent Settlement*. Paris, Mouton.

——————, 1983. *Elementary Aspects of Peasant Insurgency in Colonial India*. New Delhi, Oxford University Press.

—————— (ed.), 1982–9. *Subaltern Studies*. New Delhi, Oxford University Press.

Guha, Sumit, 1995. 'Lords of the Land Versus Kings of the Forest: Conflict and Collaboration in Peninsular India, *c.* 1500–1980'. Paper presented to the Weekly Colloquium, Program in Agrarian Studies, Yale University.

304 • Select Bibliography

—————, 1996. 'Forest Polities and Agrarian Empires: The Khandesh Bhils, *ca.* 1700–1850'. *Indian Economic and Social History Review* 33(2): 133–53.

—————, forthcoming. *Lords of the Land and Kings of the Forest: Environment, Ethnicity, and Politics in Western India, 1350–1980.* Cambridge, Cambridge University Press.

Guha-Thakurta, Prabir, 1966. *Game Laws in West Bengal.* Proceedings of Symposia Arranged in Connection with West Bengal Forest Centenary 1964, Calcutta, West Bengal Department of Forests.

Gupta, Ranjan Kumar, 1971. *The Economic Life of Bengal District, Birbhum: 1770–1857.* Burdwan, Burdwan University.

Habermas, Jurgen, 1974. 'The Public Sphere'. *New German Critique,* 3: 49.

—————, 1981. 'Modernity Versus Postmodernity.' *New German Critique* 22: 3–14.

—————, 1989. *The Structural Transformation of the Public Sphere,* Thomas Burger (trans.). Cambridge, MA, MIT Press.

Hacking, Ian, 1975. *The Emergence of Probability: A Philosophical Study of Early Ideas about Probability, Induction and Statistical Inference.* Cambridge, Cambridge University Press.

—————, 1983. *Representing and Intervening.* Cambridge, Cambridge University Press.

—————, 1992. 'The Self-Vindication of the Laboratory Sciences'. In Andrew Pickering (ed.), *Science as Practice and Culture.* Chicago, University of Chicago Press: 29–64.

Hadfield, H.B. and D. Bryant, 1895. 'Elephant Catching Operations of the Annamalai Hills'. *Indian Forester* 21: 118–20, 200–2, 75–6.

Haeuber, Richard, 1993. 'Indian Forest Policy in Two Eras: Continuity or Change?' *Environmental History Review* 17(1): 49–76.

Hagen, Joel, 1992. *An Entangled Bank: The Origins of Ecosystem Ecology.* New Brunswick, Rutgers University Press.

Haines, H.H., 1905. *Working Plan for Reserved Forests of Singhbhum, 1903–1918.* Calcutta, Bengal Secretariat Press.

Hamilton, Walter, 1815. *East India Gazetteer.* London, John Murray.

—————, 1820. *A Geographical Statistical and Historical Description of Hindustan and the Adjacent Countries.* London, John Murray.

Handelman, Don, 1990. *Models and Mirrors: Towards an Anthropology of Public Events.* Cambridge, Cambridge University Press.

Hannay, S.F., 1845. 'Observations on the Quality of Principal Timber Trees Growing in the Vicinity of Upper Assam'. *Journal of the Agricultural and Horticultural Society of India* 4: 116–33.

Haraway, Donna, 1988. 'Situated Knowledges: The Science Question in Feminism and the Privilege of Partial Perspective'. *Feminist Studies* 14(3): 575–99.

Haraway, Donna, 1991. *Simians, Cyborgs and Women: the Re-invention of Nature.* New York, Routledge.

Hardiman, David, 1987. *The Coming of the Devi: Adivasi Assertion in Western India.* New Delhi, Oxford University Press.

————, 1994. 'Power in the Forests: The Dangs, 1820–1940'. In David Arnold and David Hardiman (eds), *Subaltern Studies, Vol. VIII: Essays in Honour of Ranajit Guha*. New Delhi, Oxford University Press.

Harding, Sandra, 1986. *The Science Question in Feminism*. Ithaca, Cornell University Press.

Harley, J.B., 1992. 'Deconstructing the Map'. In Trevor Barnes and James Duncan (eds), *Writing Worlds: Discourse, Text and Metaphor in the Representation of Landscape*. London, Routledge.

Harrison, Mark, 1994. *Public Health in British India: Anglo-Indian Preventive Medicine, 1859–1914*. Cambridge, Cambridge University Press.

Harrison, Robert Pogue, 1992. *Forests: The Shadow of Civilization*. Chicago, University of Chicago Press.

Harvey, David, 1989. *The Condition of Postmodernity: An Inquiry into the Origin of Cultural Change*. Oxford, Basil Blackwell.

Hatt, C.C., 1905. *Working Plan for the Reserved Forests of the Buxa Division*. Calcutta, Bengal Secretariat Book Depot.

Haugerud, Angelique, 1995. *The Culture of Politics in Modern Kenya*. Cambridge, Cambridge University Press.

Hays, Samuel P., 1959. *Conservation and the Gospel of Efficiency: The Progressive Conservation Movement, 1890–1920*. Cambridge, MA, Harvard University Press.

Herzfeld, Michael, 1992. *The Social Production of Indifference: Exploring the Symbolic Roots of Western Bureaucracy*. Chicago, University of Chicago Press.

Hettne, Bjorn, 1990. *Development Theory and the Three Worlds*. London, Methuen.

Hill, Christopher V., 1991. 'Philosophy and Reality in Riparian South Asia: British Famine Policy and Migration in Colonial North India'. *Modern Asian Studies* 25(2): 263–79.

Hirsch, Eric and Michael O'Hanlon (eds), 1995. *The Anthropology of Landscape: Perspectives on Space and Place*. Oxford, Clarendon.

Hole, R.S., 1913. 'Useful Exotics in Indian Forests'. *Indian Forest Records* IV(3).

————, 1914–16. 'Oecology of Sal Shorea robusta, Part I: Soil Composition, Soil Moisture, Soil Aeration, Part II: Seedling Reproduction in Natural Forests and its Improvement, Part III: Soil Aeration and Water Cultures'. *Indian Forest Records* Botany Vol. 4.

————, 1919. 'Regeneration of Sal Forests'. *Indian Forester* 45: 119–32.

Holston, James, 1989. *The Modernist City: An Anthropological Critique of Brasilia*. Chicago, University of Chicago Press.

Homfray, C.K., 1936. 'Notes on Thinnings in Plantations'. *Bengal Forest Bulletin*, no. 1, Silviculture Series. Alipore, Bengal Government Press.

Hooker, Joseph, 1904. *A Sketch of the Flora of British India*. London, His Majesty's Stationery Office.

Humphreys, Sally, 1985. 'Law as Discourse'. *History and Anthropology* 1: 241–64.

Hunt, Lynn, 1989. 'Introduction'. In Lynn Hunt (ed.), *The New Cultural History*. Berkeley, University of California Press.

Hunter, W.W., 1868. *Annals of Rural Bengal*. New York, Leypoldt and Holt.

——————, 1877. *A Statistical Account of Bengal.* London, Trubner & Co.

——————, 1894. *Bengal Ms. Records: A Select List of 14,136 Letters in the Board of Revenue, Calcutta, 1782–1807, with an Historical Dissertation and Analytical Index.* London, W.H. Allen.

Hurley, Andrew, 1995. *Environmental Inequalities: Class, Race and Industrial Pollution in Gary, Indiana, 1945–80.* Chapel Hill, University of North Carolina Press.

Hussain, Shaukat, 1925. 'The Development of Sal Seedlings in Gorakhpur Taungya'. *Indian Forester* 51(2): 69–72.

Hutchins, Francis, 1967. *The Illusion of Permanence: British Imperialism in India.* Princeton, Princeton University Press.

Inden, Ronald, 1990. *Imagining India.* Oxford, Basil Blackwell.

Irvine, R.H., 1846. 'Observations on the Products and Resources of Darjeeling'. *Journal of the Agricultural and Horticultural Society of India* 5: 181–97.

Jalal, Ayesha, 1995. *Democracy and Authoritarianism in South Asia: A Comparative and Historical Perspective.* Cambridge, Cambridge University Press.

Jaundice, Jeremiah, 1834. 'Tiger Shooting from a Machan'. *The Bengal Sporting Magazine* 2: 251–7.

Jessop, Bob, 1990. *State Theory: Putting the Capitalist State in its Place.* Cambridge, Polity Press.

Jewitt, Sarah, 1995. 'Europe's "Others"? Forestry Policy and Practices in Colonial and Postcolonial India'. *Environment and Planning D: Society and Space* 13: 67–90.

Jha, J. C., 1957. 'Early British Penetration into Chotanagpur'. *Journal of Bihar and Orissa Research Society* 43(3–4): 329–33.

——————, 1961. 'British Contact with Singhbhum, 1821–31'. *Journal of Bihar and Orissa Research Society* 47(1–4): 124–8.

——————, 1960. 'Early British Penetration into the Jungle Mahals.' *Journal of Bihar and Orissa Research Society* 46(1–4): 312–17.

——————, 1964. *The Kol Insurrection of Chotanagpur.* Calcutta, Thacker & Spinck.

——————, 1966. *The Bhumij Revolt.* Delhi, Manohar.

——————, 1987. *The Tribal Revolt of Chotanagpur, 1831–32.* Patna, Kashi Prasad Jayaswal Research Institute.

Johnson, Daniel, 1822. *Sketches of Indian Field Sports with Observations on the Animals [and] also an Account of Some of the Customs of the Inhabitants with a Description of the Art of Catching Serpents as Practiced by the Conjoors, and their Methods of Curing Themselves When Bitten, with Remarks on Hydrophobia and Rabid Animals.* London, Longman.

Jones, E.L., 1981. *The European Miracle: Environments, Economics, and the Geopolitics in the History of Europe and Asia.* Cambridge, Cambridge University Press.

Joseph, Gil and Daniel Nugent, 1994. *Everyday Forms of State Formation: Revolution and the Negotiation of Rule in Modern Mexico.* Durham, Duke University Press.

Joshi, H.B., 1980. *Troup's Silviculture of Indian Trees.* Dehradoon, Forest Research Institute.

Kaplan, Martha, 1990. 'Meaning, Agency and Colonial History: Navasovakadua and the Tuka movement in Fiji'. *American Ethnologist* 17(1): 3–22.

Keane, John, 1984. *Public Life and Late Capitalism: Towards a Socialist Theory of Democracy.* Cambridge, Cambridge University Press.

——————, 1988. *Democracy and Civil Society: On the Predicaments of European Socialism, the Prospects for Democracy, and the Problem of Controlling Social and Political Power.* London, Verso.

Kellogg, Susan, 1991. 'Histories for Anthropology: Ten Years of Historical Research and Writing by Anthropologists, 1980–1990'. *Social Science History* 15(4): 417–55.

Kelly, William, 1985. *Deference and Defiance in Nineteenth-Century Japan.* Princeton, Princeton University Press.

Kerr, I.J., 1983. 'Constructing Railways in India: An Estimate of the Numbers Employed, 1850–1880'. *Indian Economic and Social History Review* 20(3): 317–39.

Khan, M.A. Waheed, H.P. Bhatnagar and A.C. Gupta, 1961. 'Physiological-Ecological Studies on Sal'. In *Proceedings of the Tenth Silvicultural Conference.* Dehradoon, Forest Research Institute.

Kloppenburg, Jack, 1991. 'Social Theory and the De/Reconstruction of Agricultural Science: Local Knowledge for an Alternative Agriculture'. *Rural Sociology* 56(4): 519–48.

Kohli, Atul, 1994. 'Centralisation and Powerlessness: India's Democracy in a Comparative Perspective'. In Joel Migdal, Vivienne Shue and Atul Kohli (eds), *State Power and Social Forces: Domination and Transformation in the Third World.* Cambridge, Cambridge University Press.

Kuklick, Henrietta, 1991. *The Savage Within: The Social History of British Anthropology, 1885–1945.* Cambridge, Cambridge University Press.

Kumar, Deepak, 1980. 'Patterns of Colonial Science'. *Indian Journal of History of Science* 15(1): 105–13.

——————, 1984. 'Science, Resources and the Raj: a Case Study of Geological Works in Nineteenth Century India'. *Indian Historical Quarterly* X(1–2): 66–89.

——————, 1990. 'The Evolution of Colonial Science in India: Natural History and the East India Company'. In J. M. Mackenzie (ed.), *Imperialism and the Natural World.* Manchester, Manchester University Press: 52–61.

Laclau, Ernesto, 1977. *Politics and Ideology in Marxist Theory.* London, New Left Books.

Latour, Bruno, 1993. *We Have Never Been Modern.* Cambridge, MA, Harvard University Press.

Leach, Melissa, 1994. *Rainforest Relations: Gender and Resource Use among the Mende of Gola, Sierra Leone.* Washington DC, Smithsonian Institution Press.

Lederman, Rina, 1986. 'Changing Times in Mendi: Notes Towards Writing Highland New Guinea History'. *Ethnohistory* 33(1): 1–30.

Lefebvre, Henri, 1991. *The Production of Space*. Oxford, Basil Blackwell.

Leopold, Aldo, 1986 (1933). *Game Management*. Madison, University of Wisconsin Press.

Long, J., 1854–6. 'The Indigenous Plants of Bengal: Notes on Peculiarities in their Structure, Functions, Uses in Medicine, Domestic Life, Arts and Agriculture'. *Journal of the Agricultural and Horticultural Society of India* 9: 399–424.

Lowood, Henry, 1990. 'The Calculating Forester: Quantification, Cameral Science, and the Emergence of Scientific Forestry Management in Germany'. In Tore Frangsmyr, J.L. Heilbron and Robin E. Rider (eds), *The Quantifying Spirit in the Eighteenth Century*. Berkeley, University of California Press.

Ludden, David, 1989. *Peasant History in South India*. New Delhi, Oxford University Press.

————, 1992a. 'India's Development Regime'. In Nicholas Dirks (ed.), *Colonialism and Culture*. Ann Arbor, University of Michigan Press.

————, 1992b. 'Anglo-Indian Empire'. In Burton Stein (ed.), *The Making of Agrarian Policy in British India, 1770–1900*. New Delhi, Oxford University Press.

————, 1993. 'Orientalist Empiricism: Transformations of Colonial Knowledge'. In Carol Breckenridge and Peter Van der Veer (eds), *Orienalism and the Postcolonial Predicament*. New Delhi, Oxford University Press.

Mackay, David, 1985. *In the Wake of Cook: Exploration, Science and Empire, 1870–1901*. London: Croom Helm.

Mackenzie, Fiona, 1991. 'Political Economy of the Environment, Gender and Resistance under Colonialism: Murang'a District, Kenya, 1910–1950'. *Canadian Journal of African Studies* 25(2): 226–56.

Mackenzie, John, 1988. *The Empire of Nature: Hunting, Conservation and British Imperialism*. Manchester, Manchester University Press.

————, 1989. 'Chivalry, Social Darwinism and Ritualized Killing: The Hunting Ethos in Central Africa up to 1914'. In David Anderson and Richard Grove (eds), *Conservation in Africa: People, Policies and Practice*. Cambridge, Cambridge University Press.

————, 1990. 'Introduction'. In J. M. Mackenzie (ed.), *Imperialism and the Natural World*. Manchester, Manchester University Press.

————, 1995. *Orientalism: History, Theory and the Arts*. Manchester, Manchester University Press.

MacPhail, James, 1922. *The Story of the Santhal: With an Account of the Santhal Rebellion*. Calcutta, Thacker, Spinck & Co.

Mactier, T.B., 1848. 'Memorandum on Teak Plantations in Bancoorah'. *Journal of the Agricultural and Horticultural Society of India* 6: 242–6.

Maithana, G.P., V.K. Bahuguna et al., 1986. 'Effect of Forest Fires on the Ground Vegetation of a Moist Deciduous Sal Shorea robusta Forest.' *Indian Forester* 112(8): 646–77.

Makins, F.K., 1920. 'Natural Regeneration of Sal in Singhbhum'. *Indian Forester* 46: 292–6.

Malcolm, C.A., 1917. 'Some Problems in Connection with Grazing in Central Provinces'. *Indian Forester* 43(1): 10–15.

Man, E.G., 1867. *Sonthalia and the Santhals*. Calcutta, George Wyman.

Mandala, Elias, 1990. *Work and Control in a Peasant Economy: A History of Lower Tchiri Valley in Malawi, 1859–1960*. Madison, University of Wisconsin Press.

Mann, Michael, 1986a. 'The Autonomous Power of the State: Its Origins, Mechanisms and Results'. In John Hall (ed.), *States in History*. Oxford, Basil Blackwell.

Mann, Michael, 1986a. *The Sources of Social Power, Vol. I*. Cambridge, Cambridge University Press.

————, 1995. 'Ecological Change in North India: Deforestation and Agrarian Distress in the Ganga-Jamna Doab, 1800–1850'. *Environment and History* 1(2): 201–20.

Markham, Clements, 1878. *A Memoir on the Indian Surveys*. London, Her Majesty's Stationery Office.

Marriott, R.G., 1928. 'Sleepers and Sleeper Woods in India'. *Empire Forestry Journal* 7: 76–83.

Martin, Montgomery, 1833. *The History, Antiquities and Topography and Statistics of Eastern India, Volume II*. London, W.H. Allen.

Martinez-Alier, Juan and Ramachandra Guha, 1997. *Varieties of Environmentalism: Essays North and South*. London, Earthscan.

Martinussen, John, 1994. 'Marx and Weber and the Understanding of Politics Within Development Studies'. In John Martinussen (ed.), *The Theoretical Heritage from Marx and Weber in Development Studies*. Roskilde University, International Development Studies.

Maser, Chris, 1988. *The Redesigned Forest*. San Pedro, R&E Miles.

Massey, Doreen, 1984. *Spatial Divisions of Labour: Social Structures and the Geography of Production*. London, Macmillan.

————, 1994. *Space, Place, and Gender*. Minneapolis, University of Minnesota Press.

McAlpin, M.C., 1909 (1981 reprinted Calcutta: Firma K.L. Mukhopadhyaya). *Report on the Conditions of Santals in the Districts of Birbhum, Bankura, Midnapore and North Balasore*. Calcutta, Government Press.

McEvoy, Arthur, 1986. *The Fisherman's Problem: Ecology and Law in the California Fisheries, 1850–1980*. Cambridge, Cambridge University Press.

McIntire, A.L., 1909. 'Notes on Sal in Bengal'. *Forest Pamphlet*. No. 5. Calcutta, Superintendent of Government Printing.

McIntosh, Robert P., 1988. *The Background of Ecology: Concept and Theory*. Cambridge, Cambridge University Press.

McLane, John, 1984. 'Revenue Farming and the Zamindari System in Eighteenth Century Bengal'. In Robert Frykenburg (ed.), *Land Tenure and Peasant in South Asia*. Delhi, Manohar.

————, 1985. 'Bengali Bandits: Police and Landlords after the Permanent Settlement'. In Anand Yang (ed.), *Crime and Criminality in Colonial India*. Tucson, University of Arizona Press.

————, 1993. *Land and Local Kingship in Eighteenth Century Bengal*. Cambridge, Cambridge University Press.

McNeile, D.J., 1866. *Report on the Village Watch in the Lower Provinces of Bengal*. Calcutta, Bengal Secretariat Press.

Menzies, Nicholas, 1992. 'Strategic Space: Exclusion and Inclusion in Wildland Policies in Late Imperial China'. *Modern Asian Studies* 26(4): 719–33.

————, 1994. *Forests and Land Management in Imperial China*. New York, St. Martin's Press.

Merchant, Carolyn, 1989. *Ecological Revolutions: Nature, Gender, and Science in New England*. Chapel Hill, University of North Carolina Press.

Metcalf, Thomas, 1979. *Landlords and the British Raj: Northern India in the Nineteenth Century*. Berkeley, University of California Press.

Migdal, Joel, 1988. *Strong Societies and Weak States: State-Society Relations and State Capabilities in the Third World*. Princeton, Princeton University Press.

————, 1994a. 'The State in Society: An Approach to Struggles for Domination'. In Joel Migdal, Atul Kohli and Vivienne Shue (eds), *State Power and Social Forces: Domination and Transformation in the Third World*. Cambridge, Cambridge University Press.

————, 1994b. 'Introduction'. In Joel Migdal, Atul Kohli and Vivienne Shue (eds), *State Power and Social Forces: Domination and Transformation in the Third World*. Cambridge, Cambridge University Press.

———— et al. (eds), 1994. *State Power and Social Forces: Domination and Transformation in the Third World*. Cambridge, Cambridge University Press.

Miliband, Ralph, 1983. 'State Power and Class Interests.' *Monthly Review* 138(37–68).

Millington, Andrew, 1989. 'Environmental Degradation, Soil Conservation and Agricultural Policies in Sierra Leone, 1895–1984'. In David Anderson and Richard Grove (eds), *Conservation in Africa: People, Policies and Practice*. Cambridge, Cambridge University Press.

Milroy, A.J.W., 1930. 'The Relations Between Sal Forests and Fire'. *Indian Forester* 56(10): 442–7.

————, 1936. 'Sal Natural Regeneration in Assam'. *Indian Forester* 62: 355–60.

Milton, Kay (ed.), 1993. *Environmentalism: The View From Anthropology*. London, Routledge.

Minchin, A.A.F., 1921. *Working Plan for Gumsur Forests, Ganjam District*. Madras, Government Press.

Mitchell, Timothy, 1988. *Colonizing Egypt*. Cambridge, Cambridge University Press.

————, 1991a. 'The Limits of the State: Beyond Statist Approaches and their Critics'. *American Political Science Review* 85(1): 77–96.

—————, 1991b. 'America's Egypt: Discourse of the Development Industry'. *Middle East Report* March–April: 18–34.

Mobbs, E.C., 1936. 'Sal Natural Regeneration Experiments in the United Provinces'. *Indian Forester* 62: 260–7.

—————, 1941. 'The Early History of Forestry in India'. *Indian Forester* 67(5): 231–4.

Mohanty, Chandra, 1984. 'Under Western Eyes: Feminist Scholarship and Colonial Discourse'. *Boundary* 2 Spring/Fall: 71–92.

Mohapatra, Prabhu, 1985. 'Coolies and Colliers: A Study of the Agrarian Context of Labour Migration from Chotanagpur, 1880–1920'. *Studies in History* new series 1(2): 247–303.

—————, 1991. 'Class Conflict and Agrarian Regimes in Chotanagpur, 1860–1950'. *Indian Economic and Social History Review* 28(1): 1–42.

Mooney, H.F., 1938. A Synecological Study of the Forests of Western Singhbhum with special reference to their Geology. *Indian Forest Records* new series II(7). Delhi, Government of India Press.

Moore, Barrington, 1909. 'Notes on the Forests of Northern India and Burma'. *Indian Forester* 35(4, 5): 213–19; 57–62.

Moore, Donald, 1996. 'Marxism, Culture and Political Ecology: Environmental Struggles in Zimbabwe's Eastern Highlands'. In Richard Peet and Michael Watts (eds), *Liberation Ecologies: Environment, Development, Social Movements*. London, Routledge.

—————, forthcoming. 'Clear Waters and Muddied Histories: Competing Claims to the Kaerezi River in Zimbabwe's Eastern Highlands'. *Journal of Southern African Studies*.

Moore, Henrietta and Megan Vaughan, 1994. *Cutting Down Trees: Gender, Nutrition and Agricultural Change in the Northern Province of Zambia, 1890–1990*. London, James Currey.

Mukherjee, S.N., 1968. *Sir William Jones: A Study in Eighteenth Century British Attitudes to India*. Cambridge, Cambridge University Press.

Murali, Atluri, 1995. 'Whose Trees? Forest Practices and Local Communities in Andhra, 1600–1922'. In David Arnold and Ramachandra Guha (eds), *Nature, Culture, Imperialism: Essays on the Environmental History of South Asia*. New Delhi, Oxford University Press.

Neumann, Roderick P., 1995. 'Colonial Recasting of African Society and Landscape in Serengeti National Park'. *Ecumene* 2(2): 149–69.

Nigam, Sanjay, 1990. 'Disciplining and Policing the "Criminals by Birth": The Making of a Colonial Stereotype—The Criminal Tribes and Castes of North India'. *Indian Economic and Social History Review* 27(2): 131–64.

Norman, A.P., 1991. 'Telling it Like it Was: Historical Narratives on their Own Terms'. *History and Theory* 30(2): 119–35.

Nugent, Daniel, 1993. *Spent Cartridges of Revolution: An Anthropological History of Namiquipa, Chihuahua*. Chicago, University of Chicago Press.

Nugent, David, 1994. 'Building the State, Making the Nation: The Bases and Limits of State Centralization in "Modern" Peru'. *American Anthropologist* 96(2): 333–69.

O'Malley, L.S.S., 1910a. *Bengal Bihar, Orissa and Sikkim*. Calcutta, Government of India Press.

—————, 1910b. *Bengal District Gazetteers: Singhbhum, Saraikela and Kharsawan*. Calcutta, Bengal Secretariat Book Depot.

—————, 1910c. *Bengal District Gazetteers: Santhal Parganas*. Calcutta, Bengal Secretariat Book Depot.

—————, 1911. *Bengal District Gazetteers: Midnapore*. Calcutta, Bengal Secretariat Book Depot.

Odum, E.P., 1971. *Fundamentals of Ecology*. Philadelphia, Saunders.

Oldham, C.E.A.W., 1930. *Introduction to Journal of Francis Buchanan Kept During the Survey of the District of Bhagalpur in 1810–11*. Patna, Government of Bihar and Orissa.

Oldham, W.B., 1894. *Some Historical and Ethnical Aspects of Burdwan District*. Calcutta, Bengal Secretariat Press.

Ortner, Sherry, 1989. *High Religion: A Cultural and Political History of Sherpa Buddhism*. Princeton, Princeton University Press.

Osmaston, F.C., 1935. 'The Effect of Burning on Medium Quality Sal Coppice'. *Indian Forester* 61(5): 311–13.

Padel, Felix, 1987. 'British Rule and the Konds of Orissa: A Study of Tribal Administration and its Legitimating Discourse'. DPhil thesis, Oxford University.

Pagden, Anthony, 1995. *Lords of All the World: Ideologies of Empire in Spain, Britain and France, c.1500 –c.1800*. New Haven, Yale University Press.

Pant, Rashmi, 1987. 'The Cognitive Status of Caste in Colonial Ethnography: A Review of Some Literature of the Northwest Provinces and Oudh.' *Indian Economic and Social History Review* 24(2): 145–62.

Parajuli, Pramod, 1991. ' Power and Knowledge in Development Discourse'. *International Social Science Journal* 127: 173–90.

Parker, R.N., 1923. 'The Indian Forest Department'. *Empire Forestry Journal* 2(1): 34–42.

Parmentier, Richard, 1987. *The Sacred Remains: Myth, History and Polity in Belau*. Chicago, Chicago University Press.

Parry, Benita, 1987. 'Problems in Current Theories of Colonial Discourse'. *Oxford Literary Review* 9(27–58).

Parsons, Talcott (ed.), 1964. *Max Weber: The Theory of Economic and Social Organization*. New York, Free Press.

Pathak, Akhileshwar, 1994. *Contested Domains: The State, Peasants and Forests in Contemporary India*. New Delhi, Sage.

Pearson, R.S., 1912. *Commercial Guide to the Forest Economic Products of India*. Calcutta, Superintendent of Government Printing.

Peel, J.D.Y, 1983. *Ijeshas and Nigerians: The Incorporation of a Yoruba Kingdom, 1890s–1970s*. Cambridge, Cambridge University Press.

—————, 1995. 'For Who Hath Despised the Day of Small Things? Missionary Narratives and Historical Anthropology'. *Comparative Studies in Society and History* 37(3): 581–607.

Peet, Richard and Michael Watts, 1996. *Liberation Ecologies: Environment, Development, Social Movements*. London, Routledge.

Pelczynski, Z.A. (ed.), 1984. *The State and Civil Society: Studies in Hegel's Political Philosophy*. Cambridge, Cambridge University Press.

Peluso, Nancy, 1992. *Rich Forests, Poor People: Resource Control and Resistance in Java*. Berkeley, University of California Press.

——————, 1993. 'Coercing Conservation? The Politics of State Resource Control'. *Global Environmental Change* 3(2): 199–217.

——————, 1995. 'Whose Woods are These? Counter-mapping Forest Territories in Kalimantan, Indonesia.' *Antipode* 27(4): 383–406.

——————, Peter Vandergeest, and Leslie Potter, 1995. 'Social Aspects of Forestry in Southeast Asia: A Review of Postwar Trends in the Scholarly Literature'. *Journal of Southeast Asian Studies* 26(1): 196–218.

Peters, Pauline, 1994. *Dividing the Commons: Politics, Policy and Culture in Botswana*. Charlottesville, University of Virginia Press.

Phillimore, R.H., 1945–50. *Historical Records of the Survey of India, Volumes I and II*. Dehradun, Survey of India.

Phillips, P.J., 1924. *Revised Working Plan for the Reserved Forests of the Saranda and Kolhan Division in the Singhbhum District, Bihar and Orissa Circle*. Patna, Superintendent of Government Printing.

Pickering, Andrew, 1992. 'From Science as Knowledge to Science as Practice'. In Andrew Pickering (ed.), *Science as Practice and Culture*. Chicago, University of Chicago Press.

——————, 1995. *The Mangle of Practice: Time, Agency and Science*. Chicago, University of Chicago Press.

Pickett, S.T.A. and P.S. White, 1985. *The Ecology of Natural Disturbance and Patch Dynamics*. Orlando, Academic Press.

Pigg, Stacy Leigh, 1992. 'Constructing Social Categories through Place: Social Representations and Development in Nepal'. *Comparative Studies in Society and History* 34(3): 491–513.

Pinchot, Gifford, 1947. *Breaking New Ground*. New York, Harcourt, Brace.

Polanyi, Karl, 1957. *The Great Transformation*. Boston, Beacon Press.

Pouchepadass, Jacques, 1995. 'British Attitudes Towards Shifting Cultivation in Colonial South India: A Case Study of South Canara District, 1800–1920'. In David Arnold and Ramachandra Guha (eds), *Nature, Culture, Imperialism: Essays on the Environmental History of South Asia*. New Delhi, Oxford University Press.

Poulantzas, Nicos, 1978. *State, Power, Socialism*. London, Verso.

Prakash, Gyan, 1990a. *Bonded Histories: Genealogies of Labour Servitude in Colonial India*. Cambridge, Cambridge University Press.

——————, 1990b. 'Writing post-Orientalist Histories of the Third World: Perspectives from Indian Historiography'. *Comparative Studies in Society and History* 32(2): 383–408.

——————, 1992a. 'Science "Gone Native" in Colonial India'. *Representations* 40: 153–77.

——————, 1992b. 'Introduction: The History and Historiography of Rural Labourers in Colonial India'. In Gyan Prakash (ed.), *The World of the Rural Labourer in Colonial India, Themes in Indian History*. New Delhi, Oxford University Press.

——————, 1995. 'Introduction: After Colonialism'. In Gyan Prakash (ed.), *After Colonialism: Imperial Histories and Postcolonial Displacements*. Princeton, Princeton University Press.

Prasad, Archana, 1994. 'Forests and Subsistence in Colonial India: A Study of the Central Provinces, 1830–1945'. PhD thesis, New Delhi, Jawaharlal Nehru University.

Pratap, Ajay, 1987. 'Paharia Ethnohistory and the Archaeology of the Rajmahal Hills: Archaeological Implications of an Historical Study of Shifting Cultivation'. PhD thesis, Cambridge University.

Pratt, Mary Louise, 1992. *Imperial Eyes: Travel Writing and Transculturation*. London, Routledge.

Pred, Alan, 1984. 'Place as Historically Contingent Process: Structuration and the Time-Geography of Becoming Places'. *Annals of the Association of American Geographers* 74: 279–97.

Price, J.C., 1876. *Notes on the History of Midnapore*. London, Smith, Elder & Co.

——————, 1953 (1874). 'Notes on the Chuar rebellion of 1799'. In A. Mitra (ed.), *Midnapore: District Census Handbook*. Alipore, Superintendent of Government Printing.

Pudup, Mary Beth, 1988. 'Arguments Within Regional Geography'. *Progress in Human Geography* 12(4): 369–80.

Pyne, Stephen, 1982. *Fire and America: A Cultural History of Wildland and Rural Fire*. Princeton, Princeton University Press.

——————, 1994. 'Nataraja: India's Cycle of Fire'. *Environmental History Review* 18(3): 21–36.

Radhakrishna, Meena, 1989. 'The Criminal Tribes Act in Madras Presidency: Implications for Itinerant Trading Communities'. *Indian Economic and Social History Review* 26(3): 269–95.

Rajan, Ravi, 1998. 'Imperial Environmentalism or Environmental Imperialism? European Forestry, Colonial Foresters and the Agendas of Forest Management in British India, 1850–1900'. In Richard Grove, Satpal Sangwan and Vinita Damodaran (eds), *Nature and the Orient: Essays on the Environmental History of South and Southeast Asia*. New Delhi, Oxford University Press.

——————, forthcoming. 'The Baconian Environmentalist Tradition, East India Company Scientists, and the Origins of Forestry in the British Empire, 1800–1850'. *Indian Economic and Social History Review*.

Ramsbotham, R.B., 1930. *Studies in the Land Revenue History of Bengal 1769–1787*. London, Oxford University Press.

Rangarajan, Mahesh, 1992. 'Forest Policy in the Central Provinces, 1860–1914'. PhD thesis, Oxford University.

——————, 1994. 'Imperial Agendas and India's Forests: The Early History of Indian Forestry, 1800–1878'. *Indian Economic and Social History Review* 31(2): 147–67.

——————, 1996a. *Fencing the Forest: Conservation and Ecological Change in India's Central Provinces, 1860–1914*. New Delhi, Oxford University Press.

————, 1996b. 'Environmental Histories of South Asia'. *Environment and History* 2(2): 129–44.

————, 1998. 'Production, Desiccation and Forest Management in Central Provinces'. In Richard Grove, Vinita Damodaran and Satpal Sangwan (eds), *Nature and the Orient: Essays on the Environmental History of South and Southeast Asia.* New Delhi, Oxford University Press.

Ranger, Terence, 1994. 'The Invention of Tradition Revisited: The Case of Colonial Africa.' In Preben Kaarsholm and Jan Hultin (eds), *Inventions and Boundaries: Historical and Anthropological Approaches to the Study of Ethnicity and Nationalism, Occasional paper no. 11.* Roskilde University, International Development Studies.

Ravina, Mark, 1995. 'State Building and Political Economy in Early Modern Japan'. *Journal of Asian Studies* 54(4): 997–1022.

Ray, Ratnalekha, 1979. *Change in Bengal Agrarian Society, 1760–1850.* Delhi, Manohar.

Raynor, E.W., 1940. 'Sal Regeneration de novo'. *Indian Forester* 66(9): 525–9.

Reid, Anthony, 1995. 'Humans and Forests in Pre-colonial Southeast Asia'. *Environment and History* 1(1): 93–110.

Reingold, N. and M. Rothenberg (eds), 1987. *Scientific Colonialism: A Cross-Cultural Comparison.* Washington DC, Smithsonian Institution Press.

Rennell, James, 1976(1793). *Memoir of a Map of Hindustan or the Mughal Empire.* Calcutta, Editions Indian.

Ribbentrop, B., 1900. *Forestry in British India.* Calcutta, Office of the Superintendent of Government Printing.

Richards, John and Elizabeth Flint, 1990. 'Long-term Transformations in the Sunderbans Wetlands Forests of Bengal'. *Agriculture and Human Values* 7(2): 17–33.

Richards, John and James Hagen, 1987. 'A Century of Rural Expansion in Assam, 1870–1970'. *Itinerario* 9: 193–209.

Richards, John and Michelle McAlpin, 1983. 'Cotton Cultivating and Land Clearing in the Bombay Deccan and Karnatak, 1818–1920'. In Richard Tucker and John Richards (eds), *Global Deforestation and the Nineteenth Century World Economy.* Durham, Duke University Press.

Richards, John, James Hagen, et al., 1985. 'Changing Landuse in Bihar, Punjab and Haryana, 1850–1970'. *Modern Asian Studies* 19(3): 699–732.

Richards, Paul, 1988. 'The Sierra Leone Department of Agriculture, 1912–1960: Lessons for Today?' In C.M. Fyle (ed.), *History and Socio-economic Development in Sierra Leone.* Freetown, SLADEA.

Ricklefs, Robert and Dolph Schluter, 1993. *Species Diversity in Ecological Communities: Historical and Geographical Perspectives.* Chicago, University of Chicago Press.

Risley, Herbert, 1981 (1891). *Tribes and Castes of Bengal, Volume II.* Calcutta, Firma K.L. Mukhopadhyay.

————, 1991 (1915, 2nd edn). *The People of India.* Delhi, Munshiram Manoharlal.

Ritvo, H., 1987. *The Animal Estate: The English and Other Creatures in the Victorian Age.* Cambridge, Harvard University Press.

Robb, Peter, 1993. 'Introduction: Meanings of Labour in Indian Social Context'. In *Dalit Movements and the Meanings of Labour in India*. New Delhi, Oxford University Press.

Robertson, George et al., 1996. *FutureNatural: Nature, Science, Culture*. New York, Routledge.

Rocheleau, Diane, Barbara Thomas-Slayter et al. (eds), 1996. *Feminist Political Ecology*. London, Routledge.

Rocher, Rosanne, 1983. *Orientalism, Poetry and the Millennium: The Checkered Life of Nathaniel Brassey Halhead, 1751–1830*. Delhi, Motilal Banarsidass.

—————, 1993. 'British Orientalism in the Eighteenth Century: The Dialectics of Knowledge and Government'. In Carol Breckenridge and Peter Van der Veer (eds), *Orientalism and the Postcolonial Predicament*. New Delhi, Oxford University Press.

Rodger, A., 1925. 'Research in Forestry in India'. *Empire Forestry Journal* 14: 45–53.

Rosaldo, Renato, 1980. *Ilongot Headhunting, 1883–1974: A Study in Society and History*. Stanford, Stanford University Press.

Roseberry, William, 1989. *Anthropologies and Histories: Essays in Culture, History and Political Economy*. New Brunswick, Rutgers University Press.

—————, 1994. 'Hegemony and the Language of Contention'. In Gil Joseph and Daniel Nugent (eds), *Everyday Forms of State Formation: Revolution and the Negotiation of Rule in Mexico*. Durham, Duke University Press.

Rowbotham, C.J., 1924. 'The Taungya System in Cachar Division, Assam'. *Indian Forester* 50: 356–58.

Royle, J. Forbes, 1839. *Illustrations of the Botany and other Branches of the Natural History of the Himalayan Mountains and of the Flora of Cashmere*. London, W.H. Allen.

Sachs, Wolfgang (ed.), 1992. *The Development Dictionary*. London, Zed.

Sack, Robert, 1986. *Human Territoriality: Its Theory and History*. Cambridge, Cambridge University Press.

Sahlins, Marshall, 1981. *Historical Metaphors and Mythical Realities: Structure in the Early History of the Sandwich Island Kingdom*. Association for Social Anthropology in Oceania, Special Publication No. 1. Ann Arbor, University of Michigan Press.

—————, 1985. *Islands of History*. Chicago, University of Chicago Press.

—————, 1993. 'Goodbye to Tristes Tropes: Ethnography in the Context of Modern World History'. *Journal of Modern History* 65: 1–25.

Sahlins, Peter, 1989. *Boundaries: The Making of France and Spain*. Berkeley, University of California Press.

—————, 1994. *Forest Rites: The War of the Demoiselles in Nineteenth Century France*. Cambridge, Harvard University Press.

Said, Edward, 1979. *Orientalism*. New York, Vintage.

—————, 1993. *Culture and Imperialism*. New York, Alfred Knopf.

Samaddar, Ranabir, 1998. *Memory, Identity, Power: Politics in the Jungle Mahals, 1890–1950*. London, Sangam Books.

Sanderson, G.P., 1879. *Thirteen Years among the Wild Beasts of India*. London, W.H. Allen.

————————, 1884–1885. 'The Asiatic Elephant in Freedom and Captivity'. *Indian Forester* 10: 533–9, 576–82; 11:32–8, 75–81.

Sangren, Steven, 1995. ' "Power" Against Ideology: A Critique of Foucaultian Usage.' *Cultural Anthropology* 10(1): 3–40.

Sangwan, Satpal, 1991. *Science, Technology and Colonization: an Indian Experience, 1757–1857*. New Delhi, Anamika Prakashan.

————————, 1994. 'Reordering the Earth: the Emergence of Geology as a Scientific Discipline in Colonial India.' *Indian Economic and Social History Review* 31(3): 291–310.

Sanyal, Hiteshranjan, 1971. 'Continuities of Social Mobility in Traditional and Modern Society in India: Two Case Studies of Caste Mobility in Bengal'. *Journal of Asian Studies* 30(2): 315–39.

Sayer, Derek and Phillip Corrigan, 1985. *The Great Arch: English State Formation as Cultural Revolution*. Oxford, Basil Blackwell.

Schlich, William, 1884. 'Review of Forest Administration for British India'. *Indian Forester* 10(8): 372–8.

————————, 1906. *Manual of Indian Forestry, Volume 1*. Calcutta, Superintendent of Government Printing.

Schwerin, Detlef, 1978. 'The Control of Land and Labour in Chotanagpur,1858–1908'. In D. Rothermund and D.C. Wadhwa (eds), *Zamindars, Mines and Peasants: Studies in the History of an Indian Coalfield and its Rural Hinterland*. Delhi, Manohar.

Scott, James, 1990. *Domination and the Arts of Resistance: Hidden Transcripts*. New Haven, Yale University Press.

————————, 1998. *Seeing Like a State: How Certain Schemes to Improve the Human Condition Have Failed*. New Haven, Yale University Press.

Sen, J.N. and T.P. Ghose, 1925. 'Soil Conditions Under Sal.' *Indian Forester* 51(6): 243–53.

Sen, Suchibrata, 1984. *The Santals of the Jungle Mahals: An Agrarian History, 1793–1861*. Calcutta, Ratna Prakashan.

Sengupta, B., 1910. 'Fire Conservancy in Indian Forests'. *Indian Forester* 36(3): 132–45.

Sharma, Sanjay, 1993. 'The 1837–38 Famine in UP: Some Dimensions of Popular Action'. *Indian Economic and Social History Review* 30(3): 337–72.

Shebbeare, E.O., 1930. 'Fire and Sal.' *Indian Forester* 56(3): 2–6.

Sheets-Pyenson, Susan, 1986. 'Cathedrals of Science: The Development of Colonial Natural History Museums during the late Nineteenth Century'. *History of Science* XXV: 279–300.

Sherwill, Walter, 1851. 'Note upon a Tour Through the Rajmahal Hills'. *Journal of Asiatic Society of Bengal* 20(7): 544–606.

————————, 1855. *Geographical and Statistical Report of the District of Bhirbhoom*. Calcutta, Bengal Secretariat Press.

Shiva, Vandana, 1989. *Staying Alive: Women, Ecology and Development*. London, Zed.

————————, 1991. *Ecology and the Politics of Survival: Conflicts over Natural Resources in India*. New Delhi, Sage.

Silverman, Marilyn and P.H. Gulliver, 1992. 'Historical Anthropology and the Ethnographic Tradition: a Personal, Historical and Intellectual Account.' In Marilyn Silverman and P.H. Gulliver (eds), *Approaching the Past*. New York, Columbia University Press.

Simson, Frank, 1886. *Letters on Sport in Eastern Bengal*. London, R.H. Porter.

Sinclair, John, 1794. *The Statistical Account of Scotland*. Edinburgh, William Creech.

Singh, Chetan, 1988. 'Conformity and Conflict: Tribes and the "Agrarian System" of Mughal India'. *Indian Economic and Social History Review* 25(3): 319–40.

Singh, K. Suresh, 1983. *Birsa Munda and his Movement, 1874–1901: A Study of a Millenarian Movement in Chotanagpur*. New Delhi, Oxford University Press.

Singha, Radhika, 1990. 'A "Despotism of Law": British Criminal Justice and Public Authority in North India, 1772–1837'. PhD thesis, Cambridge University.

———, 1993a. '"Providential" Circumstances: The Thuggee Campaign of the 1830s and Legal Innovation'. *Modern Asian Studies* 27(1): 83–146.

———, 1993b. 'The Privilege of Taking Life: Some "Anomalies" in the Law of Homicide in the Bengal Presidency.' *Indian Economic and Social History Review* 30(2): 181–214.

Sinha, Dikshit, 1984. *The Kharia of West Bengal*. Calcutta, Anthropological Survey of India.

Sinha, J.N., 1962. *Working Plan for the Reserved, Protected and Private Protected Forests of Manbhum Division*. Patna, Superintendent of Government Printing.

Sinha, Surajit, 1957. 'The Media and Nature of Bhumij-Hindu Interactions'. *Journal of Asiatic Society of Letters and Sciences* 23(1): 23–37.

Siu, Helen, 1989. *Agents and Victims in South China: Accomplices in Rural Revolution*. New Haven, Yale University Press.

Sivaramakrishnan, K., 1995a. 'Imagining the Past in Present Politics: Colonialism and Forestry in India'. *Comparative Studies in Society and History* 37(1): 3–40.

———, 1995b. 'Situating the Subaltern: History and Anthropology in the Subaltern Studies Project'. *Journal of Historical Sociology* 8(4): 395–429.

———, 1996a. 'Forests, Politics, and Governance in Bengal, 1794–1994'. PhD thesis, Yale University.

———, 1996b. 'British Imperium and Forested Zones of Anomaly in Bengal, 1767–1833'. *Indian Economic and Social History Review* 33(3): 243–82.

———, 1996c. 'Fire and the Politics of Forest Regeneration in Colonial Bengal'. *Environment and History* 2(2): 145–94.

———, 1997. 'A Limited Forest Conservancy in Southwest Bengal, 1864–1912'. *Journal of Asian Studies* 56(1): 75–112.

Skaria, Ajay, 1992. 'A Forest Polity in Western India: The Dangs, 1840s–1920s'. PhD thesis, Cambridge University.

————————, 1995. 'Hegemony and Affirmatory State Ritual in India: The Dangs Darbar, 1840s–1990s' (manuscript).

————————, 1997. 'Shades of Wildness: Tribe, Caste and Gender in Western India'. *Journal of Asian Studies* 56(3): 726–45.

————————, 1998. 'Timber Conservancy, Dessicationism and Scientific Forestry: the Dangs, 1840–1920.' In Richard Grove, Vinita Damodaran and Satpal Sangwan (eds), *Nature and the Orient: Essays in the Environmental History of South and South-east Asia.* New Delhi, Oxford University Press.

————————, 1999. *Hybrid Histories: Forests, Frontiers and Wildness in Western India.* New Delhi, Oxford University Press.

Skocpol, Theda, 1985. 'Bringing the State Back In: Strategies of Analysis in Current Research'. In P.B. Evans, D. Rueschemeyer and T. Skocpol (eds), *Bringing the State Back In.* Cambridge, Cambridge University Press.

Slade, H., 1896. 'Too Much Fire Protection in Burma'. *Indian Forester* 22(5): 172–6.

Slym, M.J., 1876. *Memorandum on Jungle Fires.* Moulmein, Tenasserim Press.

Smil, Vaclav, 1993. *Global Ecology: Environmental Change and Social Flexibility.* London, Routledge.

Smith, David, 1986. *The Practice of Silviculture.* New York, John Wiley.

Smith, Dorothy, 1987. *The Everyday World as Problematic: A Feminist Sociology.* Boston, Northeastern University Press.

Smith, Neil, 1990. *Uneven Development: Nature, Capital and the Production of Space.* Oxford, Basil Blackwell.

————————, 1994. 'Geography, Empire and Social Theory'. *Progress in Human Geography* 18(4): 491–500.

Smith, Richard S., 1985. 'Rule by Records and Rule-by-Reports: Complementary Aspects of the British Imperial Rule of Law'. *Contributions to Indian Sociology* 19(1): 153–76.

Smythies, E. A., 1931. ' Sal Regeneration in the United Provinces'. *Indian Forester* 57(6): 298–301.

————————, 1932. 'Sal Regeneration in United Provinces'. *Indian Forester* 58(4): 196–210.

————————, 1940. 'Sal Regeneration de novo'. *Indian Forester* 66(4): 193–9.

Soja, Edward, 1989. *Postmodern Geographies: The Reassertion of Space in Critical Social Theory.* London, Verso.

Soni, R.C., 1953. *A Note on the Natural Regeneration of Sal.* Proceedings of the All-India Sal Study Tour and Symposium. Dehradoon, Forest Research Institute.

Spivak, Gayatri Chakravarty, 1987. *In Other Worlds: Essays in Cultural Politics.* New York, Methuen.

Spurr, David, 1994. *The Rhetoric of Empire: Colonial Discourse in Journalism, Travel Writing and Imperial Administration.* Durham, Duke University Press.

Stafford, Robert, 1984. 'Geological Surveys, Mineral Discoveries and British

Expansion, 1835–1871'. *Journal of Imperial and Commonwealth History* 12: 5–32.

————, 1989. *The Empire of Science: Sir Roderick Murchison, Scientific Exploration and Victorian Imperialism.* Cambridge, Cambridge University Press.

————, 1990. 'Annexing the Landscapes of the Past: British Imperial Geology in the Nineteenth Century'. In J.M. Mackenzie (ed.), *Imperialism and the Natural World.* Manchester, Manchester University Press.

Stebbing, E.P, 1926. *The Forests of India, Volume III.* London, John Lane.

————, 1924. *The Forests of India, Volume II.* London, John Lane.

————, 1922. *The Forests of India, Volume I.* London, John Lane.

————, 1930. *The Diaries of a Sportsman Naturalist in India.* London, John Lane.

Stein, Burton, 1989. *Thomas Munro: The Origins of the Colonial State and his Vision of Empire.* New Delhi, Oxford University Press.

Stigloe, John, 1976. 'Jack o'Lanterns to Surveyors: The Secularization of Landscape Boundaries'. *Environmental Review* 1: 14–31.

Stoler, Ann Laura 1992. ' "In Cold Blood": Hierarchies of Credibility and the Politics of Colonial Narratives'. *Representations* 37: 151–89.

Stone, Laurence, 1979. 'The Revival of Narrative: Reflections on New Old History.' *Past and Present* 85: 3–24.

Suleri, Sara, 1992. *The Rhetoric of English India.* Chicago, University of Chicago Press.

Sundar, Nandini, 1995. 'In Search of Gunda Dhur: Colonialism and Contestation in Bastar, Central India, 1854–1993'. Ph.D. thesis, Columbia University.

Taylor, Alan, 1995. 'The Great Change Begins: Settling the Forest of Central New York'. *New York History* 76: 265–90.

————, 1996. 'Unnatural Inequalities: Social and Environmental Histories.' *Environmental History* 1(4): 6–19.

Taylor, G.F., 1981. 'The Forestry Agriculture Interface: Some Lessons from Indian Forest Policy'. *Commonwealth Forestry Review* 60(1): 45–52.

Thackeray, S. and A. Shapin, 1974. 'Prosopography as a Research Topic in History of Science: The British Scientific Community 1700–1900'. *History of Science* 12: 1–28.

Thapar, Romila and Majid Siddiqi, 1991. 'Tribals in History: The Case of Chotanagpur'. In Dipankar Gupta (ed.), *Social Stratification,* Oxford in India Readings in Sociology and Social and Cultural Anthropology. New Delhi, Oxford University Press.

Thomas, Keith, 1984. *Man and the Natural World.* Harmondsworth, Penguin.

Thomas, Nicholas, 1994. *Colonialism's Culture: Anthropology, Travel and Government.* Cambridge, Polity Press.

Thompson, E. P., 1975. *Whigs and Hunters: The Origins of the Black Act.* New York, Pantheon.

————, 1979. The Crime of Anonymity. In Douglas Hay et al. (eds), *Albion's Fatal Tree: Crime and Society in Eighteenth Century England.* New York, Pantheon.

Tilly, Charles, 1984. 'Social Movements and National Politics'. In Charles Bright and Susan Harding (eds), *Statemaking and Social Movements: Essays in History and Theory*. Ann Arbor, University of Michigan Press.

—————, 1992. *Coercion, Capital and European States, 990–1992*. Cambridge, Basil Blackwell.

Tinne, P., 1907. *Second Working Plan for the Reserved Forests of the Tista Division*. Calcutta, Bengal Secretariat Press.

Totman, Conrad, 1989. *The Green Archipelago: Forestry in Pre-Industrial Japan*. Berkeley, University of California Press.

—————, 1993. *Early Modern Japan*. Berkeley, University of California Press.

Toynbee, George, 1888. *A Sketch of the Administration of the Hooghlee District from 1795 to 1845 with Some Account of the Early English, Portuguese, Dutch, French and Danish Settlements*. Calcutta, Bengal Secretariat Press.

Tracey, P.D., 1956. 'The Development of Forestry in Assam in the Last Fifty Years'. *Indian Forester* 82(12): 619–23.

Trafford, F., 1905. *Working Plan Revised for the Reserved Forests of the Jalpaiguri Division*. Darjeeling, Bengal Secretariat Tour Press.

—————, 1911. *Working Plan for the Forests of the Sundarbans Division*. Calcutta, Bengal Secretariat Press.

Trouillot, Michel-Rolph, 1989. 'Discourses of Rule and the Acknowledgment of the Peasantry in Dominica, W.I., 1838–1928'. *American Ethnologist* 16: 704–18.

Troup, R.S., 1905a. 'Fire Protection in the Teak Forests of Burma'. *Indian Forester* 31(3): 138–46.

—————, 1905b. 'Fire Protection in the Teak Forests of Burma'. *Indian Forester* 31(9): 503–5.

—————, 1921. *Sylviculture of Indian Trees*. Delhi, Government of India Press.

Tsing, Anna Lowenhaupt, 1993. *In the Realm of the Diamond Queen: Marginality in an Out-of-the-Way Place*. Princeton, Princeton University Press.

—————, 1994. 'From the Margins'. *Cultural Anthropology* 9(3): 279–97.

Tucker, Richard, 1989. 'The Depletion of India's Forests under British Imperialism: Planters, Forests and Peasants in Assam and Kerala'. In Donald Worster (ed.), *The Ends of the Earth: Perspectives on Modern Environmental History*. Cambridge, Cambridge University Press.

Urry, John, 1987. ' Society, Space and Locality'. *Environment and Planning D: Society and Space* 5: 435–44.

Vandergeest, Peter and Nancy Peluso, 1995. 'Territorialization and State Power in Thailand'. *Theory and Society* 24: 385–426.

Varshney, Ashutosh, 1995. *Democracy, Development and the Countryside: Urban-Rural Struggles in India*. Cambridge, Cambridge University Press.

Verdery, Katherine, 1995. 'Notes Towards an Ethnography of a Transforming State: Romania'. In Jane Schneider and Rayna Rapp (eds), *Articulating Hidden Histories: Exploring the Influence of Eric Wolf*. Berkeley, University of California Press.

Vicziany, Marika, 1986. 'Imperialism, Botany and Statistics in Early Nineteenth-

Century India: The Surveys of Francis Buchanan 1762–1829.' *Modern Asian Studies* 20(4): 625–60.

Voelcker, John Augustus, 1893. *Report on the Improvement of Indian Agriculture*. London, Eyre and Spottiswoode.

Waddell, L.A., 1893. 'The Traditional Migration of the Santhal Tribe'. *Indian Antiquary* 22: 294–6.

Warner, John Harley, 1991. 'Ideals of Science and their Discontents in Late Nineteenth-Century American Medicine'. *Isis* 82(313): 454–78.

Warr, J.H., 1926. 'The Sleeper Problem in India'. *Empire Forestry Journal* 5: 235–48.

Warren, W.D.M., 1940. 'Sal Regeneration de novo.' *Indian Forester* 66(6): 334–40.

—————, 1941. 'Sal Regeneration de novo in B-3 Sal.' *Indian Forester* 67(3): 116–23.

Washbrook, David, 1976. *The Emergence of Provincial Politics: The Madras Presidency, 1870–1920*. Cambridge, Cambridge University Press.

—————, 1981. 'Law, State and Society in Colonial India'. In Chris Baker, Gordon Johnson and John Gallagher (eds), *Power, Profit and Politics*. Cambridge, Cambridge University Press.

Watts, Michael, 1992. 'Space for Everything: A Commentary'. *Cultural Anthropology* 7(1): 115–29.

—————, 1995. ' "A New Deal in Emotions": Theory and Practice and the Crisis of Development.' In Jonathan Crush (ed.), *Power of Development*. London, Routledge.

Weber, Max, 1968. *Economy and Society*. New York, Bedminster Press.

White, Hayden, 1980. 'The Value of Narrativity in the Representation of History.' *Critical Inquiry* 7: 5–25.

—————, 1984. 'The Question of Narrative in Contemporary Historical Theory'. *History and Theory* 13: 1–33.

White, Richard, 1990. 'Environmental History, Ecology and Meaning.' *Journal of American History* 76: 1111–16.

Williamson, Thomas, 1808. *Oriental Field Sports, Being a Complete, Detailed and Accurate Description of the Wild Sports of the East*. London, W. Bulmer.

Wilmsen, Edmund, 1989. *Land Filled with Flies: A Political Economy of the Kalahari*. Chicago, University of Chicago Press.

Wilson, Alexander, 1992. *The Culture of Nature: North American Landscape from Disney to the Exxon Valdez*. Oxford, Blackwell.

Wolf, Eric, 1982. *Europe and the People Without History*. Berkeley, University of California Press.

—————, 1990. 'Facing Power: Old Insights, New Questions'. *American Anthropologist* 92: 586–96.

Wood, B.R., 1922. 'The Artificial Regeneration of Sal in Gorakhpur'. *Indian Forester* 48(2): 55–67.

Woost, Michael, 1993. 'Nationalizing the Local Past in Sri Lanka: Histories of Nation and Development in a Sinhalese Village.' *American Ethnologist* 20(3): 502–21.

Worboys, Michael, 1988. 'The Discovery of Colonial Malnutrition Between the Wars'. In David Arnold (ed.), *Imperial Medicine and Indigenous Societies*. Manchester, Manchester University Press.

——————, 1990. 'The Imperial Institute: The State and Development of the Natural Resources of the Colonial Empire, 1887–1923'. In J.M. Mackenzie (ed.), *Imperialism and the Natural World*. Manchester, Manchester University Press.

Worster, Donald, 1985. *Rivers of Empire: Water, Aridity and the Growth of the American West*. New York, Pantheon.

——————, 1988. 'Doing Environmental History'. In Donald Worster (ed.), *The Ends of the Earth: Perspectives on Modern Environmental History*. New York, Cambridge University Press.

——————, 1990. 'Seeing Beyond Culture'. *Journal of American History* 76: 1142–7.

——————, 1993. *The Wealth of Nature: Environmental History and the Ecological Imagination*. New York, Oxford University Press.

——————, 1994. *Nature's Economy: A History of Ecological Ideas*. Cambridge, Cambridge University Press.

——————, 1996. 'The Two Cultures Revisited'. *Environment and History* 2(1): 3–14.

Yang, Anand, 1989. *The Limited Raj: Agrarian Relations in Colonial India, Saran District, 1793–1920*. Berkeley, University of California Press.

Young, Crawford, 1994. *The African Colonial State in Comparative Perspective*. New Haven, Yale University Press.

Young, Robert, 1995. *Colonial Desire: Hybridity in Theory, Culture and Race*. London, Routledge.

Index